U0184293

智慧城市与智能建造论文集（2021）

Proceedings of Smart City and Intelligent Construction（2021）

华中科技大学土木与水利工程学院
中国建筑学会工程管理研究分会 编

中国建筑工业出版社

图书在版编目（CIP）数据

智慧城市与智能建造论文集：2021：Proceedings of Smart City and Intelligent Construction 2021 / 华中科技大学土木与水利工程学院，中国建筑学会工程管理研究分会编. — 北京：中国建筑工业出版社，2021.11

ISBN 978-7-112-26915-0

Ⅰ. ①智… Ⅱ. ①华… ②中… Ⅲ. ①智能技术－应用－基础设施－市政工程－城市规划－文集②智能技术－应用－土木工程－文集 Ⅳ. ①TU-53

中国版本图书馆 CIP 数据核字（2021）第 248838 号

2021 智慧城市与智能建造高端论坛秉持建筑业智能发展理念，聚焦"城市智慧基础设施""建筑产业互联网与数字经济""建筑工业化与建造机器人""医养结合与健康住宅"四类前沿主题，邀请相关专家学者进行综合分析与探讨，并将优秀论文收录于本论文集中供广大学者研究参考。

希望能借助本论坛与这些优秀论文，与广大学者共同探索建筑业实现智能化转型升级的思路、任务和前景，为推动产业变革提供一定的理论依据和实践基础，更好地促进我国持续往数字化、信息化、智能化建设强国转变。

为扩大本论文集级作者知识信息交流渠道，本论文集已被《中国学术期刊网络出版总库》及 CNKI 系列数据库收录。

责任编辑：朱晓瑜
责任校对：李美娜

智慧城市与智能建造论文集（2021）

Proceedings of Smart City and Intelligent Construction（2021）

华中科技大学土木与水利工程学院
中国建筑学会工程管理研究分会　编

*

中国建筑工业出版社出版、发行（北京海淀三里河路 9 号）
各地新华书店、建筑书店经销
北京红光制版公司制版
北京中科印刷有限公司印刷

*

开本：880 毫米×1230 毫米　1/16　印张：21　字数：497 千字
2022 年 4 月第一版　　2022 年 4 月第一次印刷
定价：**79.00** 元
ISBN 978-7-112-26915-0
（38750）

版权所有　翻印必究
如有印装质量问题，可寄本社图书出版中心退换
（邮政编码 100037）

《智慧城市与智能建造论文集》编委会

顾　　问：谢礼立　柴洪峰　谢先启　钮新强　秦顺全

编委会主任：丁烈云

编委会委员（按姓氏笔画排序）：

丁荣贵	于　洁	王　帆	王广斌	王文剑	王红卫
王建廷	王雪青	方　俊	方东平	尹贻林	邓利明
邓明胜	朱　涛	朱东飞	朱宏平	朱雁飞	刘伊生
刘贵文	刘晓君	苏　成	杜　强	李　青	李　霆
李　霞	李久林	李正农	李丽红	李启明	吴佐民
何　政	佟德辉	沈元勤	宋维佳	张　宏	张　鸿
张　琨	张云波	陈　珂	陈立文	陈宝智	陈维亚
邵韦平	欧立雄	周　迎	周　诚	赵宪忠	姜　军
姚玲珍	骆汉宾	袁　烽	顾勇新	徐卫国	高　欣
高宗余	唐孟雄	黄　刚	黄　斌	黄健陵	龚　剑
盛　谦	常　锱	葛汉彬	曾志刚	薛小龙	魏　来

　　《智慧城市与智能建造论文集》编委会一共由72名来自土木与水利工程、管理科学与工程、人工智能等方面的院士、专家和学者组成，其中设顾问5人、主任1人、委员66人，研究方向涵盖城市智慧基础设施、建筑产业互联网与数字经济、建筑工业化与建造机器人、医养结合与健康住宅等多个新兴交叉领域。

前　言

随着我国经济的快速发展，传统建造模式已难以满足规模化、个性化和高质量的生产需求。智能建造作为建筑产业转型升级的核心引擎，对建筑业产业链价值提升意义重大。以"数字经济、人工智能、智慧城市、智能建造"为主题，突出政、产、学、研、金、服、用模式，围绕智慧城市与智能建造领域的技术前沿、产业应用、人才培养等热点问题进行深入研究，可以有效推动建筑产品升级并引领产业变革与创新发展。2021智慧城市与智能建造高端论坛秉持建筑业持续智能发展理念，聚焦"城市智慧基础设施""建筑产业互联网与数字经济""建筑工业化与建造机器人""医养结合与健康住宅"四类前沿主题，邀请相关专家学者就上述主题进行综合分析与探讨，探索建筑业实现建造智能化转型升级的思路、任务和前景。

以民为本的智慧城市，需要智能基础设施来保证其智能互联的生态系统。即智能基础设施是智慧城市及其高效、可持续运作能力的关键。但基础设施具备范围广、投入大、技术复杂、关联性强、不确定因素多等特点，其数字化和智能化难度较大，给工程决策和项目管理带来了巨大的挑战。智能基础设施的研究由来已久，海口经济学院秦桂芳从战略背景、发展形势、技术应用等方面，就新型智慧城市基础设施建设进行深入分析和探讨，对如何改革地方政府现有智慧城市基础设施建设的管理模式这一问题进行了深入探索；中国矿业大学许娜等人通过对有关知识情境的相关文献进行分析，提出一个新的、全面的领域知识情境维度的分类框架，从而有效识别城轨安全风险管理的一系列情境要素；福建工程学院管理学院田昌民等人从政府实施责任激励约束行为的角度构建了"政府—社会资本方"的智慧产业项目合作的演化博弈模型，并进行均衡策略分析；吉林建筑大学苗泽惠等人通过 IoT（物联网）等技术将传感器、机器、人、物连接起来，实现施工过程的智能收集、定位、监控、处理、分析和管理，并与 BIM 模型进行数据交互，从而实现智能化管理。

近年来，蓬勃发展的信息技术成为破题的关键。高清晰度测量和定位技术、数字化协同和移动技术、先进的设计和施工技术、建筑产业转型目标从原来粗犷式施工转变为工业级精细化水平，蓬勃发展的信息技术是其破题关键。物联网和高级分析技术作为引领未来建筑业变革及发展的数字化、信息化新技术，对建筑过程元素进行控制，优化建筑全过程所涉要素资源，并对建筑价值链的广度和深度进行扩展。大连理工大学王琳卓、张超等人基于制度逻辑理论和网络治理理论，通过案例研究方法，以中交集团为研究对象，识别出重大工程项目供应商网络治理的筛选进入机制、评价与分级机制、淘汰退出机制和制度拼凑过程，构建了重大工程供应商网络治理的多层次治理体系，从制度逻辑视角丰富了中国情境下重大工程项目的治理理论，同时提出相应的实践建议；南京大学陈波等人通过对近年的法律判决文书以及相关法律规范的研读，通过 Protégé 软件构建了变更的合规案例知识本体，以实现案例知识的共享和重用。同时在本体的知识管理基础上，运用案例推理算法来获取当前工程变更的相似案例，指导新变更的处理。通过自然语言处理的方法提取案例文本中的隐性特征计算文本相似度。结合时间序列数据挖掘，计算动态时间弯曲距离进行时间序列相似度量化。在此基础上，基于熵值法耦合层次分析法赋权各相似度指标权重。研究结果表明，论文的案例推理算法能在案例库中匹配出较为相似的历史案例供用户进行学习和参考；英国特许土木工程测量师学会（香港分会）理事兼义务司库余伊琪探讨建筑企业及工程咨询顾问尤其是内地与港澳三地工程承包企业及工程咨询顾问在参与融入粤港澳大湾区的过程中遇上"一带一路"的持续稳步发展，如何善用这个千载难逢的机会以满足三地建筑可持续发展的融合与共同参与国际建筑市场的机遇并符合联合国的环保、社会及管治政策 ESG 和 17 项永续发展目标 SDG。

智能建造与建筑工业化协同发展代表了建筑业高质量发展方向。大力支持建筑机器人及智能施工设备研发应用，同时融合应用大数据、云计算技术，设立建筑业大数据创新中心，实现行业数字化赋能，加快实现智能建造技术和产品的市场化应用，对推动建筑行业升级意义重大。福建工程学院土木工程学院尤志嘉等人描述了一套智能建造理论体系框架，将其划分为"基础理论、支撑技术、管理机制、参考架构、工作机理、集成方案、业务场景、运行机制、实施路径、核心目标、评价机制"11 项关键子领域，分别阐释其科学内涵并揭示其内在逻辑联系。在此基础上，展望智能建造领域未来的研究发展趋势，并识别潜在的研究方向；武汉称象科技有限公司熊彪及湖北第二师范学院宋金强等人研发了一套基于激光扫描的自动化检测系统，利用三维激光扫描仪采集毫米级精度的点云数据，并通过 4G/5G 网络快速上传到云端，在云端利用人工智能算法自动分析并生成质量报表。自动检测的指标包括平整度、垂直度、阴阳角、方正度、截面尺寸、楼板厚度等。经实验证明，该系统的测量精度完全满足当前行业规范，可有效支撑工程项目的提质增效；北京交通大学王雨桐等人以高速铁路无砟轨道板为研究对象，提出了预制轨道板多批次生产调度优化模型，优化配送批次和构件调度方案以实现按时交付及生产成本最少化。与传统生产方法相比，该模型能够降低生产成本，减少延迟、提前交付发生的可能性，推动高速铁路工程智能建造转型发展。

医养结合与健康住宅研究的序幕已拉开，国内外涌现出大量将医疗卫生服务体系与建筑业相结合的产品，这些产品可有效推动医疗卫生服务延伸至社区、家庭，真正实现以人为本的建筑理念。北京林业大学李叶等人通过 SWOT 态势分析法对智能养老家具的可行性进行分析，提出医养结合养老模式下的智能养老家具发展战略。最终总结了智能医养家具的安全性、易用性、多样性、情感化设计四项设计原则，并从建设智能控制系统、健康检测系统、信息交互适老化设计三个方面提出医养结合模式下智能养老家具的设计方法，并指导概念生成和设计实践；南昌航空大学伍昭亮等人为探索提升智慧城市韧性的路径，梳理智慧城市的构成要素和韧性城市的特征，定性分析二者之间的内在联系，并以"新冠肺炎疫情"为背景，从基础设施韧性、经济韧性、制度韧性、社会韧性等四个维度出发，对智慧城市在疫情期间的运营状况进行了综合分析，提出增强智慧城市韧性建设的对策与建议；西南石油大学刘红勇等人基于社会认知、积极组织行为、自我决定等相关理论，构建了心理韧性各子维度（情绪调节、现场应对、坚韧、自强、乐观）与安全行为之间的关系，并探讨了安全氛围的中介作用。

随着信息技术的快速发展，工程管理的教学模式也在不断创新。华南理工大学王幼松等人通过对比华南理工大学、东南大学、大连理工大学和重庆大学四大工学院工程管理专业的相关要素，对华南理工大学工程管理的专业特色进行提炼，并提出改进建议，研究结果也可为其他学校工程管理的发展提供参考；华南理工大学闫辉等人以课程的产教融合为突破口，以华南理工大学"建设工程造价管理"课程为例，对课程的教学改革与企业发展的合作模式进行探索，从前期准备、教学活动、实习实践和后续深化四个环提出课程产教融合的实施路径。研究成果也可为相关领域其他课程体系的改革提供方向和借鉴。

随着经济全球化的发展，我国建筑业逐渐走出大门。作为新型智能产业，国际视野逐渐成为我国建筑业产品发展升级的必要前提。英国诺丁汉特伦特大学的 Priyen Halai 等人通过访谈了解英国建筑业主要利益相关者对英国 2050 年净零目标的可实现性的看法，同时结合案例，分析了将英国建筑行业推向净零未来的可行性。研究结果表明，缺乏政府立法和激励措施，以及设计和生产净零建成资产的初始成本增加是该行业面临的最大障碍；英国诺丁汉特伦特大学的 X. F. Zheng 等人提出了一种专为住宅设计的热电联产系统，并在台架原型机上进行了测试以研究其性能。分析表明，提议的系统概念可提供合理的成本回收期（少于 5年）；山东交通大学的 Ying Li 等人以清澜高速公路莱泰段改扩建工程为例，系统介绍了工程安全管理制度以及该工程在疫情下采取的专项安全措施。

在智能建造已成为建筑产业转型升级的核心引擎这一背景下，建筑业的劳动密集型施工生产模式已悄然发生改变，"数字经济、人工智能、智慧城市、智能建造"依然成为研究主题。以上研究为推动建筑智能化变革提供了一定理论依据和实践基础，期望能更好地促进我国持续往数字化、信息化、智能化建设强国转变。

目 录

Contents

建筑工业化与建造机器人

医养结合与健康住宅

教学研究

海外专题

城市智慧基础设施

Smart City Infrastructure

论我国智慧城市基础设施建设管理

秦桂芳

（海口经济学院雅和人居工程学院，海南海口　571127）

【摘　要】 推进智慧城市建设，是城市提档升级、增强竞争实力的必要举措。基础设施是用于保证国家或地区社会经济活动正常进行的公共服务系统。它是社会赖以生存发展的一般物质条件，是智慧城市赖以生存和发展的重要基础，是新形势下推动城乡一体化建设的重要载体。基础设施的不断完善，对改善人民生活水平，提升智慧城市形象具有重要作用。近年来，随着经济社会的发展，智慧城市化进程的不断推进，城乡基础设施建设需求不断增加，城乡基础设施建设任务也不断加大。虽然我国智慧城市基础设施建设已经取得了很大成就，但与城市发展、人民需求相比，仍然处于相对滞后的地位，当前各地政府在城市建设和城乡基础设施建设管理方面还存在诸多与基础设施建设增长需求不相适应的地方。如何改革地方政府现有智慧城市基础设施建设的管理模式，是摆在我们面前最为突出的课题，本文从战略背景、发展形势、技术应用等方面，就新型智慧城市基础设施建设进行深入分析和探讨。

【关键词】 智慧城市；基础设施；建设；管理

1　智慧城市基础设施建设管理概述

1.1　智慧城市基础设施建设概念

智慧城市基础设施建设包括智慧城市供水、节水、供热、供气、公共交通、排水、污水处理、道路与桥梁、市政设施、市容环境卫生、垃圾处置和绿化、防洪等方面，不仅是智慧城市社会经济运行的基础，也是智慧城市现代化发展的前提。

1.2　智慧城市基础设施的分类

对于智慧城市基础设施的建设来说，可以根据不同的分类标准，划分为不同类别的基础设施。具体如表1所示。

基础设施分类　　　　　表1

设施类型	典型设施
公共事业基础设施	电力、电信、管道煤气、供水、卫生设施以及排污、固体废弃物收集与处理系统
公共工程基础设施	道路、大坝和灌溉及排水渠道工程
交通基础设施	铁路、交通、港口、水运和机场

2　智慧城市基础设施的特点

智慧城市基础设施是城市赖以生存和发展的基础。研究智慧城市基础设施建设管理，必然要了解它所固有的性质和特点。只有充分认识和掌握其性质和特点，我们才能更好地了解和运用。智慧城市基础设施的主要特点有以下几方面。

2.1　服务公共性和双重性

智慧城市基础设施的服务是面向整个城市的,是社会化的,不是面向特定的或个别的人群。因此,其具有服务的公共性。同时,按照公共产品理论界定,基础设施分为纯公共品、准公共品和非公共品。与此同时,它所提供的服务和产品,既是为了满足智慧城市各行各业生产所必需的,也是城市居民生活所必不可少的。因此,其又具有服务的双重性。

2.2　社会效益和经济效益

智慧城市基础设施建设的根本目的是保证城市各行各业生产和居民生活所需城市基础设施中的供水、能源设施能改善城市环境质量,从而带来环境效益。其交通、能源、防灾、绿化等设施,在方便生活、提高效率、美化环境、保障安全等方面都具有重要作用,具有明显的社会效益。与此同时,智慧城市基础设施中道路桥梁等设施的建设,可以加快相关企业部门的生产周转,降低运输成本,必然为这些部门和单位提高了经济效益。

2.3　建设的超前性和同步性

智慧城市基础设施项目一般规模比较大,建设周期比较长,大部分是基础工程或地下工程。随着工业化和智慧城市化的发展,城市不断扩展,城市人口逐步增长。城市大部分基础设施建设项目,如供水、供气等都具有一定的规模效应,不能随着城市的扩展而逐步建设,只能进行跳跃式、超前式建设。还有一部分基础设施,如道路桥梁等,扩宽增容费用高,难度大,也只能超前建设。只有这样才能满足城市发展需要,也才能与智慧城市其他建设同步

形成并协调发展。

2.4　垄断性和经营多样性

城市基础设施的垄断性,主要源于规模经济理论和公共产品理论。诸多经济学家认为,供水、供电等基础设施,存在规模经济和公共产品属性,由于其资源稀缺性等特点,独家垄断经营效益最高。如果允许多家企业竞争,就会造成规模经济损失,大大增加成本,甚至造成企业不能维持简单再生产。同时,将降低公共产品服务水平。一些外国学者认为,智慧城市基础设施社会最优目标是实现较高的生产效率和社会分配效率,即垄断企业以较低的成本向社会提供产品和服务,并包括正常利润在内的成本定价。而由私人企业垄断经营,难以实现社会最优目标,特别是难以保证社会分配效率。同时,在智慧城市基础设施领域也有社会化的准公共产品和非公共产品,其经营好坏、服务质量高低,并不取决于是否由政府垄断经营,这些产品也可由地方政府委托私人企业经营,只要政府给予合理的规制,即可使其朝着有益于提高经济效益的同时,达到提升社会效益的效果。因此,智慧城市基础设施又具有经营的多样性。

3　我国当前基础设施建设的现状和问题

随着经济水平的提高和城市化的发展,城市基础设施建设日益受到关注,全国各地建设项目的开展在一定程度上缓解了许多城市基础设施薄弱的状况,产生了良好的社会效益和环境效益。但是,当前我国城市基础设施建设依然存在诸多问题亟待解决。

3.1　重短期效益,轻长期效益

环境城市基础设施建设滞后。道路、供水、供电等生产性城市基础设施建设直接作

用于城市经济增长，短期经济效益明显，因此，受到城市政府的高度关注和大力投入。而污水处理、垃圾处理等环境基础设施虽关系到城市社会与生态可持续性发展，但因其经济效益的间接性和长期性使其滞后于经济社会发展需求，导致城市自然生态环境恶化、污染严重。

3.2 缺乏统一规划，综合协同能力差

在城市基础设施系统建设中，由于条块管理、规划不足等原因，给水排水、能源、交通等基础设施难以实现城市内部协调布局和区域间有效衔接，一定程度上导致发展失衡，综合效益无法得到充分发挥。

3.3 管理体制和运营机制不健全

城市基础设施效率的发挥很大程度上取决于管理。长期以来，由于我国城市基础设施实行高度集中的政府所有、垄断经营的管理体制，导致基础设施服务质量差、效率低、浪费严重，无法满足城市可持续发展对基础设施的需求。

3.4 建设资金不足

我国城市基础设施建设投资缺乏是一个长期存在的问题，城市基础设施建设投资的比例多年来均达不到合理的水平，逐年累积形成了巨额的投资欠账。同时，稳定规范的城市基础设施建设投资渠道一直未能有效建立起来，融资渠道单一，资金来源十分有限，资金严重缺乏。

3.5 设计、施工、监理招标投标不规范

在项目建设过程中存在的主要问题是项目设计、施工、监理招标投标工作不规范。许多地方建设主管部门与建设企业有着千丝万缕的联系，在对建设单位、监理单位进行公开招标的过程中难免存在着有违公开、公平、公正、诚实信用原则的行为。监理单位监管力度一旦不到位，建设企业就难以形成有效的自我约束机制，对工程质量的影响十分严重。

4 完善我国智慧城市基础设施建设管理的建议

智慧城市一直在想方设法地优化其基础设施建设管理。借助智能电网、楼宇智能化和智能交通系统等智能基础设施解决方案，助力中国城市充分挖掘城市基础设施管理的潜力。

城市化、人口增长和资源的日益减少，对城市基础设施建设管理造成的压力日渐增大，不仅是在中国。通过对欧美国家在基础设施领域改革的历程和经验做法可以看出，各个国家在推进基础设施领域改革时都是站在如何提高基础设施建设的效率和提高服务质量的立场上推进的，为我国基础设施建设管理和改革提供了借鉴。

4.1 树立智慧理念，系统谋划智慧城市基础设施实施步骤

智慧城市建设首先要树立智慧理念，只"智"缺"慧"的新城市构建模式偏重信息技术和硬件设施的组合，缺少对技术和硬件设施应用的结果进行考量，缺乏从问题导向来解决大城市病的方法。系统谋划智慧城市基础设施建设项目的实施，一是要从各个城市的实际问题出发，"一城一策"地创建智慧城市基础设施建设方案；二是做好智慧城市基础设施建设的顶层设计工作，注重网络化和公共形象平台的搭建，把能源、通信、给水排水、交通等基础设施载体建设好；三是探索智慧城市基础设施建设投融资体系，并建立智慧城市保障体系；四是强化典型智慧基础设施的应用。

4.2 提高管理水平，展开智慧城市基础设施建设规划

智慧城市的核心目标就是城市的可持续发展，如何利用信息技术以可持续发展为目标编制基础设施建设规划模式，将成为传统城市规划突破过于理想化的瓶颈、理性规划、推动城市合理发展的重要研究方向。要发挥智慧基础设施建设作为智慧城市的根基作用，必须合理布局信息基础设施，城市基础硬件包括城市公共设施、地下管线、电力布线等，城市基础软件包括平台、支撑硬件的软件，确保智慧城市基础设施科学规划、合理布局、统筹安排、分期实施。

4.3 整合行政资源，理顺智慧城市基础设施建设体制

目前智慧城市基础设施建设涉及多个管理部门，政府应组织电力、通信、管道煤气、水务、卫生设施以及排污、交通等部门建立智慧基础设施建设综合协调小组，协调解决智慧城市建设中的重点和难点问题。城市公共信息平台是智慧城市建设最重要的物理平台，要像城市的路网、能源系统一样，围绕着感知、共享、协同三个理念进行合理化改造，对公共信息的管理应当实行集中统一、信息共享、服务社会、保障安全的原则，最大限度地搭建智慧城市信息公共平台。

4.4 做好项目引进，不断推进重点领域智慧基础设施建设

一是要通过互联网、现代通信网和物联网把城市中的网络基础设施、信息基础设施、社会基础设施和商业基础设施连接起来，建设成新的智慧化基础设施；二是落实运营项目，包括智慧的交通、智慧的教育、智慧的各项事务

服务管理等；三是发展与物联网相关联的产业。

4.5 促进多元投资，探索智慧基础设施建设投融资

面对资金不足的问题，首先，拓展民间融资渠道，鼓励民间资本通过 BOT、BOO 等方式直接参与新基础设施项目的投资或通过契约管理方式参与存量基础设施项目的经营。其次，发展资本市场融资。充分利用债券融资，要推进预算法修改工作，明确城市政府市政债券的法律地位；着力发挥好企业债券和可转换债券的融资作用，扩大其融资总量。同时尽可能利用股票融资，通过对大型国有基础设施企业进行重组，在上市企业数量和质量上寻求新的突破，鼓励采用独资和股份制等多种形式的融资方式。最后，大力引进外资。我国智慧城市基础设施建设资金来源有限，基础设施产业自身又缺乏积累机制。因此，必须以促进多元化投资、多种融资渠道方式来解决城市基础设施建设所需资金。

5 结论

智慧城市基础设施建设是一项浩大的工程，不仅需要政策支持，还需要大量资金的注入。当前，在国内外的智慧城市建设中，基础设施建设主要是以政府投入为主体，辅以与实力强大的商业公司合作；战略规划与顶层设计是请商业公司、科研机构和智库进行，以确保方向的正确与实践的成效；应用领域开发是政府与商业公司水乳交融、不分上下。总之，只有社会各界的广泛参与，才能推动智慧城市基础设施建设的欣欣向荣与蓬勃发展。

参考文献

[1] 雷冰. 我国城市基础设施建设存在的问题及对策分析[J]. 山西建筑，2018(1)：238-239.

［2］　张振凯. 我国城市基础设施建设存在的问题及其对策分析［J］. 经济师，2017(3)：96-97.

［3］　李惠萍. 论我国城市基础设施建设存在的问题及对策［J］. 商业现代化，2018(7).

［4］　张颖. 新形势下城市基础设施运营模式探讨［J］. 科学观察，2018(6)：88-89.

［5］　郑吉. 中国城市市政基础设施水平综合评价［J］. 智慧城市规划，2019(4)：125-126.

［6］　杨政坤. 我国城市基础设施投融资现状分析［J］. 现代商贸工业，2018(15)：55-56.

［7］　王丽英. 我国城市基础设施建设与运营管理研究［D］. 天津：天津财经大学，2017.

［8］　殷晓梅. 我国城市基础设施建设薄弱环节的思考及对策［J］. 中国集体经济，2018（18）：76-77.

智能技术在机坪运行安全管理中的应用

丁晓欣[1]　刘力嘉[1]　乔嘉宁[2]　高林帅[3]

(1. 吉林建筑大学经济与管理学院，吉林　长春　130118；
2. 吉林省民航机场集团有限公司运行指挥中心，吉林　长春　130000；
3. 吉林建筑科技学院管理工程学院，吉林　长春　130114)

【摘　要】 随着民航旅客吞吐量的增长，航班起降架次的增多，以及"四型机场"建设的不断推进，传统的机坪运行安全管理模式已经不能满足当前需要，阻碍了机坪运行安全效率的提升。为了解决机坪运行安全管理中的人员复杂、车辆使用不规范、设施设备众多等问题，本文从机坪运行安全管理系统的视角，梳理机坪运行安全现状，分析存在的问题及成因，运用 BIM 技术、GIS 技术、大数据等智能技术，提出智能机坪运行安全管理系统解决方案，达到提高机坪运行安全管理效率的目的，并为智慧机场建设提供参考。

【关键词】 智能技术；机坪；运行；安全；管理

2020 年，我国民航局印发了《中国民航四型机场建设行动纲要（2020—2035 年）》，为实现民航强国推进建设以"安全、绿色、环保、智慧"为目标的四型机场提供了指导。各地机场旅客吞吐量持续增多，如图 1 所示。机坪运行安全作为机场运行管理中的重要环节凸显其重要性。管理人员工作效率低下、设备设施操作不规范等运行安全问题，随着智能技术的发展及其在智慧机场中的应用，将得到有效解决，同时必将促进机坪运行安全的高质量发展。

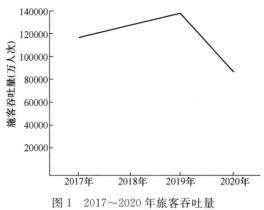

图 1　2017～2020 年旅客吞吐量

1　智能技术概述

本文借助智能技术，解决机坪运行安全管理问题。以 BIM、GIS、大数据、物联网和云技术等智能技术为载体，可实现机场机坪运行安全高效，杜绝安全隐患的发生，减少航班延误的同时减少机坪运行时出现的危险事件，提高管理水平，保证机坪运行高质量安全管理。

（1）BIM 技术

BIM（Building Information Model），又称建筑信息模型，其主要特点就是可以将二维图纸变为三维可视化，将纸质资料变成电子资料阅览，打破传统的二维思维的局限，有利于机坪运行全生命周期的安全管理。如北京大兴国际机场航站楼的运行维护管理就是运用 BIM 技术完成的[1]，同时也是首次将 BIM 技术运用到机坪航油管线当中[2]，这说明了 BIM 技术不仅能运用到航站楼的建设当中，对于管道建设和维护也可以做出突出贡献。

BIM 技术的可视化、协调性、模拟性、优化性特点可以对机场机坪运行安全进行全方位管控，便于各单位、环节查漏补缺，系统化管理。可以将多方信息整合到一起，实现机坪运行智能安全管理，提高运行效率，确保各方信息不丢失，保障机坪运行长期安全。

（2）GIS 技术

地理信息系统（Geographic Information System，GIS），有空间信息和位置信息准确的特点，且具有前瞻性和广阔的发展前景，如首都国际机场运用 GIS 技术搭建机场地理空间信息服务系统，将空间信息、基础信息、系统信息等纳入平台，实现数据共享管理[3]。

BIM 技术和 GIS 技术的有机结合，有助于机坪空间管理能力的提升，对后期运维管理有着重要作用，可以降低机坪运行安全中不安全事件的频率。

（3）大数据技术

随着物联网的迅速发展，智慧机场建设也离不开大数据，汪伟等人指出为解决运行资源的合理化，大数据在机场运行管理当中充当着重要的角色，保障飞行环境、海量数据分析、航班延误再安排、合理分配等[4]。

其数据管理和数据整合能力较强，海量的数据分析给机坪运行管理数据提供了新的管理思路，使工作人员可以更加直观地分析问题，能够及时反馈信息，减少机坪运行风险的发生。

（4）物联网技术

物联网技术可以实现物品向人类传递信息。通过无线射频识别（RFID）、传感器等可以对车辆、设备进行感知、定位、收集信息

等，并反馈给系统，通过与互联网的联系，进行实时管理。运用物联网技术可以精准化管理，可以解决航空器降落到再次起飞时，机坪上需要的车辆、设备、人员以及流程复杂的安全问题，减少危险事件的发生，智能技术给机坪运行安全带来了便利条件。

（5）云技术

云计算（Cloud Computing）是将大量的数据群体通过"云"进行处理，拥有强大的数据分析处理的能力，然后通过服务器各端传输给用户。机坪运行安全过程中有海量的数据，需要对这些数据进行分析。通过云计算技术可以打造一个云平台，减少运行成本，提高运行安全效率[5]。

2 机坪运行安全现状与存在的问题

我国机坪运行安全管理是依据其必须遵循的规章制度、法律法规来约束工作者的行为。如机坪运行安全过程中需要特种车辆进行保障作业，蒋丽就指出发生车辆事故的原因有人为操作错误、管理不到位等[6]。黄世明指出机坪运行要按照其规章准则，明确工作范围所对应的职责[7]。目前，传统的机场机坪安全管理模式都是在 SMS 安全管理体系之下，从事后分析事故的原因，容易忽略各个单位之间的协调能力且监管难度大，智能技术使用较少，已经不能适应智慧机场的发展。一方面，工作人员的业务能力低下导致工作低效[8]，设备老化、更新慢影响日常工作效率；运行保障流程复杂，保障单位较多易出现危险[9]，外来物侵犯风险大[10]，设施设备等维护成本高；另一方面，由于机坪场地面积大，出现问题时追踪溯源不及时，数据信息不准确，机坪监管部门管理难度大[11]。机坪运行安全具有特殊性，从以上来看这些都不同程度地影响着机坪运行安全，要想合理地解决这些问题，就要引入智能

技术为机坪运行安全打好坚实的基础。

3 智能机坪运行安全管理系统设计

该系统以 Revit 模型为基础，以物联网为载体，形成智能机坪运行安全管理系统，从各个模块实现对机坪运行安全的管控，旨在为机场管理机构提供高效便捷的工作方式，降低运行安全成本，减少事故的发生，让机坪更加畅快地运行工作。系统设计如图 2 所示。

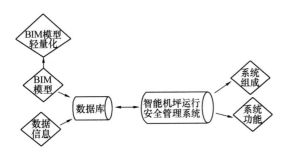

图 2　系统设计

3.1 BIM 模型的构建

根据智能机坪运行安全管理系统需求，将机场机坪这个既有建筑通过以往的 CAD 图纸、点云技术等将信息进行翻模，从而得到 BIM 模型数据库，主要包括几何信息、设施设备维护信息等。实现机坪全生命周期的信息共享，提高机坪运行安全管理水平及效率。

3.2 BIM 模型轻量化

模型轻量化处理可减少 BIM 模型中的数据过多造成浏览卡顿现象，减少智能机坪运行安全管理系统的负担，保障运行数据安全稳定。

3.3 系统的组成

智能机坪运行安全管理系统是以 BIM 技术的可视化特点为载体搭建而成的。系统通过 GIS 技术、物联网、大数据等智能技术统筹资

源最大化利用，与已有的视频监控系统等相结合，使得各个工作流程协同运作，实现对机坪人员、机坪车辆、机坪设施设备等的集成管理，解决运行当中遇到的安全难点问题，打破信息孤岛，及时了解机坪运行情况，为后期机坪运行安全的成本管理和维护管理打下坚实的基础。系统组成如图 3 所示。

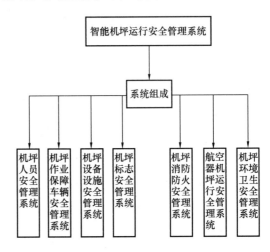

图 3　系统组成

3.3.1 机坪人员安全管理系统

机坪人员管理包括工作信息、奖惩情况、个人信息等。机坪人员管理将运用物联网技术和大数据技术的有机结合来实现，便于清晰掌握工作范围，并直接影响到机坪运行安全的保障。通过将场地 BIM 模型导入平台，可使得工作人员进行模拟作业，减少机坪运行安全隐患的发生，从而保证机坪运行安全。工作人员身份信息查看、工作信息的采集和整理汇总形成电子版信息，通过大数据可以将大量的数据整理在一起，形成数据库以便进行随时调取，为机坪运行安全的日常管理提供便利条件。

3.3.2 机坪作业保障车辆安全管理系统

机坪上的保障车辆一般分为特种作业车辆和一般车辆，其最重要的是车辆信息是否准确，保障作业分配是否合理，这一部分管理当中就要借助智能技术，通过无线射频技术、

GPS 系统等获取车辆运行信息，并且 GIS 技术对于空间管理占据优势，使工作人员实时监督车辆运行的标准化、规范化。保证车辆使用安全，减少车辆与人、车辆与车辆、车辆与航空器之间的碰撞，并将每次保障车辆的信息形成电子版文档上传到云平台以便后期调取。通过 BIM 技术可以了解车辆行驶路径是否准确，是否符合规章制度。

3.3.3　机坪设备设施安全管理系统

首先，场地 BIM 模型要进行轻量化处理；其次，与传感器和物联网进行对接，实时传送设施设备使用情况和维修巡检情况。设施设备的维护信息及时记录形成电子信息，及时反馈到管理层，有利于提高设施设备运行使用率，增强设施设备耐久性和可靠性，保证各类设施设备在工作期间始终处于适用状态，避免事故的发生，提高设施设备的使用寿命，降低运行成本。

3.3.4　机坪标志安全管理系统

《民用机场飞行区技术标准》MH 5001—2013 精准地规定了建设要求，对于机场规划、设计、建设是标准性的参考文件。机坪上的标志需要时刻保持清晰，不能处于遮蔽状态。采用 BIM 技术可以将机坪上的标志准确表达，并且工作人员可以通过三维模型提前知道工作区域位置及路线，防止产生工作错误和危险事件。

3.3.5　机坪消防防火安全管理系统

安全问题一直是每个民航工作者最关心的问题，从航站楼的消防管理到机坪运行的消防管理都是关键。通过 BIM 技术可以进行火灾模拟，让工作人员更好地了解消防通道，增强消防意识，保障机坪运行安全。

3.3.6　航空器机坪运行安全管理系统

航空器进港和离港流程都需要人员和保障车辆等的运用，可以通过人员、车辆定位进行航空器机坪运行安全管理。通过运用相应的

BIM 技术做出教学视频，进行有效学习，使工作人员熟悉保障作业流程，防止工作中出现失误。

3.3.7　机坪环境卫生安全管理系统

机坪环境卫生管理系统当中最重要的就是机场外来物（Fod）管理，按照机场日常管理规则进行机坪外来物管理。设施设备在机坪上的摆放位置规范、统一调配解决了机坪上使用设施设备杂乱无章的现象，通过工作人员的定时清理和监控视频的及时查看有助于提升对外来物的防范意识。

3.4　系统功能模块实现路径

（1）系统总体构架

智能机坪运行安全管理系统利用物联网技术，将组织架构分为数据层、应用层、展现层和用户层，如图 4 所示。

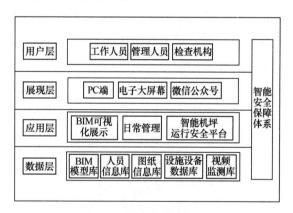

图 4　系统总体构架

数据层：包括 BIM 模型数据库、人员信息、设施设备信息、视频数据、图纸信息和民航局颁布的规章、制度、标准及法律法规。主要通过传感器等手段对机坪运行需要的基础信息进行全面感知。利用物联网进行数据交换收集。数据层的建立便于信息共享和日常管理的文件查找。设施设备和车辆数据通过接口处理对接传感器、监控设备等获取数据。

应用层：包括各个子系统的管理以及通过

BIM 可视化软件做出的场景展示管理、对 BIM 模型的浏览查询，满足基层人员学习需求，为机坪运行安全管理带来更大的价值。

展现层：通过物联网、大数据等智能技术，将系统各个模块通过终端进行展示，通过微信公众号和 PC 端进行实时管理，实时提供所需信息。充分展示智能技术的优势，便于监管部门工作人员监管，细化管理细则。

用户层：方便从上级到下级工作人员检查、信息查找，有效提供数据信息，管理系统全面安全可靠。在机坪运行管理当中，人员管理占据重要一部分，运用智能技术将信息收集整合到一起，用可视化贯穿整个运行周期，有效提高管理水平，最终达到统一协调管理，保证机坪安全运行。

（2）系统实现功能模块

1）数据信息的集成化管理：

对人、设备、车辆等信息的收集整理，建立电子信息库。对设施设备运行情况进行记录，细化管理，解决以往信息数据不准确的问题。机坪运行安全贯穿着机场运行的全生命周期，智能技术能准确高效地将数据信息进行梳理，便于管理，促进机坪运行安全畅通，减少航班延误。平台收集各方资料进行内部整合，数据信息清晰明了，查找问题时方便快捷，利于日后的整改。对于设备管理，可以实现数据信息共享，快速进行运行决策，资源合理配置，资源最大化利用，利于掌控全局。

2）可视化的运维管理，可视化展示提高工作人员安全意识：

通过数据信息模型进行可视化表达，并结合 VR、AR 进行展示，清晰直观全方位地表述，有效降低未来发生风险的可能性。减少机坪内事故发生造成的经济损失和时间成本，进而有效提高机坪运行的安全性。

3）提高工作人员工作效率，实行视频培训课程：

将机坪上发生的违反规章制度、不安全事件等通过草图大师软件建模，导入 Lumion 当中进行渲染，做成三维动画，使用 PR 等软件进行后期处理，指派专人讲解，如图 5 所示。既能避免类似不安全事件的发生，又能让员工熟记工作内容的规章制度，也可作为新员工入职培训案例使用。能有效提高工作人员认真学习的积极性；并且通过视频案例讲解，能够预防此类事件的再度发生，培养智能型机坪运行人员，有利于机坪运行安全的文化建设。

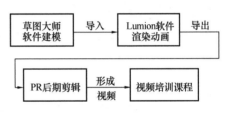

图 5　视频培训流程

4　案例分析

4.1　项目介绍

以某机场项目为例，探讨本文提出的智能技术在机坪运行安全可行性中的应用。该机场有 T1 和 T2 两个航站楼，占地 20.1 万 m^2，跑道材质为水泥，长 3200m、宽 45m。21 世纪初期机场通航，随着经济的增长、人流量的增多，机场也在不断扩建，2018 年 T2 航站楼投入使用，设计容量 1600 万人次。机场机坪人员、车辆、设备多，流动性大，流程复杂，对机坪运行安全管理提出挑战。课题组利用草图大师（Sketchup）、Autodesk Revit 等软件绘制机场模型形成数字化的机场建筑，如图 6 所示。

4.2　项目智能运行安全管理

以 Revit 模型为基础载体，根据三维模型和各种属性信息构建数据库，形成数字驾驶

图 6　航站楼 Revit 模型

舱，通过云计算、大数据等储存数据、分析数据，打造可视化管理平台，使机坪运行更加智能化、信息化，提高综合应用价值。

4.3　项目应用

（1）以三维可视化模型为基础，可以涉及智能机坪运行安全管理系统的各个子系统，模型与实际相符，工作准确率高。

（2）将最终信息汇总到模型当中，让模型始终保持与实际运行状态数据同步。设备设施、车辆的维护信息通过传感器上传到平台进行汇总，更加便捷的记录方式为后期的运行维护提供保障。

（3）为了提高机坪运行效率，降低事故发生率，通过将三维模型导入 Lumion 软件，剪辑生成动画视频，用于安全教育培训、风险防范培训、规章制度安全学习培训等。提高了机坪运行安全效率，保证机坪运行安全可靠，如图 7 所示。

图 7　机坪场景图

5　结语

建设智慧机场，将智能技术运用在机坪运行中是大势所趋。在实际工作当中，面对高消耗人力、低效率问题，能够充分发挥智能技术作用，构建机坪运行安全管理系统，形成正确的工作方式，打造高质量、高品质、高运行的机坪运行管理模式。提高工作人员的积极性和学习性，降低机坪运行中的资源浪费，从而打造机坪运行智能安全管理平台，实现运行安全、智能，达到机坪运行长治久安。

参考文献

[1]　徐洋，王志浩，冷婷婷，等. BIM 技术在机场航站楼项目全生命周期管理中的应用研究[J]. 项目管理技术，2020，18(8)：21-26.

[2]　田文钊，刘志文，等. 浅谈 BIM 在机坪航油管线中的应用[J]. 安装，2020(1)：48-52.

[3]　李涛. 地理信息系统在机场领域的应用研究[J]. 软件，2014，35(3)：43-44，51.

[4]　汪伟，李向明. 浅谈大数据在机场运行管理中的运用[J]. 民航管理，2018(1)：60-62.

[5]　马原. 云计算技术发展分析及其应用[J]. 电子技术与软件工程，2017(18)：177.

[6]　蒋丽. 特种车辆机坪运行安全研究[J]. 民航管理，2014(8)：73-74.

[7]　黄世明. 机坪安全运行的规章标准以及监管职责探讨[J]. 民航管理，2013(11)：22-28.

[8]　王云龙. 民用机场停机坪事故分析和预防[J]. 中国民用航空，2009(5)：59-61.

[9]　陈明亮，张元，陈艳秋. 基于安全绩效的机坪运行风险评估研究[J]. 民航学报，2018，2(6)：90-94.

[10]　王厚苏，丁甜. 航空器地面运行安全问题及应对措施探讨[J]. 民航管理，2020(6)：58-59.

[11]　张严峰. 中小机场安全运行管理[J]. 北京航空航天大学学报(社会科学版)，2019，32(6)：98-102.

城市轨道交通施工安全风险管理领域
知识情境维度研究

许　娜[1]　赵丽丽[1]　常　弘[1]　谢　颖[1]　李　解[2]

（1. 中国矿业大学力学与土木工程学院，江苏徐州　221000；

2. 内江师范学院建筑工程学院，四川内江　6410002）

【摘　要】[目的/意义]知识情境能够提升人们对知识的理解和减少知识过载，在知识管理中发挥着重要的作用。而目前情境维度的研究没有考虑到知识流动全过程所涉及的所有情境要素，进而导致无法对知识存储时的状态进行有效而全面系统的描述。当前，鲜有学者研究知识情境对城市轨道交通施工安全风险管理（以下简称"城轨安全风险管理"）的影响。[方法/过程]文章通过对有关知识情境的相关文献进行分析，旨在提出一个新的、全面的领域知识情境维度的分类框架；通过创建的领域知识情境分析方法，有效识别城轨安全风险管理的一系列情境要素；提出由频次、被引证次数和期刊影响因子三者共同构成的可信度值指标，用以判断领域知识情境要素分析结果的客观性和可信度。[结果/结论]文章提出了一个较为全面的领域知识情境维度的分类模型，并探讨了分类框架各个层级及维度特征。该模型包含三层：一级情境、一级情境衍生出来的16个维度的知识情境要素以及用于描述各个情境要素状态值的构成项。领域知识情境分类框架描述了知识流动全过程所涉及的情境维度，该框架对于提升城轨安全风险管理决策水平具有重要参考价值。与此同时，城轨安全风险管理领域知识情境分析方法也可应用到其他领域，并为其提供情境要素分析的方法支持。

【关键词】城市轨道交通；安全风险管理；知识情境；情境要素；内容分析法

1　引言

我国城市轨道交通建设规模和速度逐年增大，但是由于城市轨道交通项目施工作业技术复杂，且总是处于高密度的发展地区，施工难度较大，城市轨道交通事故率难以降低[1~3]。因此要推动快速、安全、稳定的城市轨道交通网络的建设，必须做好施工阶段的安全风险管理[4]。目前，关于风险评估理论和方法的研究日渐成熟，但是风险决策领域尚有许多问题亟待解决[5]。

从安全风险动态管理过程来看，信息检索是获取知识的核心，是知识管理系统的重要组成部分，安全风险管理者需要从各种渠道获取的大量非结构化数据中提取有用的信息，并在综合分析这些信息后作出及时、准确的安全风险决策，这一过程要求管理者具有丰富的专业知识和理论、实践深度。当前，项目组织内的知识普遍存在非结构化、无序的存储状态和储存位置不明等问题[6]，信息检索过程中，内容描述和查询技术主要是基于关键词而非模型。因此，有效捕捉和利用用户需求和内容含义的水平有限[7]。下面的场景说明了这些困难：一个建筑项目工程师在为加利福尼亚的一个项目场地选择合适的泥沙控制措施，他使用搜索引擎检索了有关泥沙控制实践的实例化方法。除此之外，他需要花费更多的时间和精力去搜寻在特定的情境中（即参与者、项目位置、地形、天气、施工顺序）支持决策的具体信息，这种信息的缺乏会导致决策效率低下、工作重复和成本增加[8]。临时性的知识收集所花费的时间与安全风险动态管理的快速决策形成矛盾，因此，绝大多数管理者都依靠项目经验和主观的判断进行风险决策，而不是依据知识进行理性、客观地分析，决策结果难免会发生偏差。专业人员需要相关和可靠的信息来支持有效的决策[9]。

城轨安全风险管理过程中，当组织内部人员在面临一些亟待解决的问题时，会通过知识获取、共享、创造、利用、储存等过程在组织内部搜索知识，促使知识流动才能解决问题。知识产生于特定的情境，并且知识价值的实现高度依赖其所属的情境[10]。知识所包含的情境不仅从过程的角度对知识进行了一次扩展性描述，而且也刻画了知识的个性化特征[11,12]，使得即使拥有同样内容的知识也会因为情境的不同而有所区别[13,14]，这样就更有利于我们区分和选择知识。因此，有必要将知识情境引入城轨安全风险管理过程中，在解决风险决策问题时就能把当前问题情境和知识情境进行对接和比较，择优选择将知识结构中能解决当前安全风险管理问题的知识提供给组织子系统中各参与方，从而实现安全风险管理的知识支持过程。

本文构建了一个城轨安全风险管理领域知识情境维度的分类框架，并通过对其中领域知识情境要素以及构成项的分析与识别，揭示了安全风险管理过程中项目参与方的知识需求，希望能够指导项目各参与方进行风险决策；除此之外本文的贡献还在于建立了一个广泛而灵活的知识情境要素识别方法，该方法可以根据研究领域的不同进行调整，具有普适性。

后续章节如下，第 2 节对目前国内外有关知识情境维度研究的现状进行了分析；第 3 节详细介绍了本文提出的知识情境框架和采用的分析思路及方法；第 4 节解释了如何利用文献分析和内容分析法得到领域知识情境要素的组成；最后一节我们探究如何将提取出的情境要素应用到一个程序中去解决问题。

2 国内外知识情境研究现状

情境是任何可用于描述实体、其条件或者周围情况的信息，并且这些信息与用户和活动之间的交互有关[15~18]。关于知识情境维度的研究得到了学者们的广泛关注[19]，大多数学者将情境信息大致分为个人、位置、时间、活动和关系五个类别[14, 20~22]。例如，Gary[23]在情境的两层次分析法中，用 Who、What、When、Where、Why 五个维度描述综合情境；Villegas[24]、Norha[16] 将情境描述为个人、位置、时间、活动和关系五个一般范畴，并指出其他的情境特征都可以从这些一般类别中实例化；潘旭伟[25]基于"5W"也进行了情境建模的研究。但是在进行知识选择的过程中，仅考虑那些与任务本身特性以及业务流程有关的情境要素，无法描述出知识的全部特征。因此，迫切需要对现有的情境维度进行扩展，定义一个能够覆盖多种情况的通用模型[26]。

在这种需求下，Florent[27]围绕企业目标建立了一个包含任务、任务活动、目标、对象状态和资源五个维度的概念模型。强调知识的多个参数，如活动、角色等都应该明确表述出来，并基于嵌套图的形式设计了多尺度知识库，用于情境知识的查询。Dhuieb[15]提出了处理制造业务问题的三维度知识情境模型：由操作情境、组织情境和用户情境三个维度组成，操作情境是用来确定在执行时间内工作人员所涉及的过程、活动和任务；组织情境是从组织的角度来确定工人的角色即相关信息；用户维度是由一组描述用户知识能力概况的参数组成。用其构建工作环境中可能发生的所有情况的情境存储库。上述研究虽然提出了与知识选择有关的情境要素，但是学者们对这些情境要素并没有达成广泛的共识，现有的结论也无法直接对情境要素进行整合。因此，我们的工作有必要对其加以整合以发挥知识情境最大的价值。

为了不断适应内外部环境变化以及市场的需求，情境本体应该具有可重复性并且能够被细化[28]，从而保证在复用过程中能够不断变化更迭来适应新的知识。Gary[23]指出离散情境是嵌套在综合情境中的，综合情境的效果是由离散情境变量或者他们之间的相互作用决定的。离散的情境适用于任何层次的分析。Dey[29]提出将情境分为初级和次级两个类型。实践、身份、地点和活动等主要情境完全可以捕获任何给定的情况，而其他情境都是次要的，可以从主要的情境派生出来。基于这种观点，周明建对影响企业知识管理的目标、资源、人员、组织、领域及时间五个维度进行建模，将其细化为若干子类别[30]。邬益男在基于领导者胜任力、团队协作氛围、现场管理、

绩效管理的需求四维度知识情境模型中明确将维度转化为可测量的问项[31]。Patrick[18]提出了适用于工业应用中情境类型的分类框架，包含用户、环境、系统、信息检索、模式识别在内的五个最重要的情境类别，将情境类别作为一个高层次的分类框架，在此基础上派生出特定于工业应用程序的上下文类型。由此可见，对不同维度的情境要素进行更精确的描述，使得知识更易于管理。

基于文献分析可以得出，情境维度框架至少应该分为三层，并且每层逐步细化对知识的描述。其中执行者、对象、时间、地点、目标、资源是描述知识必不可少的情境要素。与此同时，从知识流动的角度来看，研究大多关注的是知识获取、知识共享和知识创造的过程，忽视了知识储存时的组织内部知识数量的现状以及储存后知识数量的变化，如知识的类别是什么、知识数量有多少或者知识质量的好坏。如Dhuieb指出可以通过构建多尺度知识库来存储和标准化不同形式的知识，通过专家知识和显性知识的形式化来包含隐性知识。Adrian S[32]研究了一个包含基本情境和方法要素的质量框架来促进学习和知识创造，指出情境要素与探索性学习和隐性知识有关、方法论要素与开发性学习和显性知识有关。张发平[33]构建了包含任务和用户两个主维度以及衍生出来的一系列第二层维度的多维层次情境模型，并提出一个集成知识存储、管理、索引等的知识库。但是仅用知识编号、知识名称等来描述知识存储时的状态，显然无法描述出知识的全部特征。Dhuieb（2016）强调目前的知识结构化方法没有考虑参与者的认知能力这一维度，每个工人的能力水平应该分为新手、中级和专家三个层次，以便根据专业水平为其提供适当的知识集。Heisig（2009）和Rodrigo认为人力资源管理中最重要的是员工的数

量和质量，知识管理的主动性取决于人们愿意分享他们的知识和技能。没有合格的员工，组织就无法生产知识，群体决策可以使得问题易于解决[35]。文献[8]提出知识存量对知识的流动起到促进作用，当知识存量越多，解决问题时，我们就能查询到更多知识。

在某些情况下，缺乏情境信息会导致管理者无法选择恰当的知识，显然这些用来描述知识储存状态的要素也应该纳入情境维度模型中。但是目前并没有一个完整的知识情境框架包含所有上述知识储存要素。

3 领域知识情境要素分析思路及方法

3.1 领域知识情境维度模型建立

为了完整地表示知识利用的整个过程，本文提出了一个情境维度模型，如图1所示。该模型主要由三个描述知识流动过程的一级情境组成，每个一级情境都可以用若干维度的情境要素来描述。为了更精确地描述不同维度的情境要素，模型中将情境要素细化到了构成项级别。

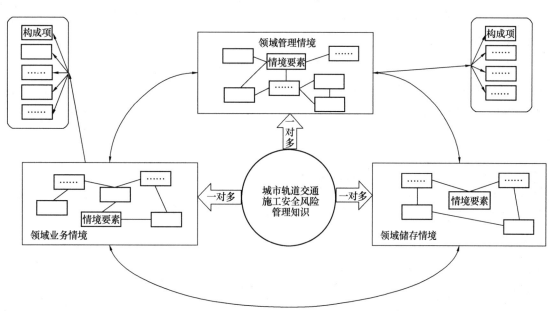

图1　领域知识情境维度模型

知识情境维度模型包含三个一级情境，用来描述知识产生前状态的领域管理情境，用于描述知识利用过程的领域业务情境以及记录组织内知识数量以及知识持续积累过程的领域储存情境。每个一级情境都是由若干与该过程相关的情境要素组成，即一级情境维度可以用若干个情境要素来刻画知识的个性化特征，相当于对知识进行了一次扩展性的描述。

此外，鉴于每种情境要素存在由多个状态值构成的状态域，用于阐述情境要素所包含的

所有可能性，如知识质量的高、中、低三个状态。因此可以利用构成项来明确表示情境要素的状态值，即用有限个相互区别的条目说明某一特定情境要素所包含的所有可能存在的情况及取值。知识产生于特定的情境中，明确了情境要素的构成项后，我们才能清楚地了解和记录知识究竟产生于这些情境中的哪种具体情况，更有利于知识支持的实现。

情境要素与知识之间是一对多的映射关系，一个知识可能与多个不同的知识情境对

应，例如，知识应用于某一执行过程，实现某一目标，面向某一对象，由某个执行者负责。但是知识不可能同时处于某个情境的两种状态中，因此知识与具体的情境要素构成项间呈现一对一的映射关系。初步构建的领域知识情境维度框架如图1所示。

3.2 情境要素分析思路及方法

在城市轨道交通施工安全风险管理过程中，各参与方管理者关注的是能够帮助其快速筛选领域知识的情境要素。为了能够快速找到适合特定用户的知识，需明确和领域知识内容有直接关系且被管理者关注的情境要素类别，才能将其纳入知识支持结构中。这一节将提出领域知识情境要素的分析方法，并用该方法识别和确定领域知识情境要素。分析思路及方法如图2所示。

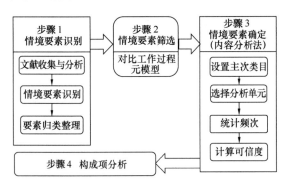

图 2　城市轨道交通施工安全风险管理领域知识情境要素分析思路及方法

领域知识场分析过程实质上为领域知识情境要素识别、筛选和确定的过程，其分析思路和方法如下：

（1）情境要素识别及归类：大量收集和阅读国内外研究知识场和情境的文献资料，识别文献中给出的各个情境要素；随后借鉴文献中的描述，根据领域管理情境、领域业务情境和领域储存情境的内涵对其进行归类[6]。

（2）情境要素初步筛选：工作过程模型是用于对工作流过程定义的基本元素和规则进行抽取，并加以一般性描述，用来指导工作过程建模、重组和执行。组织内管理者在其工作过程中最关心的是其工作范围所涉及的事务，为了精确地找到这一部分要素，我们将第一步中提取的情境要素与工作过程模型中涉及的有关工作过程的相关要素进行对比，一旦重叠则意味着该要素是管理者工作过程所关注的，可以用来区分领域知识，完成对领域知识情境要素的筛选。

（3）情境要素确定：借助内容分析法处理文献资料以确定领域管理情境、领域业务情境和领域储存情境所包含的各情境要素，利用内容分析法处理文献资料的内容详见3.3节。

（4）构成项分析：结合城市轨道交通施工安全风险管理实际情况确定领域知识情境要素的具体构成项。

3.3 内容分析法

内容分析法是一种从文本中做出可复制和有效推断的研究技术[36]。通过对文献资料内容所包含的信息或事实进行分析，并把研读结果借助频次量化指标反映在事先确定的研究范畴内，以完成对事实的认定与判断[37]，内容分析法分为四个步骤：设置主次类目、选择分析单元并编码、编码员统计频次、计算可信度。

（1）设置主次类目：主类目与次类目可参照领域管理情境、领域业务情境和领域储存情境的情境要素筛选结果确定。

（2）选择分析单元：确定能表达各个情境要素明确含义的语句、符号、句子或段落作为分析单元。

（3）对分析单元进行编码：为尽量避免主观误差，邀请三位工程管理专业从事安全风险管理相关研究的博士研究生担任编码者研读和编码所有的分析单元，研读完所有分析单元后即可统计各类目出现的频次并提出对于事实的判断。

（4）判别可信度：借助相互判别信度公式[38]判断结果的客观性和可靠性，如公式（1）所示。

$$R = \frac{n \times k}{1 + (n-1) \times k} \qquad (1)$$

式中，R 为相互判别信度；n 为参加内容分析的人数；k 为所有参与者平均相互同意度。

2 名参与者相互同意度如公式（2）所示。

$$K_{AB} = \frac{2M_{AB}}{N_A + N_B} \qquad (2)$$

式中，K_{AB} 为参与者 A 与参与者 B 的相互同意度；M_{AB} 为两者分析结果完全相同的分析单元数；N_A 和 N_B 分别代表两者分析和研读的分析单元数。

本文选取了三名编码员，这里只需按照公式（2）计算两者相互同意度的算数平均值即可得到三者的平均相互同意度。当相互判别信度 $R > 0.8$ 时，可认为分析的结果是可靠和相对客观的。

目前，内容分析法仅采用频次作为事实分析的指标存在较大缺陷，有可能忽略频次小但非常重要的类目。同样的道理，选定学术期刊文献为分析单位，要素 1 在水平较低且在未实行同行评议的期刊中出现了 5 次，而要素 2 则只在世界顶级期刊中出现过 1 次，显然要素 2 的重要性及可信度远大于要素 1。基于以上考虑，为弥补内容分析法面临的问题，需加入新的指标使分析结果更客观。丁佐奇[39]提到期刊影响因子和论文的被引证次数已被视为衡量期刊、论文质量、科技成果的学术意义甚至学者学术水平的"金指标"，是当前科研绩效评价体系中具有决定性作用的重要量化指标之一。因此，在频次的基础上加入期刊影响因子和被

引证次数用于判断分析单元的质量和确定分析结果的客观性和可信度，如公式（3）所示。

$$S = F \times C \times I \qquad (3)$$

式中，S 表示内容分析结果的可信度；F 表示频次；C 表示以文献资料为分析单位的被引证次数；I 表示文献资料来源期刊的影响因子。

4　领域知识情境要素分析过程与结果

4.1　领域知识场情境要素识别及归类

4.1.1　文献资料检索及统计

关于知识情境要素国内外学者已经进行了充分的讨论与研究，检索与研读相关文献资料可识别目前已被广泛认可和达成共识的知识情境要素。

Nonaka and Konno 曾提出"知识是否像其他资源一样能被管理"的问题，为进一步讨论该问题，他们提出"Ba"的概念，认为"Ba"是知识活动参与者共享的"Context"，包括知识产生和应用的背景、条件和环境等。国内学者在从事相关研究时则把"Ba"和"Context"翻译为"知识场"和"情境"，由此可把"Ba""Context""知识场"和"情境"作为检索关键词收集文献资料，使用的数据库包括：Sciencedirect，Emerald，CNKI。其中，从 1990 年 1 月到 2019 年 12 月在 Sciencedirect 上总计搜集 22561 篇文章，在 Emerald 上总计搜集 17023 篇文章，在 CNKI 上总计搜集 19963 篇文章。

为了限制检索的论文数量并找出与研究最相关的文章，本文基于下列检索原则：

（1）由于不同的行业特征和背景，导致所涉及的情境要素会有较大的区别，而本文所研究的城市轨道交通施工安全风险管理是以具体项目为依托并由项目组织通过协作共同完成的一项临时性努力和任务，阅读和筛选所搜集到

的文献资料,仅保留题目与摘要中提到的"项目""组织""任务""流程"和"业务"等关键词的文献。

(2)由于被引证次数与影响因子两个指标可以用来判断显性知识质量的高低[40]。这里将期刊的影响因子和文章的被引证次数作为筛选标准来进行筛选。

检索并筛选后共获得文献资料145份:其中中文文献49份,分别来源于EI、CSSCI、中文核心,如《计算机学报》《管理科学学报》《现代管理科学》等期刊;英文文献96份,来源于SSCI、SCI、EI,如 Journal of Knowledge Management、International Journal of Project Management、Creativity & Innovation Management 和 Knowledge Management 等期刊。大部分期刊的影响因子在0~6之间,其中最高达到11.25,并且这145份参考文献的平均被引证次数为165,最高的达到4172次。

4.1.2　领域知识场情境要素识别

研读145份文献资料,共识别知识情境要素28个,各情境要素都以不同的途径影响着组织内知识流动,如表1所示。根据领域管理情境、领域业务情境以及领域储存情境内涵进行归类,其中领域管理情境包括组织特征(规模、结构)、领导支持、组织文化/氛围、组织交流/沟通、组织位置、组织制度、组织激励、组织关系、参与方、组织角色/责任、角色关系任务、任务活动、时间、资源、任务目标和信息环境/技术17个情境要素,领域业务情境要素为产品/服务/对象1个。还有与知识本身有密切关系并从不同的角度描述了知识储存特征的情境要素,包括说明知识总体状况的知识存量、知识结构、储存位置、载体形式;用来判断隐性和显性知识差异性和质量的工作经验、技能/能力、教育程度、兴趣爱好和知识评价,以及简要说明只是内涵的内容描述。

领域知识情境要素　　　　　　　　　　　　　　　　　　　表1

序号	知识情境	知识情境要素	对组织知识流动的影响
1	知识管理情境	组织特征(规模、结构)	组织规模决定了员工数量,员工数量与电脑使用情况及采纳信息技术意愿呈现正相关关系,其在推动组织内信息技术的变革中发挥着关键作用,使知识流动获得信息技术的支持而对其产生积极正向影响
……		……	……
18	知识业务情境	产品/服务/对象	因为产品/服务/对象其中一个或者多个不同会给知识流动带来不同的影响
……		……	……
26		内容描述	精简的语句概括知识内容以提高知识获得效率
27	知识储存情境	储存位置	决定了知识获取的成本、时间和难易程度,相比于组织外专家的隐性知识而言,组织内数据库中的文档更易获取
28		载体形式	

4.2　领域知识场情境要素筛选

表1所识别的28个情境要素有些仅能促进知识流动,在描述知识的差异性上没有突出作用,因此,有必要对其加以筛选。本文将领域知识情境要素识别结果与Millie提出的工作

过程元模型进行对比并完成要素的筛选，突出情境要素中有关参与方所担任的角色为实现组织目标，在组织资源的支持下根据任务对象的

特征去执行具体任务的要素[42]，工作过程元模型如图3所示。

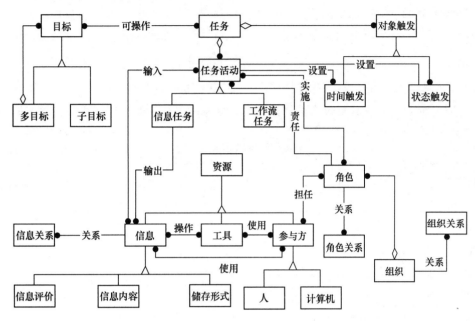

图3 工作过程元模型

对比图3和表1筛选知识场情境要素，如表2所示。

领域知识情境要素筛选结果　　表2

领域知识情境	情境要素	工作过程元模型要素	是否与知识情境要素一致
领域管理情境	目标	目标	是
	任务	任务	是
	任务活动	任务活动	是
	资源	资源	是
	信息环境/技术	工具	是
	参与者	参与者	是
	组织角色/责任	角色/责任	是
	角色关系	角色关系	是
	组织关系	组织关系	是
领域业务情境	产品/服务/对象	对象	是

续表

领域知识情境	情境要素	工作过程元模型要素	是否与知识情境要素一致
领域储存情境	储存形式	储存形式	是
	知识结构	信息关系	是
	技能		是
	工作经验		是
	教育程度	信息评价	是
	知识存量		是
	知识评价		是
	内容描述	信息内容	是

表2表明，共有18个领域知识场情境要素与工作过程元模型要素内涵一致。

4.3　领域知识情境要素确定

为消除文献分析和与工作过程元模型对比

过程中主观判断对筛选结果的影响及误差,利用内容分析法定量化处理145份文献资料,借助可信度量度判断所有研究学者对筛选结果是否持有支持的态度和一致性的认知,最终确定已形成共识且适用城市轨道交通施工安全风险管理的情境要素。

1) 设置领域管理情境、领域业务情境和领域储存情境三个主类目,其所包含的情境要素为次类目;2) 从上述145份参考文献中选择除了能够代表学者主要态度和观点的结论性句子522项,将其作为分析单元;3) 邀请3位工程管理专业人士对提取的分析单元进行独立编码,记录编码结果和出现不一致认识的分析单元数量,并统计各类目出现的频次,提出对事实的判断;4) 利用式(1)~式(3)计算情境要素的可信度和相互判别信度,如表3所示。

领域知识情境要素可信度　　　　　　　　　表3

领域知识情境 (主类目)	情境要素 (次类目)	可信度值	可信度级别	相互判别信度	处理结果
领域管理情境	目标	17299	万级	0.959	保留
	任务	39432	万级	0.953	保留
	任务活动	22197	万级	0.945	保留
	资源	27984	万级	0.962	保留
	信息环境/技术	38671	万级	0.948	保留
	参与方	20460	万级	0.943	保留
	组织角色/责任	38461	万级	0.982	保留
	角色关系	974	百级	0.934	删除
	组织关系	284	百级	0.951	删除
领域业务情境	产品/服务/对象	28725	万级	0.972	保留
领域储存情境	储存形式	777	百级	0.938	删除
	知识结构	873	百级	0.927	删除
	技能	39290	万级	0.985	保留
	工作经验	37128	万级	0.984	保留
	教育程度	15386	万级	0.984	保留
	知识存量	11726	万级	0.973	保留
	知识评价	20550	万级	0.941	保留
	内容描述	16412	万级	0.966	保留

内容分析法的研究和应用中对于效度的探讨和分析较少。内容分析法多为描述性质的研究,因此进行内容效度研究是有必要的[43]。首先,本文构建的领域知识场情境要素是建立在大量已有的内容紧密相关的研究基础之上的。无论是从理论上还是实践中都具有充分的依据;其次,主类目和次类目的概念界定十分清晰且界限分明,都来源于大量的以往研究;再次,编码人员均为工程管理专业从事安全风险管理相关研究的博士研究生,对涉及类目的确切含义有清晰的认识且能够有效地进行判别。最后,本文为了弥补内容分析法有可能忽略频次小但非常重要的类目,在频次的基础上加入期刊影响因子和被引证次数用于判断分析单元的质量

和确定分析结果的客观性和可信度。表3中互相判别信度全部都大于 0.8，也佐证了分析过程基本排除了编码者的主观影响，因此，该内容分析结果具有较高的内容效度水平。

4.4　领域知识情境要素构成项

由表2确定的所有领域知识情境要素中，目标、任务、任务活动及参与方等要素结合城市轨道交通施工安全风险管理实际实施情况即可明确其构成项，如参与方为施工方、设计方、监理方等。类似地，可将参与方要素分为建设单位、勘察单位、监理单位等8个部分；安全管理角色责任可以为负责、监管、参与三大类，参与方按各自的角色来执行责任，其构成项为：建设单位监管勘察单位的初步设计、详细勘察；项目管理单位监管勘察单位的施工勘察等50项责任；安全风险管理工作任务目标构成项由安全风险识别、估计、评价、控制和跟踪5项构成；任务构成项包含初步勘察、初步勘察阶段安全风险管理等22项任务内容；将任务参与方与对应的活动进行逻辑搭配可明确任务活动构成项由初步勘查阶段组织设计单位提出勘察的技术要求、建设单位提出初步勘察方案审查意见等424项；知识资源要素有制度类、标准与规范类等7个构成项。根据我国相关法规规定，将员工受教育程度分为初中及以下、高中/中专、本科/专科等5个构成项；将技能掌握情况分为无职称、初级职称等4个构成项。

在城市轨道交通施工安全风险管理背景下，产品/服务/对象表现为施工业务及其相对应的安全风险因素状态。施工业务构成项参考WBS分解准则和命名规则将其描述为明挖车站、矿山法区间、盾构法区间3个单位工程和10个分部工程、36个子分部工程以及84个分项工程。借鉴现场和质量管理的方法，把

4M1E法引入安全风险管理中来明确事故发生过程和确定安全风险因素分类及关系，可以将事故的发生解释为人、机、料、法、环五种安全风险因素中某些因素共同作用的结果[42]。从这五个角度去分析收集的158份事故报告（来自国家安全生产监督管理总局、住房和城乡建设部、各省市相关的行政管理部门及安全管理行业学习交流等平台），分析其结果并统计事故的安全风险因素类型组合，即可确定安全因素的构成项是由人法、人法环、人料环等31种情况组成。类比安全风险因素构成项的确定方法，通过对住房和城乡建设部工程质量安全监管司的26个城市轨道交通在建项目调研，可将项目参建人员的配置情况划分为9个区间以确定员工数量构成项的组成；通过对项目中各单位员工从业年限的统计将员工工作年限构成项分为3个区间；载体数量构成项借助洛特卡定律来估计，可以分为小于10、10～20等9个构成项；关键词分为1～9等9个构成项，用以简要地表达物理载体的核心内容。被引证次数与影响因子两个指标可以用来判断知识质量的高低[39]，通过对收集的145篇"情境"论文的被引证次数的统计将其划分为9个区间，也即9个构成项。影响因子由被引证次数和物理载体总数量之比求得，因此其构成项也可分为9个。

知识产生于特定的情境中，明确了情境要素的构成项后，我们才能清楚地了解和记录知识究竟产生于这些情境中的哪种具体情况，更有利于知识支持功能的实现。

4.5　领域知识情境维度模型

确定了城市轨道交通施工安全风险管理领域知识情境要素及其构成项之后，将3.1节提出的情境维度框架补充完整。如图4所示，图中的双箭头表示领域知识与领域管理情境、领

域业务情境和领域储存情境所包含的知识情境要素之间的集成关系。当管理者获取、共享、创造和储存知识时，即可利用给定的构成项明确地记录其所处的具体情境状态。当知识复用

时，各种情境构成项都有助于区分和筛选知识，相当于为知识提供了除内容分类外的其他分类维度，从而满足安全风险管理过程不同的知识需求。

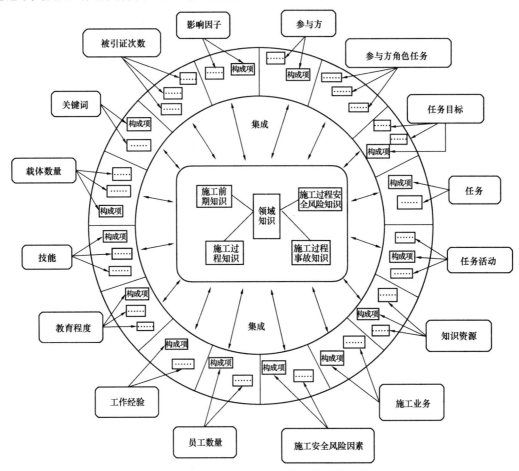

图 4　城市轨道交通施工安全风险管理领域知识情境维度模型

5　领域知识情境维度模型应用案例

徐州地铁三号线，项目全长约 18.13 km，共设 16 座车站，全部为地下车站，3 号线是徐州规划开通的第一批地铁线路。为了保证其施工过程中能够安全高效地进行安全风险管理，基于本文的知识情境维度模型（本体中知识类的概念和术语都来源于城市轨道交通施工安全风险管理知识支持结构中的领域管理场、领域业务场、领域储存场的情境要素及要素构成项），结合本体、数据库和 Java 等技术为地

铁三号线施工过程开发了城市轨道交通施工安全风险支持系统，设计的城市轨道交通施工安全风险管理知识支持系统人机交互界面如图 5 所示，包括用户登录、项目概况、新建项目、隐性知识储存、专家黄页、专家主页、知识问题、显性知识储存、知识仓库、知识引证、安全风险识别、安全风险评估、安全风险控制、安全风险跟踪和安全风险查询 15 个界面。

当管理者实施施工安全风险管理时，管理者不同的个体特征和任务特征将催生出不同的知识需求，此时正是施工安全风险领域知识区

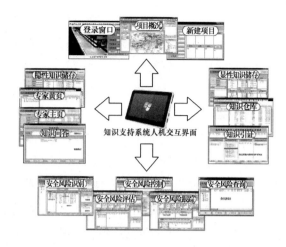

图5　知识支持系统人机交互界面示意图

别性复用的时机。首先通过查看项目概况，了解目前施工安全风险管理工作的实施情况、工作成果及安全风险基本信息（与每个具体的施工业务对应）；然后通过风险识别、风险评估、风险控制及风险跟踪等过程，厘清安全风险所在的施工业务、安全风险因素构成、风险类型、可能导致的事故、发生概率、事故后果。知识支持系统是按照"任务目标"情境要素分别设置风险识别、风险评估、风险控制和风险跟踪四个独立界面，同时从数据库中匹配与当

前系统用户（管理者）有相似"工作经验""教育程度"和"技能"的专家及其所储存的显性知识载体，保证每个界面中系统只推送与当前具体任务目标有关且能让管理者理解和接受的知识。其中施工安全风险识别界面划分为工作区域和知识支持两个区域，在工作区域内设置选择和操作"施工业务"和"施工安全风险因素"两个知识情境要素的树形图和下拉菜单；在知识支持区域设置如知识资源类型、任务、任务活动等知识情境要素和领域知识的下拉菜单，其他如"隐性因子"和"被引证次数"等则和知识一同显示，以便引导管理者选择与当前其所处情境相似的历史情境要素，及时获取能辅助其做出安全风险决策的显性知识与隐性知识，以实现知识对安全风险识别的支持。最后提出控制措施和应急措施等，以保证施工处于相对安全的状态。

以风险识别过程为例，说明该系统是如何利用情境维度框架模型去调用储存在数据库中的各种知识。风险识别界面如图6所示。

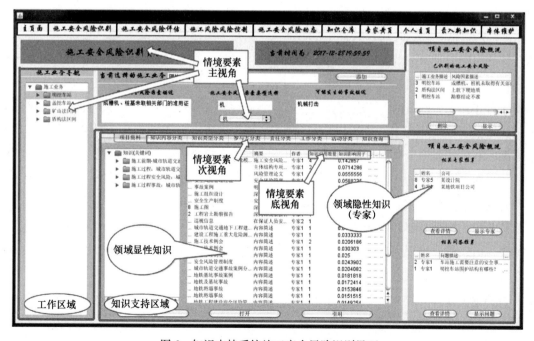

图6　知识支持系统施工安全风险识别界面

风险识别是风险管理的第一步，该步工作主要内容是明确风险发生的位置、存在哪些风险及风险类型，确定风险有可能导致的事故。当管理者开始识别某项施工业务下具体类型的施工安全风险时，如识别明挖车站基坑围护中施工机械可能存在的安全风险，在"任务目标"对应的树形图和下拉菜单中选择同样的内容，系统自动调用领域知识与领域知识情境要素集成（本体属性表示）关系，在知识支持区域显示与之对应的能被管理者理解且帮助其了解当前施工业务和判断机械安全风险的隐性知识与显性知识，如返回基坑围护施工的技术工艺原理、施工技术参数、质量控制关键点、机械故障及故障排除、机械安全防护、机械技术性能、机械安全操作及机械设备管理等领域知识所关联的各种显性知识载体，熟悉这些领域知识的专家资料及相关问答，然后管理者结合具体的个人情况及工作特征，进一步利用知识支持区域的情境要素下拉框选择最为符合的情境要素构成项，如管理者为施工单位，期望了解施工机械管理任务中机械设备安全检查任务活动环节所产生的案例类知识，此时知识载体被再次筛选与过滤，最后管理者仅需查看剩余知识载体的影响因子和被引证次数即能挑选出满足其知识需求且质量较好的知识载体。

管理者在知识支持条件下依次完成安全风险识别、评估、控制及跟踪，可在安全风险界面中查询所有的施工安全风险。查看施工安全风险时可以在安全风险选择方框内选择按照风险类型查询、按照施工业务查询或者查询某个具体施工业务的某种类型的安全风险。当确定好查询条件后，符合要求的施工安全风险就会显示在安全风险显示区域，显示内容包括安全风险所在的施工业务、安全风险因素描述、风险类型、事故发生概率、事故后果、控制措施和应急措施等，点击对应的按钮查看对应风险

的控制措施和应急措施具体内容，以指导管理者对施工过程做出合理安排和有效监控，确保安全风险管理目标的实现。

在知识使用时，管理者在开展安全风险管理工作时把工作中从组织内部与外部不同来源的知识都通过显性、隐性知识储存界面储存在数据库中，后续在处理相似情境的安全风险决策问题时，就能借助系统推送或者主动查看的方式在核心工作界面的知识显示区、专家黄页界面、专家主页界面、问答界面、知识仓库界面和知识引证界面中获取和复用有助于风险决策判断的显性知识与隐性知识。而领域知识情境保证了领域知识能通过人机交互界面准确且及时地提供给安全风险管理者，使其在适合的时候获得合适知识的支持，帮其完成风险决策，实践证明本文提出的框架模型相对完整且具有较强的实用性。

6　结论及展望

文章将知识情境的概念引入城市轨道交通施工安全风险管理领域中，填补目前城市轨道交通施工安全风险管理领域知识情境的研究空白，为后续研究奠定基础。

（1）本文从知识流动的角度提出领域知识情境三层维度框架模型，包含了知识从产生到储存全过程所涉及的所有情境。借助领域知识情境要素分析方法和参考城市轨道交通施工相关国家标准、制度、书籍及项目原始资料等文件，确定的16个情境要素及其构成项充分描述了领域知识的背景和环境，有益于管理者在特定的施工安全风险管理情境中获取所需的知识去支持其风险决策，为做出风险决策提供了有效的提升路径。

（2）文章基于文献分析和内容分析法创建的情境要素分析方法及构成项分析方法可应用于知识推送、业务问题求解等方向，并为其提

供情境要素分析的方法论支持，研究人员可以基于此思路方法构建相应的系统。

（3）提出由频次、被引证次数和期刊影响因子三者共同构成的可信度值指标，弥补了以往内容分析法在结果客观性和可信度判别上的不足。

领域隐性知识是高度个性化的知识，很难以规范的格式和内容传递给其他人，如了解工程地质情况的管理者凭借多年工程经验和个人直觉得出施工过程有可能会发生地面沉降风险，但让其把所应用到的知识明确地表达和有逻辑地说明却相当困难。未来可设置一个更为全面的专家系统，在结构化领域隐性知识的同时，将对应的专家信息进行存储，以便管理者结合自身情况和具体工作进行隐性知识与情境的集成并存储。

参考文献

[1] Yu Q Z, Ding L Y, Zhou C, et al. Analysis of factors influencing safety management for metro construction in China[J]. Accident Analysis & Prevention, 2014(68)：131-138.

[2] Qian Q, Lin P. Safety risk management of underground engineering in China：progress, challenges and strategies[J]. Journal of Rock Mechanics and Geotechnical Engineering, 2016, 8(4)：423-442.

[3] Ding X, Yang X, Hu H, et al. The safety management of urban rail transit based on operation fault log[J]. Safety Science, 2017(94)：10-16.

[4] Yan H, Gao C, Elzarka H, et al. Risk assessment for construction of urban rail transit projects[J]. Safety Science, 2019(118)：583-594.

[5] 熊自明，卢浩，王明洋，等. 我国大型岩土工程施工安全风险管理研究进展[J]. 岩土力学，2018, 39(10)：3703-3716.

[6] 李解. 城市轨道交通施工安全风险管理知识支持机制及方法研究[D]. 北京：中国矿业大学，2018.

[7] Fernández M, Cantador I, López V, et al. Semantically enhanced Information Retrieval：An ontology-based approach[J]. Web Semantics Science Services & Agents on the World Wide Web, 2011, 9(4)：452.

[8] Zhang Lu, El-Gohary, Nora M. Epistemology-Based Context-Aware Semantic Model for Sustainable Construction Practices[J]. Journal of Construction Engineering & Management, 2016.

[9] Incorporated. Improving Management of Transportation Information[M]. 2013.

[10] Hong J-y, Suh E-h, Kim S-J. Context-aware systems：a literature review and classification[J]. Expert Systems with Applications, 2009, 36(4)：8509-8522.

[11] Belkadi F, Anis Dhuieb M, Vicente Aguado J, et al. Intelligent assistant system as a context-aware decision-making support for the workers of the future[J]. Computers & Industrial Engineering, 2019.

[12] Waqar A, 邵杰, Aman K A, 等. 上下文感知推荐系统：挑战和机遇[J]. 电子科技大学学报，2019(5)：655-673.

[13] 秦亚欧，刘岩. 虚拟社区知识转移的情境研究[J]. 情报科学，2015, 33(3)：41-44.

[14] Verbert K, Manouselis N, Ochoa X, et al. Context-aware recommender systems for learning：a survey and future challenges[J]. IEEE Transactions on Learning Technologies, 2012, 5(4).

[15] Dhuieb M A, Laroche F, Bernard A. Context-awareness：a key enabler for ubiquitous access to manufacturing knowledge[J]. Procedia CIRP, 2016(41)：484-489.

[16] Villegas N M, Sánchez C, Díaz-Cely J, et al. Characterizing context-aware recommender sys-

tems：a systematic literature review［J］. Knowledge-Based Systems，2018(140)：173-200.

［17］ Pradeep P，Krishnamoorthy S. The MOM of context-aware systems：a survey［J］. Computer Communications，2019(137)：44-69.

［18］ Rosenberger P，Gerhard D. Context-awareness in industrial applications：definition，classification and use case［J］. Procedia CIRP，2018 (72)：1172-1177.

［19］ Assefa Tsehayae A，Robinson Fayek A. Developing and optimizing context-specific fuzzy inference system-based construction labor productivity models［J］. Journal of Construction Engineering ＆ Management，2016：04016017.

［20］ Villegas N M，Sánchez C，Díaz-Cely J，et al. Characterizing context-aware recommender systems：a systematic literature review［J］. Knowledge-Based Systems，2017：S0950705117305075.

［21］ Xu M，Liu S. Semantic-enhanced and context aware hybrid collaborative filtering for event recommendation in event-based social networks ［J］. IEEE Access，2019(7)：17493-17502.

［22］ Sundermann C，Domingues M，Sinoara R，et al. Using opinion mining in context-aware recommender systems：a systematic review［J］. Information (Switzerland)，2019，10(2).

［23］ Johns G. The essential impact of context on organizational behavior ［J］. The Academy of Management Review，2006，31(2)：386-408.

［24］ Villegas N M，Müller H A. Managing dynamic context to optimize smart interactions and services［C］. 2010.

［25］ 潘旭伟，顾新建，程耀东，等. 集成情境的知识管理模型［J］. 计算机集成制造系统，2006 (2)：225-230，263.

［26］ Wolff，Karin. Engineering problem-solving knowledge：the impact of context［J］. Journal of Education ＆ Work，2017，30(8)：1-14.

［27］ Laroche F，Dhuieb M A，Belkadi F，et al. Accessing enterprise knowledge：a context-based approach［J］. CIRP Annals，2016，65（1）：189-192.

［28］ 密阮建驰，战洪飞，余军合. 面向企业知识推荐的知识情境建模方法研究［J］. 情报理论与实践，2016，39(4)：78-83，59.

［29］ Dey A K. Understanding and using context［J］. Personal ＆ Ubiquitous Computing，2001.

［30］ 周明建，赵建波，李腾. 基于情境相似的知识个性化推荐系统研究［J］. 计算机工程与科学，2016，38(3)：569-576.

［31］ 邬益男，战洪飞，余军合. 基于需求层次分析的知识情境模型构建及实证研究［J］. 工业工程，2019，22(2)：73-81.

［32］ Choo A S，Linderman K W，Schroeder R G. Method and context perspectives on learning and knowledge creation in quality management［J］. Journal of Operations Management，2007，25 (4)：918-931.

［33］ 张发平，李丽. 基于多维层次情境模型的业务过程知识推送方法研究［J］. 计算机辅助设计与图形学学报，2017，29(4)：751-758.

［34］ Dominguez Gonzalez R V，De Melo T M. The effects of organization context on knowledge exploration and exploitation［J］. Journal of Business Research，2018，90(9)：215-225.

［35］ Zeng J，Zhang W，Matsui Y，et al. The impact of organizational context on hard and soft quality management and innovation performance［J］. International Journal of Production Economics，2017，185(3)：240-251.

［36］ Krippendorff K. Content analysis：an introduction to its methodology. sage，thousand oaks krippendorff K (2011) principles of design and a trajectory of artificiality［J］. J Prod Innov Manage，2004(28).

［37］ Moldavska A，Welo T. The concept of sustainable manufacturing and its definitions：a content-analysis based literature review[J]. Journal of Cleaner Production，2017（166）：744-755.

[38] 杨菊萍，贾生华. 企业迁移的动因识别——基于内容分析法的研究[J]. 地理科学，2011(1)：15-21.

[39] 丁佐奇，郑晓南. 期刊影响因子、论文被引证次数与学术质量评价的矛盾分析[J]. 中国科技期刊研究，2009，20(2)：286-288.

[40] 盖斯勒. 科学技术测度体系[M]. 北京：科学技术文献出版社，2004.

[41] Kwan M M，Balasubramanian P R. Dynamic work flow management：a framework for modeling workflows[J]. System Sciences，1997.

[42] 刘国愈，雷玲. 海因里希事故致因理论与安全思想因素分析[J]. 安全与环境工程，2013，20(1)：138-142.

[43] Holsti O R. Content analysis for the social sciences and humanities[J]. Addison-Wesley Pub.

基于责任激励约束下的智慧产业项目中
社会资本方参与合作的演化博弈研究

田昌民[1]　陈　群[1,2]　陈　哲[1]　乔丽丽[1]　李佳昕[1]

（1. 福建工程学院管理学院，福建福州　350118；2. 宁德职业技术学院，福建宁德　355099）

【摘　要】　政府实施责任激励是促使社会资本方自发履行责任合作，有效避免合作中出现道德风险机会主义行为的有效举措，对社会资本资源发挥最大效用和助力智慧城市建设与发展具有重要作用。针对社会资本方在合作行为中易出现道德风险的机会主义行为问题，本文提出从政府实施责任激励约束行为的角度构建了"政府—社会资本方"的智慧产业项目合作的演化博弈模型，并进行均衡策略分析。研究表明：政府只要确保监督成本和激励成本低于合作共赢带来的额外收益，就可保证实施责任激励约束的动力，通过采取加大对社会资本方机会主义行为的惩罚力度和提高履行责任奖励的责任激励约束实施手段，可引导社会资本方避免机会主义行为的出现，有利于促进社会资本方提高自身履行责任意识，在项目合作中表现较强责任感。

【关键词】　智慧产业项目；责任激励约束；演化博弈；智慧城市

1　引言

　　智慧城市是物联网、云计算、大数据等先进信息技术与城市化的高度融合[1]，其建设过程依托于智慧产业项目来输送血液为智慧城市发展提供动力，有助于智慧城市更好地发展。政府为了大力推广智慧城市建设，给企业起到良好引领作用，开创了许多关于智慧产业的政府性项目，为该城市的智慧化发展提供动力，为此，如何激励社会资本方积极参与智慧产业政府性项目建设是值得关注的问题。因为智慧城市的建立需要智慧产业的发展，而智慧产业相关项目仅靠政府单方面财政投入容易导致政府财力不支。在我国当前智慧城市建设中，智慧产业项目的资金投入问题成为当前制约智慧产业发展规模扩大的一大瓶颈，也是智慧城市建设过程中的重大问题[2]。因此，解决智慧产业政府项目资金投入问题，是有效促进智慧产业发展，为智慧城市建设提供动力的有力措施。

　　在相关研究中，众多学者认为吸引社会资本参与资金投入是解决其资金问题的有效途径，其中，侯羽[3]认为利用政府科技资金在城市智慧产业的发展中起着重要的先导作用和杠杆效应，一定程度的财政资金可以撬动多倍的社会资本投入智慧产业的发展中来。钱斌华[4]认为加大企业所得税、个人所得税、增值税等税种的税收优惠，提高社会资本对智慧产业发展进行风险投资的动力。杨俊杰[5]认为通过特许经营、委托经营等方式可吸引社会资本投资建设智慧产业项目。满青珊和孙亭[6]认为通过PPP方式可为智慧城市建设提供充足的资金支持。张静[7]认为利用完善基础环境、拉动重点应用以及政策指导等不同方式,可引导社会资本流向智慧产业。

基于以上文献梳理发现,吸引社会资本参与来解决智慧产业建设主要依靠政府管理或税收政策等方式,能缓解智慧产业相关项目的资金短缺问题,使政府有限的财政资源发挥最大效用。但是为了吸引社会资本参与其中,利用政府的财政优惠倾斜政策在后期合作中将会存在一些问题,尤其是由于社会资本方的盈利天性以及合作信息的不对称性,导致社会资本方在与政府合作的利益博弈问题上,易出现道德风险和机会主义行为。因此,政府分析和设置合理约束机制来规范社会资本方的合作行为最为关键。在相关研究中,高华[8]等基于风险偏好和公平偏好的分析,认为政府部门的激励水平与社会资本方的公平偏好系数呈反向关系,与其风险规避度呈正向关系;黄思韵[9]认为政府激励对社会资本在提高技术创新和减少机会主义行为方面有促进作用。Li和Cai[10]认为在需求不确定的情况下政府激励可以促使社会资本以高能力和低价格来参与项目。刘平[11]认为建立社会资本的风险分担与风险补偿制度是完善智慧产业风险投资机制和加快构建“投、保、贷”一体化的智慧产业投融资体系的有效途径。

基于以上文献梳理发现,在对社会资本方易出现的道德风险和机会主义合作行为问题的研究中,大部分都是在考虑政府如何设置风险分担与风险补偿激励制度来更好地应对社会资本方的风险偏好和公平偏好,提高社会资本方合作积极性。但很少有文献对社会资本方责任意识激励进行考虑,而社会资本方参与智慧产业项目建设的努力程度受到政府激励措施外在影响的同时,还受到自身责任意识的内在激励影响[12]。因此,本文针对社会资本方在与政府合作中易出现道德风险和机会主义行为的问题上构建了考虑政府对责任合作行为激励约束的演化博弈模型来提高社会资本方在智慧产业建设项目上的合作动力,避免社会资本方出现道德风险的机会主义合作行为。

2 模型构建与求解

2.1 模型假设

假设1:在本研究的模型中,博弈的双方分别为政府部门和社会资本方,均为有限理性“经济人”。假定不考虑其他因素,将模型简单化,在智慧产业建设项目的合作中政府有两种行为选择:实施责任激励约束和不实施责任激励约束。同理,对于社会资本方来说也存在两种行为选择:履行责任合作行为和出现道德风险的机会主义合作行为。因此,博弈的合作机制为:(实施责任激励约束,履行责任合作)、(实施责任激励约束,机会主义合作)、(不实施责任激励约束,履行责任合作)、(不实施责任激励约束,机会主义合作)。

假设2:在有限理性的情况下,政府选择不实施责任激励约束的概率为 x,实施责任激励约束的概率为 $1-x$,其中 $x \in [0,1]$,社会资本方选择积极合作的概率为 y,机会主义行为的概率为 $1-y$,其中 $y \in [0,1]$。

假设3:智慧产业建设项目合作的政府基本收入为 S_1,基本支出为 B_1,社会资本方履行责任时政府的额外收益为 D_1,政府对社会资本方进行责任激励为 F,社会资本方采取机会主义行为时政府的机会主义损失为 W,政府对社会资本方的责任惩罚为 T,责任监督成本为 C。

假设4:与智慧产业建设项目合作的社会

资本方基本收入为 S_2，基本支出为 B_2。社会资本方履行责任的收益为 D_2，履行责任成本为 E，社会资本方采取机会主义行为时额外收益为 W，机会主义成本为 J。

基于以上假设，构建博弈收益矩阵，见表1。

演化博弈模型的支付矩阵　表1

		社会资本方	
		履行责任合作 $H_1(y)$	机会主义合作 $H_2(1-y)$
政府	不实施责任激励与约束 $G_1(x)$	$S_1-B_1+D_1$ $S_2-B_2+D_2-E$	S_1-B_1-W S_2-B_2+W-J
	实施责任激励与约束 $G_2(1-x)$	S_1-B_1-F $-C+D_1$ S_2-B_2+F $+D_2-E$	$S_1-B_1-W+T-C$ S_2-B_2+W $-J-T$

在政府和社会资本方的博弈支付中，当社会资本方采取机会主义合作时，会造成政府 W 的损失，而社会资本方会得到不合理收入 J，这会使得利益分配不合理，利益不均衡。当社会资本方履行责任合作时，能使得双方利益均衡。当政府实施责任激励约束，社会资本方采取履行责任合作时能使项目利益最大化，在保证智慧产业建设项目收益的基础上，为智慧产业建设项目带来了 D_1 的额外收益。所以，当双方的合作采取（实施责任激励约束、履行责任合作）时能形成双方利益均衡，智慧产业建设项目利益最大化的良性合作机制。为了使智慧产业建设项目能在良性合作下达到双赢，让利益相关者利益均衡，智慧产业建设项目收益最大。下面将通过对政府和社会资本方的行为和合作机制进行演化博弈分析，系统全面地分析其演化均衡策略，进而为智慧产业建设项目实现利益均衡、项目整体收益最大化的良性演化提供理论依据。

2.2 演化模型构建

2.2.1 政府收益及复制动态方程

政府采取"不实施责任激励约束"的期望效益为：

$$U_{(G_1)}=y(S_1-B_1+D_1)+(1-y)(S_1-B_1-W) \quad (1)$$

政府采取"实施责任激励约束"的期望效益为：

$$U_{(G_2)}=y(S_1-B_1-F-C+D_1)+(1-y)(S_1-B_1-W+T-C) \quad (2)$$

政府平均期望效益为：

$$U_{(G)}=xU_{(G_1)}+(1-x)U_{(G_2)} \quad (3)$$

政府选择"实施责任激励约束"策略的复制动态方程为：

$$F(x)=\frac{\mathrm{d}x}{\mathrm{d}t}=x[U_{(G_1)}-U_{(G)}] \quad (4)$$

政府策略的复制动态方程表现了实施责任激励约束策略和不实施责任激励约束策略的群体演化过程，反映出复制动态方程的意义：假设实施责任激励约束策略的效益比平均效益高，那么群体中选择策略将随博弈的开展比例上升。

将式（1）和式（3）代入政府选择"实施责任激励约束"策略的复制动态方程（4），可以得到：

$$F(x)=\frac{\mathrm{d}x}{\mathrm{d}t}=x[U_{(G_1)}-U_{(G)}]$$
$$=x(1-x)[y(F+C)+(1-y)(C-T)] \quad (5)$$

2.2.2 社会资本收益及复制动态方程

社会资本方采取"履行责任合作"的期望效益为：

$$U_{(H_1)}=x(S_2-B_2+D_2-E)+(1-x)(S_2-B_2+F+D_2-E) \quad (6)$$

社会资本方采取"机会主义合作"的期望效益为：

$$U_{(H_2)} = x(S_2 - B_2 + W - J) + (1-x)(S_2 - B_2 + W - J - T) \qquad (7)$$

社会资本方平均期望效益为：

$$U_{(H)} = y U_{(H_1)} + (1-y) U_{(H_2)} \qquad (8)$$

社会资本方选择"履行责任合作"策略的复制动态方程为：

$$
\begin{aligned}
F(y) = \frac{dy}{dt} &= y[U_{(H_1)} - U_{(H)}] \\
&= y(1-y)\{-x(F+T) \\
&\quad + [F + D_2 - E - (W - J - T)]\}
\end{aligned}
\qquad (9)
$$

2.3　演化模型求解

式（5）与式（9）组成的方程组为该博弈的动态复制系统。令式（5）、式（9）分别等于 0，即

$$
\begin{cases}
F(x) = \dfrac{dx}{dt} = 0 \\[2mm]
F(y) = \dfrac{dy}{dt} = 0
\end{cases}
\qquad (10)
$$

得到两组稳定状态的解为：

$$x_1 = 0, x_2 = 1,$$
$$x_3 = \frac{(D_2 - E) - (W - J) + (F + T)}{F + T} \qquad (11)$$

$$y_1 = 0, y_2 = 1, y_3 = \frac{T - C}{F + T} \qquad (12)$$

3　模型的均衡策略分析

3.1　复制动态方程分析

3.1.1　政府复制动态方程分析

（1）当 $y(F+C) + (1-y)(C-T) = 0$，即 $y = \dfrac{T-C}{F+T}$。因为 $T > C$，$F + C > 0$，所以 $0 < \dfrac{T-C}{F+T} < 1$。此时始终等于 0，所有 x 都

处于稳定。也就是说，当社会资本方采取"履行责任合作"的比例为 $y = \dfrac{T-C}{F+T}$，政府采取"实施责任激励约束""不实施责任激励约束"两种策略收益相同没有差别。如图 1 所示。

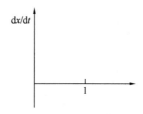

图 1　$0 < y = \dfrac{T-C}{F+T} < 1$ 时，政府的复制动态相位图

（2）当 $y < \dfrac{T-C}{F+T}$ 时，$U_{(G_1)} - U_{(G)} < 0$，这时若想 $\dfrac{dx}{dt} = 0$，则需 $x_1 = 0$ 或 $x_2 = 1$。所以，x_1 和 x_2 是 x 的两个动态稳定点。又因为 $U_{(G_1)} - U_{(G)} < 0$，所以，政府采取"实施责任激励约束"的效益低于平均水平，所以，此时 $x_1 = 1$ 即政府采取"不实施责任激励约束"是此复制动态情况下的演化稳定策略，在此状态下即使少数的个体选择"实施责任激励约束"，也会在学习中改变，从而具有抗突变能力。而如果 $x_2 = 0$ 的情况下，政府选择采取"不实施责任激励约束"，就会发现此行为选择的效益高于原有策略，就会通过不断的博弈、试错、学习而采用"不实施责任激励约束"的策略。如图 2 所示。

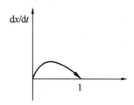

图 2　$0 < y < \dfrac{T-C}{F+T} < 1$ 时，政府的复制动态相位图

（3）当 $y > \dfrac{T-C}{F+T}$ 时，$U_{(G_1)} - U_{(G)} > 0$ 与上述情况相同，需 $x_1 = 0$ 或 $x_2 = 1$。又因为

$U_{(G_1)} - U_{(G)} > 0$，所以，政府采取"实施责任激励约束"的效益高于平均水平，所以此时 $x_2 = 0$ 即政府采取"实施责任激励约束"是此复制动态情况下的演化稳定策略。如图 3 所示。

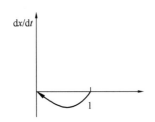

图 3　$0 < \dfrac{T-C}{F+T} < y < 1$ 时，政府的

复制动态相位图

综上所述，智慧产业建设项目合作的演化稳定策略中，$y = \dfrac{T-C}{F+T}$ 是 $x_1 = 0$ 和 $x_2 = 1$ 这两个演化稳定的分界点，所以政府对社会资本在智慧产业建设项目中的责任监督成本 C 越小，对社会资本方在合作过程中采取机会主义合作行为处以责任惩罚力度 T 越大，政府则采取"实施责任激励约束"的动力越大。而对于责任激励成本 F 来说，由于责任惩罚力度 T 同时存在于分子和分母位置，又因为 $C < T$，所以 F 的变化会引起分子产生更多变化，即激励成本 F 变小会让政府采取"实施责任激励约束"的动力增大，但是它产生的影响因素相对较小。而在实际项目中这样的博弈可以表现为，当政府责任监督成本 C、责任激励成本 F 越小，对机会主义采取的机会主义惩罚力度 T 越大，获得的额外收益越大，政府就会在不断的博弈学习中采取"实施责任激励约束"策略，而政府责任监督成本 C、责任激励成本 F 越大，对机会主义采取的机会主义惩罚力度 T 越小，获得的额外收益越小，政府会在不断的博弈学习中采取"不实施责任激励约束"策略。

3.1.2　社会资本复制动态方程分析

（1）当 $-x(F+T) + [F + D_2 - E - (W - J)$

$-T)] = 0$，$x_3 = \dfrac{(D_2 - E) - (W - J) + (F + T)}{F + T}$

时，因为 $(D_2 - E) < (W - J)$，$(W - J) - (D_2 - E) < (F + T)$，所以 $0 < \dfrac{(D_2 - E) - (W - J) + (F + T)}{F + T} < 1$，此时始终等于 0，此时所有 y 都处于稳定。也就是说，当政府采取"实施责任激励约束"的比例为 $x = \dfrac{(D_2 - E) - (W - J) + (F + T)}{F + T}$，社会资本方采取"履行责任合作""机会主义合作"两种策略收益相同没有差别。如图 4 所示。

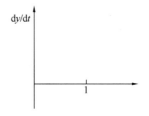

图 4　$0 < x = \dfrac{D_2 - E - (W - J)}{F + T} < 1$ 时，社会

资本方的复制动态相位图

（2）当 $x < \dfrac{(D_2 - E) - (W - J) + (F + T)}{F + T}$

时，$U_{(G_1)} - U_{(G)} > 0$，这时若想使 $\dfrac{dy}{dt} = 0$，则需 $y_1 = 0$ 或 $y_2 = 1$。所以 $y_1 = 0$ 和 $y_2 = 1$ 是 y 的两个动态稳定点。又因为 $U_{(G_1)} - U_{(G)} > 0$，所以，社会资本方采取"履行责任合作"的效益高于平均水平，所以此时 $y_1 = 1$ 即社会资本方采取"履行责任合作"是此复制动态情况下的演化稳定策略。如图 5 所示。

（3）当 $x > \dfrac{(D_2 - E) - (W - J) + (F + T)}{F + T}$

时，$U_{(G_1)} - U_{(G)} < 0$ 与上述情况相同，需 $y_1 = 0$ 或 $y_2 = 1$。又因为 $U_{(G_1)} - U_{(G)} < 0$，所以社会资本方采取"履行责任合作"的效益低于平均水平，所以此时 $y_2 = 0$ 即社会资本方采取"机会主义合作"是此复制动态情况下的演化稳定策略。如图 6 所示。

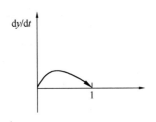

图5 $0 < x < \dfrac{D_2 - E - (W - J)}{F + T} < 1$ 时，
社会资本方的复制动态相位图

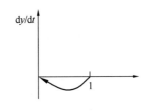

图6 $0 < \dfrac{D_2 - E - (W - J)}{F + T} < x < 1$ 时，
社会资本方的复制动态相位图

综上所述，智慧产业建设项目的演化稳定策略 $x = \dfrac{(D_2 - E) - (W - J) + (F + T)}{F + T}$ 是 $y_1 = 0$ 和 $y_2 = 1$ 这两个演化稳定的分界点，所以社会资本方在智慧产业建设项目中履行责任合作的收益越大，机会主义收益 W 越小，机会主义成本 J 越高，以及机会主义受到的惩罚 T 和履行责任合作获得的奖励 F 越大，则社会资本方采取"履行责任合作"的动力越大。在实际项目中这样的博弈可以表现为：当履行责任合作的收益越大，机会主义收益 W 越小，机会主义成本 J 越高，机会主义惩罚 T 和责任激励收益 F 越大时，社会资本方会在不断的博弈学习中采取"履行责任合作"策略。而履行责任合作的收益越小，机会主义收益 W 越大，机会主义成本 J 越低，机会主义惩罚 T 和责任激励收益 F 越小时，社会资本方会在不断的博弈学习中采取"机会主义合作"策略。

3.2 稳定性策略分析

当 $0 < \dfrac{(D_2 - E) - (W - J) + (F + T)}{F + T} < 1$，$0 < \dfrac{T - C}{F + T} < 1$，该博弈的动态复制系统有五个平衡点，分别为 $Q_1(0,0)$、$Q_2(0,1)$、$Q_3(1,0)$、$Q_4(1,1)$、$Q_5\left(\dfrac{(D_2 - E) - (W - J) + (F + T)}{F + T}, \dfrac{T - C}{F + T} \right)$，其中 $F + T \neq 0$。

依据以上的博弈动态关系和复制动态方程组，运用雅克比矩阵对演化博弈的渐进稳定性进行分析，雅克比矩阵如下：

$$J = \begin{bmatrix} \dfrac{dF(x)}{dx} & \dfrac{dF(x)}{dy} \\ \dfrac{dF(y)}{dx} & \dfrac{dF(y)}{dy} \end{bmatrix}$$

$$= \begin{bmatrix} -(2x-1)(C-T+Fy+Ty), & -x(x-1)(F+T) \\ y(y-1)(F+T), & -(2y-1)(D_2 - E + F + J + T - W - Fx - Tx) \end{bmatrix}$$

根据矩阵局部分析法，对均衡点进行稳定性分析，其判断依据为是否满足 $DetJ > 0$ 以及 $TrJ < 0$，从而判断均衡点是否为局部稳定状态，结果如表2所示。

演化博弈模型均衡点的 $DetJ$ 和 TrJ　表2

均衡点	$DetJ$	TrJ
$Q_1(0,0)$	$[(C - T)(D_2 - E) - (W - J) + (F + T)]$	$C + F + (D_2 - E) - (W - J)$
$Q_2(0,1)$	$-(C + F) \times [(D_2 - E) - (W - J) + F + T]$	$-[(D_2 - E) - (W - J) + (T - C)]$
$Q_3(1,0)$	$-(C - T) \times [(D_2 - E) - (W - J)]$	$(D_2 - E) - (W - J) - (T - C)$
$Q_4(1,1)$	$(C + F)[(D_2 - E) - (W - J)]$	$-[(D_2 - E) - (W - J) + (C + F)]$

续表

均衡点	$DetJ$	TrJ
$Q_5(Q_x^*,Q_y^*)$	$(C-T)(C+F)[(D_2-E)-(W-J)]\times[(D_2-E)-(W-J)+(F+T)]/(F+T)^2$	0

由假设可知，$(D_2-E)<(W-J)$，$(W-J)-(D_2-E)<(F+T)$，$F+C>0$，$C<T$。

(1) 当 $(C+F)>(W-J)-(D_2-E)$，$T-C>(W-J)-(D_2-E)$ 时，根据矩阵局部分析法，对五个均衡点进行稳定性分析，其判断依据是看是否满足 $DetJ>0$，$TrJ<0$，从而判断均衡点是否为局部稳定状态，结果如表3所示。

演化博弈的均衡点及稳定性　　表3

均衡点	$DetJ$	TrJ	结果
$Q_1(0,0)$	−	+	不稳定
$Q_2(0,1)$	−	−	不稳定
$Q_3(1,0)$	−	−	不稳定
$Q_4(1,1)$	−	−	不稳定
$Q_5(Q_x^*,Q_y^*)$	+	0	鞍点

(2) 当 $(C+F)>(W-J)-(D_2-E)$，$T-C<(W-J)-(D_2-E)$ 时，根据矩阵局部分析法，对五个均衡点进行稳定性分析，其判断依据是看是否满足 $DetJ>0$，$TrJ<0$，从而判断均衡点是否为局部稳定状态，结果如表4所示。

演化博弈的均衡点及稳定性　　表4

均衡点	$DetJ$	TrJ	结果
$Q_1(0,0)$	−	+	不稳定
$Q_2(0,1)$	−	+	不稳定
$Q_3(1,0)$	−	−	不稳定
$Q_4(1,1)$	−	−	不稳定
$Q_5(Q_x^*,Q_y^*)$	+	0	鞍点

(3) 当 $(C+F)<(W-J)-(D_2-E)$，$T-C>(W-J)-(D_2-E)$ 时，根据矩阵局部分析法，对五个均衡点进行稳定性分析，其判断依据是看是否满足 $DetJ>0$，$TrJ<0$，从而判断均衡点是否为局部稳定状态，结果如表5所示。

演化博弈的均衡点及稳定性　　表5

均衡点	$DetJ$	TrJ	结果
$Q_1(0,0)$	−	−	不稳定
$Q_2(0,1)$	−	−	不稳定
$Q_3(1,0)$	−	−	不稳定
$Q_4(1,1)$	−	+	不稳定
$Q_5(Q_x^*,Q_y^*)$	+	0	鞍点

(4) 当 $(C+F)<(W-J)-(D_2-E)$，$T-C<(W-J)-(D_2-E)$ 时，根据矩阵局部分析法，对五个均衡点进行稳定性分析，其判断依据是看是否满足 $DetJ>0$，$TrJ<0$，从而判断均衡点是否为局部稳定状态，结果如表6所示。

演化博弈的均衡点及稳定性　　表6

均衡点	$DetJ$	TrJ	结果
$Q_1(0,0)$	−	−	不稳定
$Q_2(0,1)$	−	+	不稳定
$Q_3(1,0)$	−	−	不稳定
$Q_4(1,1)$	−	+	不稳定
$Q_5(Q_x^*,Q_y^*)$	+	0	鞍点

综上4种情况分析可知，当 $(D_2-E)<(W-J)$，$(W-J)-(D_2-E)<(F+T)$，$F+C>0$，$C<T$，满足 $0<\dfrac{D_2-E-(W-J)}{F+T}<1$ 和 $0<\dfrac{T-C}{F+T}<1$ 要求，依据矩阵局部分析法，对五个均衡点进行稳定性分析，其判断依据是看是否满足 $DetJ>0$，$TrJ<0$，从而判断均衡点是否为局部稳定状态，结果如表7所示，演化博弈相位图如图7所示。

均衡点	$DetJ$	TrJ	结果
$Q_1(0,0)$	−	±	鞍点
$Q_2(0,1)$	−	±	鞍点
$Q_3(1,0)$	−	−	不稳定点
$Q_4(1,1)$	−	±	鞍点
$Q_5(Q_x^*,Q_y^*)$	+	0	鞍点

演化博弈的均衡点及稳定性　　表7

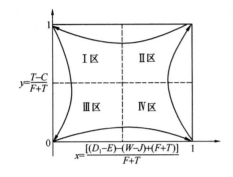

图7　演化博弈相位图

根据相位图和均衡稳定性分析可以看到该博弈根据初始状态存在不同的均衡情况，但不存在演化稳定策略：①当政府和社会资本方的最初选择情况落在Ⅰ区域时，博弈结果将向（0，0）演化。即政府采取实施责任激励约束，社会资本方采取机会主义合作，也就是合作机制（不实施责任激励约束、机会主义合作）将成为群体参与者所有人的最终决策。②当政府和社会资本方的最初选择情况落在Ⅱ区域时，博弈结果将向（0，1）演化。即政府实施责任激励约束，社会资本方采取履行责任合作，也就是合作机制（实施责任激励约束，履行责任合作）将成为群体参与者所有人的最终决策。③当政府和社会资本方的最初选择情况落在Ⅲ区域时，博弈结果将向（1，0）演化。即政府采取不实施责任激励约束，社会资本方采取机会主义合作，也就是合作机制（不实施责任激励约束、机会主义合作）将成为群体参与者所有人的最终决策。④当政府和社会资本的最初选择情况落在Ⅳ区域时，博弈结果将向（1，1）

演化。即政府采取不实施责任激励约束，社会资本方采取履行责任合作，也就是合作机制（不实施责任激励约束、履行责任合作）将成为群体参与者所有人的最终决策。也就是说如果政府用于责任监督和责任激励的成本高于政府采取责任激励约束和社会资本方达到双赢产生的高效运转的收益，就会使得政府的责任激励约束的动力不足。也会使得双方的博弈不存在恒定的演化稳定策略，从而落入了一个循环。这也是在智慧产业建设项目合作过程中最容易出现的问题，即政府责任激励约束力度不够—社会资本方产生机会主义合作—政府加大责任激励约束力度—社会资本方履行责任合作—政府责任激励约束减弱—社会资本方增加机会主义。

所以，政府在实行责任激励约束的过程中，减少责任激励约束的成本 C 和 F 是确保政府做好责任激励约束的前提条件。只有通过将责任激励约束所产生的成本转化成为社会资本方履行责任合作创造的额外收益，才能推动智慧产业建设项目合作的高效运转，创造社会效应，才能走出循环的怪圈，发挥社会资本的最大效用。

在此情况下，虽然不能形成稳定的演化博弈策略，但初始状态Ⅰ能让政府和社会资本方达成良性合作机制，在利益均衡的基础上创造更多的项目价值和社会收益。为了扩大Ⅰ的范围，则需分界线 $x = \dfrac{(D_2-E)-(W-J)+(F+T)}{F+T}$ 向右偏移，$y = \dfrac{T-C}{F+T}$ 向下偏移。则初始选择落在Ⅰ区域的机会也将变大。为了使 $x = \dfrac{(D_2-E)-(W-J)+(F+T)}{F+T}$ 变大，就需要政府的责任激励 F 和社会资本方履行责任合作带来的超额收益 D_2 提高，同时让社会资本方采取机会主义合作行为的收益变小，其机会主义成本增大。$y=\dfrac{T-C}{F+T}$ 变小,则需要政府加大对社会资本方的责任惩罚力度 T,提升责任激励成本 F。

3.3 博弈均衡结果分析

根据演化博弈模型的演化路径分析,要想使得政府和社会资本方的合作达成最优演化稳定策略(实施责任激励约束、履责合作)。一方面政府首先要确保监督成本 C、激励成本 F 低于合作共赢带来的额外收益 D_1,保证实施激励约束的动力,其次要利用责任激励约束手段,加大对社会资本方投机行为的打击力度,并给予履行责任合作的奖励,引导社会资本方的积极合作,避免机会主义行为的出现。另一方面,社会资本方要增强自身责任意识,具有较强的社会责任感,才能刺激政府的责任激励约束行为。只有通过核心利益双方的积极配合,才能使得智慧建设项目利益均衡,整体利益最大化。

4 案例验证

前文针对智慧产业建设项目的合作的演化博弈分析,分析了政府与社会资本方的决策过程,并求解出公私部门演化稳定策略的条件及影响因素,为政府和社会资本方达成良好均衡合作策略提供了思路。下面通过数值模拟对上述理论结果进行验证。

A 市智慧产业园作为 A 省政府"861"项目,该园区规划占地 384 亩,建设面积约 88 万 m^2。其中一期占地 169 亩已全面启动建设,建设面积约 32 万 m^2,其中针对云计算、物联网、电子商务、服务外包、工业设计和研发设计、文化创意、节能环保等高端服务业的社会资本方,若投资额在 100 万元以上的项目,在项目竣工投产纳税后,另给予固定资产投资额

5% 的一次性补助。以其中的节能环保服务类建设项目为例,社会资本方对此项目投资 500 万元,正常履行责任合作可收益 1000 万元,对此履行责任成本 40 万元,若对该项目合作存在机会主义,可从中获利 1200 万元,机会主义成本为 20 万元,政府对此惩罚 300 万元,监督成本为 100 万元。若社会资本方对项目合作履行责任义务,政府按固定资产投资额 5% 的补助再给予 25 万元的奖励补助。相关参数见表8。

模型参数设定(单位:万元)　　　表8

D_1	D_2	E	W	J	F	T	C
500	500	40	700	20	25	300	100

根据公式 $x=\dfrac{(D_2-E)-(W-J)+(F+T)}{F+T}$

和 $y=\dfrac{T-C}{F+T}$,从中可计算出:

$$x=\dfrac{(500-40)-(700-20)+(300+25)}{300+25}$$

$$=0.32$$

得:$1-x=0.68$

$$y=\dfrac{T-C}{F+T}=\dfrac{300-100}{300+25}=0.62$$

此时,$F+C=100+25<500$,即政府的监督成本、激励成本低于双方合作共赢带来的额外收益,保证了政府实施激励约束的动力。此时,政府实施责任激励约束的概率为 0.68;另外 $T>F>0$,即提高对社会资本方投机行为的打击力度,并给予一定的责任奖励,此时社会资本方履责合作的概率为 0.62。可见,政府实施责任激励约束对社会资本方避免出现机会主义行为具有一定积极作用。该模型得到了验证。

5 总结

本文从社会资本方在参与智慧产业建设项目合作中履行责任角度,构建了考虑政府采取

责任激励约束策略的演化博弈模型，并对双方均衡策略进行分析，结果表明：①政府只要确保监督成本和激励成本低于社会资本方履行责任合作带来的额外收益，就可保证实施责任激励约束的动力。②通过加大社会资本方采取机会主义合作行为的惩罚力度和给予社会资本方履行责任合作的一定激励，可引导社会资本方避免机会主义行为出现。③政府采取责任激励约束策略可提升社会资本方自身责任意识，在项目合作中表现较强责任感，对合作项目整体效益提升具有促进作用。

参考文献

[1] 卫慧. 智慧产业园区建筑运维系统设计与开发[J]. 电脑知识与技术，2020，16(35)：71-73.

[2] 梁昀. 智慧城市建设与城市经济发展分析[J]. 中国商论，2021(10)：116-119.

[3] 侯羽. 基于文本研究的广州市智慧产业发展政策分析[D]. 广州：华南理工大学，2013.

[4] 钱斌华. 助推智慧产业发展的财政政策研究[J]. 未来与发展，2013，36(9)：77-81.

[5] 杨俊杰. 银行服务智慧城市建设研究[J]. 科技创新导报，2017，14(15)：157-159.

[6] 满青珊，孙亭. 新型智慧城市理论研究与实践[J]. 指挥信息系统与技术，2017，8(3)：6-15.

[7] 张静. 广州市智慧城市建设的政府作用研究[D]. 广州：华南理工大学，2017.

[8] 高华，杨红伟，朱俊文. 基于风险偏好和公平偏好的PPP项目社会资本选择研究[J]. 工程管理学报，2020，34(5)：109-113.

[9] 黄思韵. 基于公私博弈的海绵城市PPP项目动态补偿机制的研究[D]. 合肥：安徽建筑大学，2020.

[10] Li S，Cai H B. Government incentive impacts on private investment behaviors under demand uncertainty [J]. Transportation Research，2017(3)：115-129.

[11] 刘平. 引导社会资本参与宁波智慧城市建设的研究[J]. 时代金融，2014(23)：216-217.

[12] 刘润楠. PPP模式在政府购买公共文化服务中的应用[J]. 学理论，2021(1)：72-74.

基于 BIM＋GIS＋IoT 的智慧楼宇集成管理平台的研发与应用

赵　璐　许磊磊　殷　超　龚　清

（中交武汉智行国际工程咨询有限公司，湖北武汉　430056）

【摘　要】　当前，在国家新基建、智慧城市、数字中国的发展战略下，以智慧楼宇等为代表的融合性基础设施作为城市的微观缩影和功能载体，空间范围较小、业务场景丰富、技术落地性强，成为各智慧城市产业链企业争先投入和实践的领域。本文从建筑用户体验、能耗管理、设施管理、安全管控和可持续发展的角度出发，通过项目调研、案例对标、需求分析、技术研发、试运行评估等步骤，研发了一套基于 BIM、GIS、IoT、大数据等技术的智慧楼宇集成管理平台，并在华中·中交城项目进行应用示范，实现年均运营人力成本节省约 80 万元/a，能耗成本预测节省约 44.6 万元/a，故障定位与响应时长、应急事件处置效率、资产盘点效率、用户体验大幅提升的目标。

【关键词】　BIM；GIS；IoT；智慧楼宇；集成管理平台

1　前言

当前，以智慧建筑、园区等为代表的融合性基础设施，作为城市的微观缩影和功能载体，空间范围较小、业务场景丰富，利于 BIM[1]、IoT[2]、GIS、大数据等新一代信息技术的集成应用落地，成为各地政府和智慧城市产业链企业争先投入和实践的领域。在智慧城市[3]顶层规划下，智慧建筑[4]作为城市生态的重要组成部分，正处于蓬勃发展的态势。但是，超高层建筑的复杂性决定了其运营管理的难度远高于普通建筑[5]，本文提出的智慧楼宇集成管理平台，是依托某集团区域性总部建设项目，研发了一套"建筑大脑"，并开发多个应用，使建筑逐步具备感知、记忆、判断和决策等综合能力，提高建筑运行效率，降低使用能耗，延长使用寿命[6]。

2　平台建设目标与内容

2.1　总体目标

建成稳定可靠、灵活兼容、互联互通、保障有力的智慧楼宇集成平台，融合大楼各类软硬件子系统，打通系统间壁垒，整合各系统功能，将楼宇、人、设备无缝连接，形成综合运维、智慧安防消防、节能降耗、智慧办公等应用场景，实现数据全融合，状态全可视，业务全可管，事件全可控，服务楼宇管理者、企业、员工、访客等，实现智慧楼宇一体化应用生态链以及功能随需定制的建设目标。

2.2 平台架构

整体架构上，从"一套物联网平台感知层、一套数字孪生展示层、一套基础功能+多个智慧场景应用层"三个层面，构建感知层+展示层+应用层的"1+1+N"总体架构（图1）。分两期实施，一期搭建平台底座，包括感知层、数字孪生和业务层基础功能；二期基于平台底座基础能力实现智慧办公、停车、安防、会议、能耗、消防、餐厅等系列智慧化应用场景。围绕大屏端、PC端、移动端三大主题，面向大楼管理者和使用者，实现精细化管理、智慧化运营、人性化服务、数字化展示，使建筑智能化、智慧化。

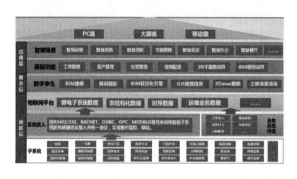

图1 系统架构图

2.3 关键技术

基于总体架构，确定 BIM+GIS+IoT 的技术路线，并命名为智行·海川 EOS 平台，主要研究并应用了以下技术：

（1）基于物联网平台的接口程序开发。

（2）BIM 模型轻量化技术。

（3）基于 SpringCloud+Spring+Jooq 微服务框架，确保平台弹性可扩展和易维护。各微服务应用之间通过 Kafka 消息队列进行通信，性能高，稳定可靠。

（4）数据库采用 MySQL+Redis，结合数据库连接池和数据库分片技术，弹性可扩展。

（5）上层应用统一通过 API Gateway 接

入授权和管理。

3 平台感知层建设

根据国家《智能建筑设计标准》GB 50314—2015，楼宇建筑针对智能化子系统尤其在硬件配置方面，普遍建设得很全面，通常会规划设计 18~30 个子系统，涵盖建筑内机电设备各方面。若按照传统楼宇建设模式，必然存在"子系统众多，各设备分散管理，难以将相关设备进行集中、关联控制与管理，信息透明化程度欠佳；运营维护基本依赖传统人工，出现问题被动响应，费时耗力，管理成本高；管理粗放，设备闲置率较高，节能环保的建设与运行机制难以发挥作用；人性化体验设计难以落地，智慧能力相对虚化"等一系列问题。

为了更好地解决上述问题，应用物联网平台建设感知层是最好的选择（图2）。项目启动时我们对所有子系统厂商一一对接协议，目前已将 26 个弱电子系统统一接入物联网平台（图3），通过数据采集、反向控制与场景联动控制基本实现了楼宇数据全融合[7]，并为上层展示和应用提供了统一数据接口。

图2 物联网平台

4 平台展示层建设

本案例楼宇建筑在设计阶段就引入 BIM 技术，提前对设计方案、材料选型、设备采购等提供优化建议，并将设计、建设阶段的 BIM 模型成果移交运维阶段[8]，再利用 GIS

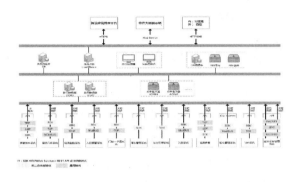

图3　子系统接口协议

技术将建筑内所有设备的地理信息数据进行存储和处理，并与物联网平台进行数据融合[9]，一定程度上实现了数字孪生[10]，真实还原物理空间。在平台上，可以全览整个园区布局，掌握楼层空间规划，了解各设备设施动静态数据信息，实现一屏总览，达到直观、方便、快捷地了解大楼一切情况的目的。展示层主要实现功能有：

（1）模型轻量化转换，以模型流的方式，用户可以实时看到已经下载的部分，对显示影响较大的部分先下载先显示，细节部分可以后显示。下载过程，用户不需要等待，可以进行其他操作。

（2）模型管理，统一管理全局的模型和附属的各类文档，通过权限控制、变更控制、版本控制管理模型、文档的上传、浏览、标识。

（3）版本管理，模型每一次修改会造成许多的历史版本，不同版次的版本之间可进行差异比较，支持随时恢复到指定的历史版本，提供增量文档传输功能，以提高发布速度。

（4）业务绑定，将BIM模型与业务系统中的设备进行关联与绑定，业务绑定是三维模型展示业务状态及相关详情的前提与基础，主要功能包括业务绑定、业务解绑、业务绑定变更。展示功能是对提供的BIM模型轻量化、规范化及模块化处理，完成建筑、设备等对象的三维立体动态展现与操作，实现建筑可视

化、环境可视化、结构可视化、区域可视化（图4）。

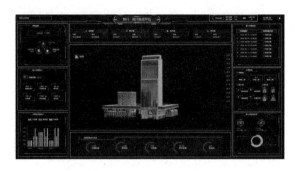

图4　PC端展示层

5　平台应用层建设

应用层首先实现了业务基础功能，包括项目管理、数据概览、2D平面图应用、BIM模型应用、运行管理、故障告警、工单管理、资产管理、系统管理、配置中心。各模块使用微服务的框架搭建，各模块之间独立运行互不影响。有新的需求时，只需添加新的模块即可，不影响其他模块的使用。应用层基于微服务架构还持续扩展了以下应用场景。

智慧运维是为设施设备运维提供故障告警和预测性维保，确保楼宇的数万个设备稳定可用，降低停机带来的损失。在设备发生硬件故障告警时，系统自动形成维修工单，以及预防性维保工单，派发给系统设置的对应维护人员。维修后，同时记录维护信息以及运行状态情况。可以快速生成运维报表，并完整保存过程数据。在不断积累运维数据之后，对未来的运维计划与费用测算提供决策支持。

智慧安防消防连接楼宇传统消防主机系统，并根据需求添加智慧消防传感设备，不影响传统消防设施结构，可实时监测楼宇内各类消防设备（消防水泵、水箱、喷淋、电气、防排烟、防火门等）运行状态。当采集数据判断触发火灾告警时，自动上报，可结合BIM模型快速定位告警设备，了解灾情所在位置，联

动周边摄像头，快速调出视频监控画面。

节能降耗基于人工智能和楼宇大数据为楼宇物业提供能源系统的实时监测、能耗优化。对接智能电表和BA系统，实时获取楼宇能耗计量数据，根据空间和设备类型分析展示在大屏和Web端。并利用时间、人数、环境等数据差分，让工作机组的设定变动尽量平滑，实现真正的节能降耗。智慧办公是实现与交建通的对接，将平台集成的子系统服务嵌入交建通移动端APP和PC端，向入驻单位的员工提供更便捷的办公体验。

智慧会议是会议子系统在预约会议之后，与相关的第三方系统对接，将会议的各个环节串联得更加智能化，提升会议体验，实现会前、会中、会后的全过程智能管理。包括会前人员获得授权进入会议室，会议开始时，调整好舒适的会议环境。会议结束后，自动关闭各类会议设施。

智慧餐厅通过交建通在手机上查看员工食堂的排队情况监控视频，避开用餐高峰期就餐。

一脸通行，在众多智慧场景的使用中，为了提升用户体验，安装了人脸识别设备，包括大楼门禁、各楼层门禁、会议室、食堂刷脸支付等，人脸信息通过移动端小程序统一接入平台后下发到各子系统实现一脸通行。

6 平台应用及实施效果

6.1 平台应用

该平台历时8个月的建设，目前已在华中·中交城成功应用近半年，实现了集成化和智能化，并探索实现智慧化（图5）。相关成果已获得软件著作权1项，集团课题1项，智慧企业案例1项，正在办理智慧场景的软件著作权，正在申报科技进步奖等相关奖项。

图5　华中·中交城指挥中心大屏

平台应用主要创新点有：

（1）项目从规划设计阶段引入BIM技术，并将设计、建设阶段的模型成果移交运维阶段，与物联网平台进行融合实现数字孪生，真正实现了BIM的全过程项目应用（图6）。

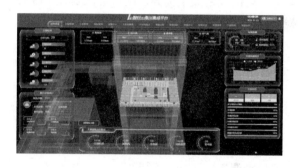

图6　华中·中交城数字孪生

（2）基于物联网平台技术打通各垂直智能系统之间的信息壁垒，破除"烟囱"现象，实现了上千种设备的统一接入，建立了建筑BIM、GIS、物联网等多源数据的融合应用标准，为后续适配各类项目需求，快速定制和开发智慧化应用场景提供了统一标准和平台基础功能支撑（图7）。

（3）面向办公、访客、物业等不同用户的

图7　消防联动安防

智慧、绿色运维场景需求，从策划咨询与功能设计再到研发应用全过程，提升运维效率、改善运维环境、增强用户体验。

（4）大数据和人工智能赋予建筑分析和学习能力。平台支持工单、设备、人员、能耗等多维度数据的整理归纳，自动生成按照日、周、月、季、半年、年度分类的指标数据报表，基于设计的节能降耗算法、忙/闲时人员调度与优化算法，输出优化改善策略，如基于工单与设备维修大数据统计，优化维修策略。

6.2　管理效益

运维效率大幅提升。项目案例实现了故障定位、应急事件处置、设备设施资产盘点数字化、在线化处理。故障定位与响应时长、应急事件处置效率、资产盘点效率，从传统处置时间天、小时跨越到分钟、秒，运维管理效率大幅提升。

项目用户服务体验评价满意率达到预期。项目案例投入运行 8 个月后，对入驻人员从便捷通行、安保措施、就餐体验、办公环境、会议体验、运维管理等维度开展用户满意度访谈和问卷调查，收集用户体验。现场访谈人员 75 人次，线上 842 有效人次参与问卷调查（其中办公人员 802 人次，运维人员 22 人次，访客 18 人次），项目智慧体验服务、绿色运维评分满意率达到 95%。

6.3　经济效益

运营人力成本节省。华中·中交城项目基于本设计方案，楼宇强电、暖通、弱电、综合维修等运营人员较常规楼宇数量减少 22 人，按照招聘市场楼宇维修技工、安保等人员平均 4000 元/月的薪资水平，预计节省人力成本约 80 万元/a。

能耗成本节省。按照我国大型公共建筑的单位建筑面积年用电量约为 100～300kW·h/(m^2·a)，采用楼宇智能化集成平台，使用节能灯具、进行节能控制策略设置等措施后，预测节约 6%～15% 的能耗[11]。本项目运行能耗监测数据统计约 80.6kW·h/(m^2·a)。按照项目建筑面积 13 万 m^3，武汉商业综合电价均值 0.7 元/(kW·h) 考虑，该项目年电费约 900 万元，水费约 20 万元。电费、水费分别取 6%、3% 进行测算，预期每年能耗成本节省约 44.6 万元。

项目出售效益明显。项目案例作为某区域智慧楼宇典型代表，项目出售部分（公寓）售价与周边同类型项目整体持平（约 15000 元/m^2），但是出售部分（公寓）去化率达到 87%，较周边同类项目去化率 48% 提升 39%。（相关数据统计截至 2021 年 8 月 1 日）

6.4　社会效益

本项目通过数字化、智慧化的技术手段，实现了建筑的绿色运维、智慧服务与体验，为"新基建""绿色建筑""智慧城市"等方面的应用落地及推广发挥了示范作用，助力国家碳达峰、碳中和战略。

7　结论

平台探索 BIM、GIS、IoT、AI 等新一代信息技术与建筑楼宇建设的深度融合，提出了可复制迭代的智慧楼宇集成管理平台的建设方案，在华中·中交城项目的建设与使用中进一步论证了本文提出的集成管理平台建设路径的可行性，通过数字化、智慧化的技术手段，实现了建筑的绿色运维、智慧服务与场景体验，打造了智慧楼宇典型示范。该平台仍需持续完善，后续将重点围绕节能降耗算法、自动控制策略、绿色运维体系和各入驻单位的个性化场景定制等多个方面开展迭代升级。相信，随着数据的不断累积，会在大数据应用、人工智能等前沿技术方面实现应用突破。

参考文献

[1] 杨军志. 基于BIM与人工智能技术结合的智慧建筑综合管理平台[J]. 智能建筑与智慧城市, 2020(2)：10-14.

[2] 薛玮. 物联网技术在智能建筑中的应用研究[D]. 济南：山东建筑大学, 2018.

[3] 郑喆. 智慧建筑视野下的我国建筑策划探索[J]. 智能建筑与智慧城市, 2020(5)：37-38, 42.

[4] 张月珍. 智慧建筑推动智慧城市发展[J]. 智能建筑电气技术, 2020, 14(1)：1-4.

[5] 陈兴海, 丁烈云. 基于物联网和BIM的建筑安全运维管理应用研究——以城市生命线工程为例[J]. 建筑经济, 2014, 35(11)：34-37.

[6] 周瑜, 刘春成. 雄安新区建设数字孪生城市的逻辑与创新[J]. 城市发展研究, 2018, 25(10)：60-67.

[7] 黄颖豪. 数字化解决方案——打造更加智慧的建筑环境[J]. 智能建筑与智慧城市, 2020(2)：17-18.

[8] 胡北. 基于BIM核心的物联网技术在运维阶段的应用[J]. 四川建筑, 2016, 36(6)：89-91.

[9] 王玲莉, 戴晨光, 马瑞. GIS与BIM集成在城市建筑规划中的应用研究[J]. 地理空间信息, 2016, 14(6)：75-78, 8.

[10] 刘万斌. 基于BIM和GIS的三维建筑信息管理系统研究[D]. 郑州：华北水利水电大学, 2019.

[11] 江亿. 我国建筑耗能状况及有效的节能途径[J]. 暖通空调, 2005, 35(5)：31-32.

基于盾构数据驱动的地质条件动态预测

夏　阳　高名岳　王朋艳

（华中科技大学土木与水利工程学院，湖北武汉　430074）

【摘　要】 隧道盾构施工过程中掌子面前方的地质条件对开挖稳定性有较大影响。当前地质条件主要依赖开挖前的钻孔勘探试验，但受限于试验条件及成本制约，只能获取稀疏条状的样本数据，难以满足盾构开挖的实时需要。对此，提出基于数据驱动的地质条件动态预测模型，将地表沉降视作盾构参数与地质条件共同作用下的土体响应行为，同时考虑到地质条件在空间上的连续性，通过 Logistic 多元回归算法建立地质条件、盾构参数与地表沉降的关系，基于观测的地表沉降响应量、盾构参数和已知钻探地质条件来动态预测掌子面前方的地质条件。将模型应用于武汉地铁某盾构区间，通过模型输出值与真实值的比较验证了模型的精度，相关结论为盾构施工提供了决策支持。

【关键词】 数据驱动；动态预测；地质条件；Logistic 多元回归算法；隧道盾构

1　引言

盾构法施工凭借其安全、快速、机械化程度高等优点在隧道工程中广泛应用。由于在地下施工，开挖面前方地质条件难以实时获取，而盾构施工过程中掌子面前方地质条件对开挖稳定性有较大影响，因此，研究如何预测施工前方开挖面地质条件对控制土体变形、指导施工、评估风险等意义重大。

隧道围岩通常由多层土体构成，具有典型的非匀质性[1]，不同施工区段地质条件也有差异，地质条件的复杂性增加了施工决策难度。传统获取地质情况的方法是通过钻孔获取数据，然后依经验对地质条件进行判断。但受成本以及实验条件等因素的限制，获取的钻孔数据通常是稀疏的，且经验判断存在精度低、可靠性差等问题，地质条件的精准预测成了隧道施工面临的难题。

随着人工智能技术的快速发展，基于数据驱动的地质建模受到了广泛关注。比如 Zhao，Wang[2] 使用贝叶斯压缩感知根据稀疏测量值对多层土质特性剖面进行插值和分层，用贝叶斯监督学习从稀疏训练数据集中插值多层土质性质剖面，然后，利用插值结果和改进的 k—均值聚类方法进行土质分层；Einstein 等[3] 基于贝叶斯网络将领域知识与数据相结合，对变量之间的依赖关系进行编码，提出地质预测模型和施工策略决策模型，帮助施工决策者进行风险评估；邓志平等[4] 提出基于钻孔资料的广义耦合马尔可夫链水平两方向转移概率矩阵极大似然估计方法，对地层不确定性进行分析。这些模型的建立主要是基于钻孔试验这一单一数据，如圆锥贯入试验获取的锥尖阻力，属于施工前的静态分析。但隧道工程开挖过程中也会产生大量与地质条件相关的监测数据，如何综合利用这些监测数据动态判断开挖面前方地质条件对指导施工意义更大。地层响应受地质条件、盾构参数等多因素的影响[5]，因此可以依据地表变形在不同地质条件和盾构参数作用下的响应机理推测盾构施工所处地质条件；另一方面，地质条件在空间上具有一定连续性，如层状土体在水平方向上的变异性较低，所以盾构施工所处地质条件与开挖上一环所处地质条件相关。

本文基于盾构参数、地质条件和地层响应之间的非线性关系，根据观测的地表沉降响应量、盾构参数和已知钻探地质条件来动态预测掌子面前方的地质条件，提出基于盾构数据驱动的地质条件动态预测模型。运用 Logistics 多元回归，将 70% 的数据样本作为训练数据集训练模型。剩下 30% 数据样本作为测试数据集验证模型。最终建立盾构参数、地质条件与地表沉降三者之间的关系，基于当前地表沉降、盾构参数和上一处土层类型，分析出当前所处土层，为盾构施工提供决策依据。

2 Logistic 回归模型

Logistic 回归可看作是线性回归的一种推广，通过引入 Sigmoid 连接函数将分类问题转换为回归问题，其函数图像如图 1 所示。设 n 维输入数据 $x = (x_1, x_2, \cdots, x_n)$，$x$ 属于某类

别的概率 P_i 为：

$$P_i = \frac{1}{1 + e^{-Z_i}} \qquad (1)$$

其中

$$Z_i = w_0 + w_1 x_1 + \cdots + w_n x_n \qquad (2)$$

w_{ij} 为逻辑回归系数。

多元 Logistic 回归模型是将多分类问题转化为若干二分类问题再采用上述方法进行求解，即对总类别为 k 的分类问题，转换为 $k-1$ 个二元回归模型，以第 k 类作为对照组，对于给定数据，建立 $k-1$ 个二元回归模型，该样本属于第 i 类的概率为

$$\begin{cases} p(y=0 \mid x;w) = \dfrac{1}{1 + \sum\limits_{i=1}^{k-1} e^{xw_i}} \\ p(y=i \mid x;w) = \dfrac{e^{xw_i}}{1 + \sum\limits_{i=1}^{k-1} e^{xw_i}}, i=1,\cdots,k \end{cases}$$

$$(3)$$

$w_i = [w_{i1}, \cdots, w_{in}]$ 为第 i 个二分类模型每个因子的逻辑回归系数。

3 地质预测模型建立

3.1 地质预测模型

设 P 为盾构参数，S 为地层响应，G 为地质条件，当前施工里程为 x，$P(x)$ 表示在里程 x 处的盾构施工参数，$S(x)$ 表示在里程 x 处的地层响应，$G(x)$ 表示在里程 x 处的地质条件。地质条件在空间上存在一定连续性，因此，$G(x)$ 表示一马尔科夫过程；盾构参数的设置是考虑与地质条件相适应的，所

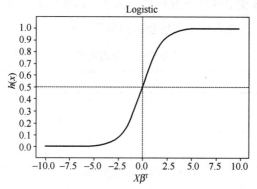

图 1　Sigmoid 函数

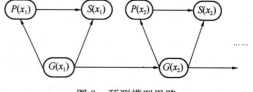

图 2　预测模型思路

以节点 $P(x)$ 受节点 $G(x)$ 影响；而盾构参数和地质条件共同影响地表沉降，所以 $G(x)$、$P(x)$ 共同指向 $S(x)$。本文目的是预测下一里程处的土质类型，故利用上一处的地质条件 $G(x_{i-1})$，以及当前位置的挖掘处盾构参数 $P(x_i)$ 和地层响应 $S(x_i)$ 作为模型的输入，分析当前开挖面的地质条件 $G(x_i)$，然后再次将该处的地质条件的实际测量值作为下一次预测的前一地质条件（图2），即：

$$G(x_i) = f[G(x_{i-1}), P(x_i), S(x_i)] \quad (4)$$

本文基于盾构施工过程中的数据来预测前方地质条件，其中包括开挖面到监测点的距离和埋深等几何特征；掘进速度、刀盘转速、刀盘扭矩、中部土压力、上部土压力和泥浆进出量等泥水盾构参数。表1中列出了变量、种类及数据类型。

以武汉某段地铁隧道为例来验证上述模型思路，其垂直剖面如图3所示。其中沉降监测点设置在隧道顶部，间距5～50m。每天对地表沉降量、地质条件等参数进行监测，最终获得157个数据点用于模型的建立与检验。

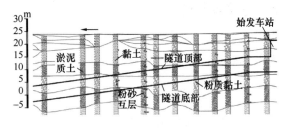

图3　隧道地质剖面图

3.2　影响地表响应的隧道几何特征

隧道直径和隧道深度是影响盾构施工过程中地面沉降和开挖面稳定性的重要因素。Shi etal[6]考虑到埋深（H），因为埋深与隧道直径的比值（$H/2R$）决定了隧道上部的荷载分布状况，所以将其作为一个关键因素。在盾构掘进隧道中，隧道直径通常是一个常数，因此可

使用隧道埋深来表征[7]。另一个重要因素是开挖面到监测点的距离，该距离对隧道纵向沉降槽有一定影响[8]。

3.3　地质条件

隧道穿越的土质类型有4种，因此用1-4来表示土质类型（即淤泥质土、粉砂互层、黏土、粉质黏土），但土质类型之间本身不存在任何序数关系。地下水位在拱底下方，因此影响较小，不做考虑。本文构建的Logistic模型将粉质黏土作为对照组。

3.4　泥水盾构参数

在盾构掘进隧道中，工作面压力通常是掘进过程中的主要控制参数，因为它对保持开挖工作面的稳定性和减少沉降至关重要。在地铁隧道中，刀盘顶部和中部均安装有压力传感器。因此，为了更好地理解工作面压力的影响，模型中考虑了上部土压力和中部土压力两个因素。盾构施工中，浆体流入和流出是关键。泥浆进出差表明了可用于估计是否发生超挖的挖土体积[9]。因此，泥浆进出差也被认为是一个因素。盾构参数中，刀盘转速、刀盘扭矩和推进速度最能反映刀盘所面对地质条件的变化[10]，其中刀盘扭矩主要是刀盘与土的相互作用引起的，切削扭矩约占总扭矩的99%[11]，所以刀盘转速、刀盘扭矩和推进速度也是反映地质条件的重要因素。因此，模型中考虑了6个因素作为泥浆盾构运行参数（表1）。

变量分类及数据类型　　表1

变量种类	变量	数据类型
几何参数	开挖面到监测点距离（m）	连续变量
	埋深（m）	连续变量
地质条件	隧道顶部的土质类型（1-4）	分类变量（1-4）

续表

变量种类	变量	数据类型
泥水盾构参数	掘进速度（mm/min）	连续变量
	刀盘转速（rpm）	连续变量
	刀盘扭矩（kN·m）	连续变量
	中部土压力（MPa）	连续变量
	上部土压力（MPa）	连续变量
	泥浆进出差（m³）	连续变量

4 数值分析

4.1 变量相关性检验

在进行分析之前，首先对各变量之间进行相关性检验。其中采用卡方检验分析两个分类变量之间的相关性，采用单因素方差分析多分类变量和连续变量之间的相关性。结果如表 2 所示，当 $F > F(0.05) = 10.128$，认为隧道顶部的地质条件与该变量有极大的显著相关性，当 $F(0.05) = 10.128 > F > F(0.1) = 5.538$，认为该变量与隧道顶部地质条件有显著相关性。当 $\chi^2 > \chi^2(0.05) = 9.348$，认为该变量与隧道顶部地质条件有极大的相关性。

变量相关性检验　表 2

变量	分析结果
开挖面到监测点距离（m）	$F = 5.602$
埋深（m）	$F = 1158.823$
隧道顶部的土质类型（m）	$\chi^2 = 256.107$
掘进速度（mm/min）	$F = 28.297$
刀盘转速（rpm）	$F = 38.028$
刀盘扭矩（kN·m）	$F = 19.724$
上部土压力（MPa）	$F = 636.863$
中部土压力（MPa）	$F = 654.955$
泥浆进出差（m³）	$F = 151.276$

4.2 多元 Logistic 回归分析

当类型 1 与类型 4 对比时，计算得到 $w_{10} = 29.134$，$w_{11} = 2.551$，$w_{12} = 40.967$，$w_{13} =$

1.684，$w_{14} = 121.356$，$w_{15} = -141.409$，$w_{16} = -35.201$，$w_{17} = -264.503$，$w_{18} = 25.023$，$w_{19} = -17.311$，$w_{110} = 23.741$，该预测模型为：

$$Z_1 = 29.134 + 2.551 x_1 + 40.967 x_2 + 1.684 x_3 + 121.356 x_4 - 141.409 x_5 - 35.201 x_6 - 264.503 x_7 + 25.023 x_8 - 17.311 x_9 + 23.741 x_{10} \quad (5)$$

当类型 2 与类型 4 对比时，计算得到 $w_{20} = -592.154$，$w_{21} = 0.371$，$w_{22} = 4.169$，$w_{23} = -13.639$，$w_{24} = -12.757$，$w_{25} = 241.717$，$w_{26} = -111.269$，$w_{27} = 277.469$，$w_{28} = -30.075$，$w_{29} = -29.618$，$w_{210} = 266.869$，该预测模型为：

$$Z_2 = -592.154 + 0.371 x_1 + 4.169 x_2 - 13.639 x_3 - 12.757 x_4 + 241.717 x_5 - 111.269 x_6 + 277.469 x_7 - 30.075 x_8 - 29.618 x_9 + 266.869 x_{10} \quad (6)$$

当类型 3 与类型 4 对比时，计算得到 $w_{30} = -90.313$，$w_{31} = 0.595$，$w_{32} = -0.118$，$w_{33} = 5.789$，$w_{34} = -47.304$，$w_{35} = 45.501$，$w_{36} = -105.457$，$w_{37} = -13.904$，$w_{38} = 2.857$，$w_{39} = 21.699$，$w_{310} = 165.732$，该预测模型为：

$$Z_3 = -90.313 + 0.595 x_1 - 0.118 x_2 + 5.789 x_3 - 47.304 x_4 + 45.501 x_5 - 105.457 x_6 - 13.904 x_7 + 2.857 x_8 + 21.699 x_9 + 165.732 x_{10} \quad (7)$$

将训练集中的数据分别代入式（1）、式（2）中得到 Z_1，Z_2，Z_3，再将 Z_1，Z_2，Z_3 经过式（4）处理，得到该处的隧道顶部土质条件为 i 的概率 $P_i(i=1,2,3,4)$，概率最大值所对应的 i 值即为土质类型。

4.3 模型检验

将训练后得出的模型用测试集进行检验，

其结果见表3。可以看到，该模型对地质类型1的预测准确率为96.6%；对类型2、类型3的准确率为100%；对类型4的准确率为90.7%；总体样本预测的准确率为95.5%。表明模型精度较高，能够较为准确地预测施工掌子面前方的地质条件。

测试集土质条件分类混淆矩阵　　表3

实测	预测				
	1	2	3	4	正确百分比
1	28	0	0	1	96.6%
2	0	1	0	0	100.0%
3	0	0	73	0	100.0%
4	0	5	0	49	90.7%
总体百分比	17.8%	3.8%	46.5%	31.8%	95.5%

此外，通过似然比检验来衡量该模型的准确程度。相对于混淆矩阵检验，似然比检验是根据变量之间的相关性进行检验，模型拟合信息中，$\chi^2 = 240.220 > \chi^2_{0.99}$（36）$= 58.619$，在99%的置信区间内，可证明拟合线性效果明显，结果如表4所示。

模型拟合信息　　表4

模型	模型拟合条件	似然比检验		
	−2 对数似然	卡方	自由度	显著性
仅截距	240.227			
最终	0.007	240.220	36	0.000

伪 R^2 检验中，伪 R^2 是决定系数，反映自变量与因变量变化的关系特征。本次模型的伪 R^2 的值为 0.878>0.5，当伪 R^2 的值越趋近于1时，变量之间的相关性越高，因变量随自变量变化相关程度越好，拟合后的模型效果越好，结果如表5所示。

伪 R^2　　表5

考克斯-斯奈尔	0.878
内戈尔科	1.000
麦克法登	1.000

5　结论

本文提出一种基于盾构数据驱动的地质条件动态预测模型，通过 Logistic 多元回归算法建立地质条件、盾构参数与地表沉降的关系，基于观测的地表沉降响应量、盾构参数和已知钻探地质条件来动态预测掌子面前方的地质条件，并通过对各类变量的相关性检验验证了多元回归分析的合理性。通过比较预测值和实测值来验证模型。最终，模型预测准确度可以达到95%以上，对于预测地质条件具有良好的效果。结果表明，可依据既有钻孔数据和盾构数据等多源信息判断地质条件。随着盾构机的前进，可以在训练集中加入新的数据来更新模型，这可以作为隧道实时控制系统的基础。然而，本文在数据分析时，变量之间的关系被简单处理，只用了线性回归的方式建立模型，没有考虑变量之间可能存在的非线性关系。因此，如何识别变量间非线性关系并建立预测模型可能需要进一步研究。

参考文献

[1] 邓志平，李典庆，曹子君，等. 考虑地层变异性和土体参数变异性的边坡可靠度分析[J]. 岩土工程学报，2017，39(6)：986-995.

[2] Zhao T Y, Wang Y. Interpolation and stratifycation of multilayer soil property profile from sparse measurements using machine learning methods[J]. Engineering Geology, 2020：265.

[3] Rita L Sousa, Herbert H Einstein. Risk analysis during tunnel construction using Bayesian Networks：Porto Metro case study[J]. Tunnelling and underground space technology, 2012, 27(1)：86-100.

[4] 邓志平，李典庆，祁小辉，等. 基于广义耦合马尔可夫链的地层变异性模拟方法[J]. 岩土工程学报，2018，40(11)：2041-2050.

［5］ 吴贤国. 盾构近接施工对既有隧道影响参数分析及近接度分区研究[J]. 土木工程与管理学报，2021，38(4)：96-109，114.

［6］ Shi J，Ortigao J A R，Bai J. Modular neural networks for predicting settlements during tunneling. J. Geotech. Geoenviron. Eng. ，1998，124(5)：389-395.

［7］ Thomas Kasper，Günther Meschke. A numerical study of the effect of soil and grout material properties and cover depth in shield tunnelling[J]. Computers and Geotechnics，2006，33(4).

［8］ 吉越亘. Prediction of ground settlements associated with shield tunnelling.[J]. 土质工学会论文报告集，1978，18(4)：47-59.

［9］ Zhang F，Zhu H，Fu D. Shield tunnelling method. Beijing：China Communications Press，2004.

［10］ 邢彤，赵阳. 盾构掘进土层识别及刀盘转速控制策略研究[J]. 浙江工业大学学报，2010，38(6)：649-654.

［11］ 潘茁，张晋勋，江玉生，等. 盾构刀盘扭矩与位移相关性分析[J]. 科学技术与工程，2016，16(10)：238-242，247.

建筑产业互联网与数字经济

Industrial Internet and Digital Economy

智能建造变革背景下建筑业电子商务发展路径研究

严小丽　李　桃

（上海工程技术大学管理学院，上海　201620）

【摘　要】 在建筑业掀起"智能建造"变革背景下，建筑业电子商务产业业务迅猛发展，涌现出一大批建筑业电子商务平台公司，为推进建筑业全产业链协同发展起到了重要作用。立足于目前建筑业电子商务产业发展的现状，深入挖掘其发展的动因，分析电子商务目前发展的行业背景，进而基于产业升级理论提出建筑业电子商务发展路径，将其分为萌芽及雏形、形成与成熟三个阶段，并根据建筑业电子商务产业目前所处的阶段，从国家、行业、企业三个层次提出了促进建筑业电子商务产业发展的对策，助力建筑业转型升级。

【关键词】 智能建造；建筑业电子商务；发展路径；产业升级

1　引言

习近平总书记指出：世界正在进入以信息产业为主导的经济发展时期。我们要把握数字化、网络化、智能化融合发展的契机，以信息化、智能化为杠杆培育新动能。要推进互联网、大数据、人工智能同实体经济深度融合，做大做强数字经济。为贯彻落实习近平总书记重要指示精神、推动建筑业转型升级、促进建筑业高质量发展，2020年7月住房和城乡建设部等13部门联合印发了《关于推动智能建造与建筑工业化协同发展的指导意见》（以下简称《指导意见》）[1]，明确提出了推动智能建造与建筑工业化协同发展的指导思想、基本原则、发展目标、重点任务和保障措施。其中，关于建筑产业互联网的发展提出两点重要意见，一是引导大型总承包企业采购平台向行业电子商务平台转型，实现与供应链上下游企业间的互联互通，提高供应链协同水平；二是鼓励企业建立工程总承包项目多方协同智能建造工作平台，强化智能建造上下游协同工作，形成涵盖设计、生产、施工、技术服务的产业链，意味着以加快打造建筑产业互联网平台为重点，促进全产业链发展，推进建筑业数字化转型是推进智能建造转型的重点和突破口之一。建筑业电子商务是建筑业产业互联网的重要组成部分，《指导意见》为建筑业电子商务的发展指明了方向。

随着互联网与经济社会各领域融合发展的进一步深化，网络经济协同实体经济共同发展成为经济增长新动力，同时形成推动经济和社会发展的新范式[2]。建筑行业把握住"互联网＋"这一历史时机，加快发展电子商务。建筑业电子商务是以建筑项目为核心，将所有参与项目的各个交易方联结在一起的复杂的电子交易系统[3]，发展电子商务能够降低信息壁垒，有效节省交易时间和交易成本，提高整个建筑行业的生产效率，缩短建筑产品生命周期，促进建筑业企业的竞争[4,5]。近年来引起了广泛关注，相关研究主要集中在：

（1）建筑业电子商务模式及创新探讨。贺亚冬[6]分析了电子商务体系在建筑行业的主要商业模式及业务形式；李敏[7]对互联网＋建筑的商业模式展开研究；李晨洋[8]主要分析研究了建筑业电子商务 B2B 模式；邵鹏等[9]对不同发展阶段的建筑电子商务商业模式创新进行分析研究。

（2）建筑电子商务平台的构建研究。李晓东等[10]、胡晓明[11]、杨静[12]等均对构建建筑业电子商务框架体系进行了研究，于颜君[13]系统地设计了建筑业电子商务平台架构；王红卫等[14]在电子商务平台基础上，创新性地提出了工程建造平台模式。

（3）建筑电子商务的发展及现状评价研究。罗振华等[15]、李晨洋[16]分别对建筑企业发展电子商务的必要性和可行性进行了论证分析；李忠富等[17]对建筑电子商务的发展现状进行了分析并提出改进对策；张春林[4]提出建筑业发展电子商务遇到的难点，并提出解决思路。

（4）建筑电子商务未来发展畅想。提出了"智慧城市""智慧建筑"[18,19]等概念，倡导数字城市与智能建筑、电子商务、智能社区的统筹规划和协调发展。

综上，"互联网＋"下建筑业电子商务新模式蓬勃发展，部分学者对建筑业电子商务的开展、应用现状与模式，以及未来形态进行研究分析，但目前尚未有建筑业电子商务发展路径规划方面的研究。由于建筑业的特性，建筑业电子商务有其特有的发展规律。本文基于《指导意见》，立足于当前建筑业电子商务发展的现状，深入挖掘促进其发展的动因，分析目前发展的行业背景，进而基于产业升级理论合理规划其发展路径，促进建筑业产业互联网平台健康发展，助力建筑业转型升级。

2　建筑业电子商务产业发展的动因

建筑业电子商务的产生与发展是宏观、中观、微观三个层面共同作用的结果。

（1）宏观——互联网创新环境及国家政策支持

建筑业电子商务起源于"互联网＋"创新环境。"互联网＋"浪潮以排山倒海之势席卷各传统行业，"互联网＋"创新环境及国家政策支持是建筑电商平台发展的首要驱动力。2016 年由商务部、中央网信办、发展改革委联合发布《电子商务"十三五"发展规划》，明确提出电子商务的发展目标和主要任务以及具体的保障措施。由住房和城乡建设部颁布的《2016—2020 年建筑业信息化发展纲要》明确指出积极探索"互联网＋"优势，发展新的商业模式，增加建筑业企业的竞争力。

（2）中观——产业和企业的需求

从建筑产业及企业发展来看，"互联网＋"工程建造将重构整个建筑行业生态圈。"互联网＋建筑业"所形成的建筑业电子商务平台可将所有参与方和相关资源进行整合，实现线上、线下资源共享，提高资源配置效率和能力，促进行业主体全面合作和有效协同，有效解决工程建造领域普遍存在的材料集中采购、工厂化建造、工程质量管理、信用管理等现实问题。

（3）微观——平台自身需求

建筑业已出现众多提供不同类型的服务的电子商务平台，实现了一定程度上的资源整合，但功能相对单一，从平台现实内在发展需求出发，将信息共享、服务交易、业务协同、监督管理等功能进行有机整合、扩展和完善，向综合化、集成化方向发展，形成功能强大的综合性服务平台是建筑业电子商务发展的内生原动力。

3 建筑业电子商务发展的行业背景

基于文献提取与现状分析，对建筑业电子商务目前发展的背景进行 SWOT 分析（表1），明确影响其发展的内外部环境，为其发展路径设计提供依据。

建筑业电子商务发展 SWOT 分析模型　表1

优势	劣势
1. 国家政策支持 2. 科技进步	1. 建筑行业技术水平较低 2. 建筑业电子商务专业人才较少 3. 传统观念转变困难
机会	威胁
1. 市场需求 2. "互联网＋"计划的实施推动 3. 建筑行业转型升级	1. 相关法律不完善 2. 监管体系不健全 3. 信息安全与隐私问题突出

（1）优势

1）政策支持。《电子商务"十三五"发展规划》，明确提出电子商务的发展目标和主要任务以及具体的保障措施。由住房和城乡建设部颁布的《2016—2020 年建筑业信息化发展纲要》也明确指出积极探索"互联网＋"优势，发展新的商业模式。《指导意见》则为建筑业电子商务发展指明了方向。

2）科技进步。①信息化水平提高。随着云计算、大数据、智能化、移动通信、物联网等信息技术发展，建筑业的信息化水平显著且全面提高，对建筑业发展电子商务提供了巨大的推动力。②新技术对建筑产业模式的革新。人工智能、区块链、虚拟现实及 BIM、3D 打印等新技术已进入建筑产业链，并从根本上改变了传统的设计、采购、施工、销售、运维模式，引发了一系列变革。

（2）劣势

1）观念转变困难。相较其他行业，建筑业相对落后，信息化水平低，创新能力弱，思想固化，转型升级举步维艰，转变企业现有保守思想非常困难，需要大量时间和成本投入。

2）技术水平较低。首先，信息技术不完善。建筑软件之间的信息融合和交互存在问题，影响建筑业电子商务的发展；电子商务平台数据巨大、相关数据维度复杂，且没有被行业认可的数据存储标准；线上支付尚无完善解决方案。其次，工程技术不成熟。建筑业供应链体系未建立，核心技术不成熟，无法发挥电子商务的优势；建筑业装配式标准化低，体系化不足，拖慢了建筑电商应用的步伐。最后，管理技术不适应。建筑业传统的商务模式，无法直接照搬到平台，新的电商模式需要探索；相关管理模式制度不规范，导致电商运作存在混乱。

3）有电子商务经验的管理与技术人员相对缺乏。建筑业电子商务缺乏具有丰富管理经验的管理者，电商系统实施过程中也缺乏具有相应技术背景的技术开发人员；员工对电子商务系统的接受能力也是阻碍电子商务发展的原因。

（3）机遇

1）市场需求。"互联网＋"下，电子商务成为交易的重要方式，制造业、零售业等行业电子商务已发展较为成熟，但建筑业电商尚处于发展初期，市场环境不断成熟过程中孕育着无限商机和可能。

2）"互联网＋"行动计划的实施及推动作用。"互联网＋"行动计划已经得广泛应用，身处"互联网＋"浪潮的建筑行业，外界环境有利于电子商务的蓬勃发展。

3）建筑行业转型升级所需。利用"互联网＋"技术使建筑行业从传统向信息化的方向转化，打破原来的传统发展模式是一次重要机遇，也是建筑产业结构调整的重要步骤。

（4）威胁

1) 现有行业法律法规的制约。目前法律对工程总承包模式，还缺乏明确的条款来规范主体的交易行为，这可能会给平台的运营管理带来一定的法律障碍。

2) 信用监管体系的不完善。我国还未建立起健全的信用管理体系，缺乏有效的失信、违规行为监督惩罚机制，这给平台运作带来了一定困难。

3) 信息安全与隐私问题。建筑电子商务需要采集有关工程建造服务交易和实施的信息，目前缺乏有关保护隐私权和责任的法律法规。如何既实现信息共享，又保障隐私安全，也是建筑电子商务所面临的难题。

4 基于产业升级的建筑业电子商务发展路径研究

"互联网＋"下，基于建筑业电子商务发展的宏观、中观、微观动因及目前发展的行业背景，其发展应遵循产业发展的内在规律，实现由低级到高级、简单到复杂、由功能单一向功能综合发展的渐进过程。图1所示为外界环境作用下的电子商务发展路径图。

具体的，建筑业电子商务将以客户需求为目标，将信息共享、服务交易、业务协同、监

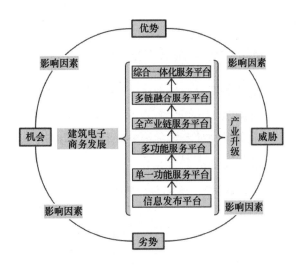

图1　建筑电子商务发展路径

督管理等功能进行有机整合，在已有的信息发布平台与服务交易平台的基础上，规划建筑全产业链的业务，实现线上建造服务，并整合价值链、创新链，向综合化、集成化的多链融合方向发展；在此基础上，进一步深度整合大数据、云计算、物联网等新兴技术，在智能建造模式逐渐形成的环境下，向智能综合一体化建造平台发展。

在建筑业电子商务目前发展背景分析的基础上，根据建筑电子商务发展路径中不同阶段的共同特点，将建筑业电子商务发展路径分为三个大的阶段，如图2所示。

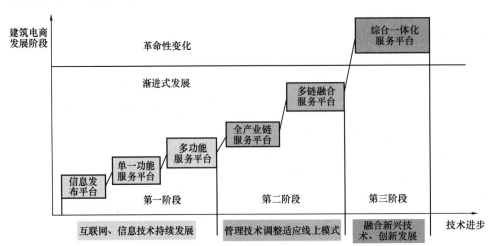

图2　基于产业升级的建筑电子商务发展路径图

4.1 萌芽及雏形阶段——信息发布平台与服务交易平台

第一阶段是萌芽及雏形阶段，目前建筑业电子商务基本处于此阶段。

"互联网＋"背景下，信息技术快速发展，建筑业信息化水平大大提高。特别是随着网络经济快速发展，建筑企业积极顺应时代发展电子商务平台。按其功能可分为信息发布平台、单一建造功能服务交易平台以及少数多功能服务平台。

建筑业电子商务起源于工程信息发布平台，其基础是信息资源整合，即根据用户需要，对不同类型的网络信息资源进行归类并整理，使用户能够通过统一的检索平台查找和浏览。工程信息发布平台主要有两种类型：一种是工程建造的基础数据资源整合和发布平台，另一种是工程项目的 BIM 数据整合平台。前者的典型案例，如全国建筑市场监管与诚信信息发布平台——四库一平台；后者如 BIM 云服务平台、BIM 中国网等。但是，目前的信息发布平台信息资源数量、类型众多，无法做到互联互通，易造成信息孤岛等问题。此阶段属于建筑业电子商务发展的萌芽期。

服务交易平台是在信息发布功能的基础上，支持服务交易功能。这个阶段的电商平台提供单一类型或者几种交易的工程建造服务。按照服务交易的内容，可以将服务交易平台分为建筑材料电商平台、招标投标平台、网络房地产市场平台、新兴的装修与设计众包平台等单一功能的电子商务平台，部分建筑电商平台也具有一部分综合功能。按照电商平台的出资方分类，包括大型企业（集团）投资成立的主要为集团服务的电商公司，有大中型民企联合投资的电商平台，也有多家创业团队联合投资基金创建的第三方电子商务平台等。此阶段属

于建筑业电子商务发展的雏形期，服务业务较为单一，呈现碎片化状态，未整合建筑业产业链的功能。

在"互联网＋"下建筑业的转型升级发展、企业自身发展等内外部因素的共同作用下，建筑业电子商务不断整合产业链资源，逐渐进入下一阶段。

4.2 形成阶段——全产业链协同平台

第二阶段是建筑业电子商务的形成阶段。进入此阶段，除了需要依据信息技术的发展突破平台建设瓶颈，还需转变工程建造传统管理技术以适应线上管理模式，打破线下线上的空间壁垒，将碎片化、间断式的服务功能统筹规划，在电商平台上实现建筑行业全产业链服务，集成投资、咨询、规划、设计、施工、制造、安装、运营、维护等全产业链企业，实现全产业链一体化联动，实现多业务、全产业的整体升级发展，即进入"全产业链协同平台"阶段。

在此基础上，通过进一步优化、理顺管理链，融通产业链，打通价值链，保证供应链，发展创新链，进入"多链融合服务平台"阶段。在此阶段，通过电子商务平台整合多条链和配置产业资源，实现业务之间的相互拉动，实现多链系统集成，从而促成行业内信息共享、资源共享、经营成本降低、利润效益最大。例如，在电子商务平台上向前后延伸业务，实现建筑产业链向两端延伸发展拉动建筑业创新链形成；反之，建筑业创新链通过为建筑产业各参与单位提供新思想、新方法、新技术、新产品和新服务等，创造建筑产业的新生态，有效推动建筑产业链的发展；进而，创新链创新传导效应作用于产业价值活动中，实现创新体系下的价值转移和创造，从而增值价值链等。

此阶段末期，随着新兴科学技术、工程技术、管理技术的进一步发展，我国建筑业发展为智能建造模式，智能建造与建筑工业化协同发展，建筑电子商务与新兴技术进一步融合，形成行业新生态，使行业内部技术结构、组织结构和管理水平升级，使建筑全产业协调发展，生产要素优化组合与合理配置，产品质量得以提高，从而推动建筑电子商务进入更高阶段。

4.3 成熟阶段——智能建造一体化平台

第三阶段是成熟阶段，也是建筑业电子商务发展的高级阶段。进入此阶段将进行革命性的变化，建筑电子商务平台将不仅仅服务于工程建造产业，更是要打破建筑行业生态模式，进行价值重构。

进入此阶段，平台化发展成为企业的常态。建筑业电子商务深度融合新兴技术如大数据、云计算等，逐步向智能综合一体化建造平台发展。智能综合一体化平台是基于新兴技术建立的虚拟平台，通过对海量数据的积累和挖掘，以及新兴技术的应用，如大数据服务、人工智能服务、BIM 云平台、VR 体验、3D 打印等，并与新兴商业模式紧密结合，形成虚拟空间的产业运营模式，为合作伙伴、客户、建筑管理者等开发出众多的业务服务，达到数据驱动的智能管理模式。同时，构建具有无限发展空间的科技生态圈，吸引技术合作伙伴、应用合作伙伴和各行业用户，逐渐形成以建筑行业为主、纵横多行业的虚拟产业云生态，实现众多产业的利益共享、资源共享的相互协助，合作共赢。

此阶段，以综合一体化电子商务平台为载体，建筑业数字化、网络化、智能化得以实现，充分满足以人为本的服务需求，实现建筑业可持续发展；同时，建筑业与其他产业实现良好协同发展。

5 促进建筑业电子商务发展的对策

基于建筑业电子商务的发展路径，目前，建筑业电子商务正处于从第一阶段向第二阶段发展的阶段，应从国家、行业、企业三个不同层次采取有力措施，促进建筑业电子商务的深入发展。

（1）国家支持

1）制定、完善建筑业电子商务相关法律法规。如《中华人民共和国招标投标法》对建筑业电子商务的众包交易与代理交易模式缺乏明确的法律条款以规范主体交易行为，给电商运营管理带来一定的法律障碍。完善相关法律是目前政府的主要工作，也是电商企业合理合法运营的基础。

2）建立有效严明的监管机制和审查机制。我国尚缺乏健全的信用管理体系，也缺乏有效的失信、违规行为监督惩罚机制，给平台的运营管理带来一定困难。完善监管和审查机制，解决交易过程中的信用问题，打击不正当竞争，促进电子商务企业间的合作交流，扩大电子商务业务领域的开放，实现电子商务市场的公平和有序。

3）加大对建筑行业转型升级的支持力度。制定相关优惠政策，鼓励企业对电子商务的投资；创造有利的市场环境，使得企业切实通过电子商务获得价值。

（2）行业促进

1）顺应时代潮流，把握市场需求。充分把握时代优势与机会，攻克现存发展困难，把握好国家政策的支持，积极适应互联网背景下电子商务平台的发展模式。

2）创新组织和管理模式。促进建筑行业内企业开发特定的商业策略，保证资金、时间和人力资源的合理调用，适应建筑工程与电子商务结合带来的改变。

3）推进行业新发展。推进建筑工程标准化进程，完善供应链体系，充分融合产业链资源，拓展业务范围，激发行业新的盈利点，构建行业新生态，探索行业新价值，才能进一步实现在平台上构造建筑业产业生态。

（3）企业建设

对于建筑业电子商务企业而言，其发展应顺应时代潮流、把握机遇、发挥优势、警惕预防威胁、补缺劣势及短板，全面、灵活地把控发展动向。

1）加紧平台建设落地。建筑电子商务企业应完善电子商务平台设计及业务功能，提高系统效率，保证平台功能切实可行，保证良好的用户体验。只有在企业系统支持下，电子商务才能真正发挥效益。

2）加快新技术、专业技术应用。企业应积极结合新兴技术，加大对新技术应用的投入力度，专业人员应深入研究、努力突破技术难关，保障电子商务平台运作流畅，并为扩大平台业务能力提供强有力的支持。

3）转变建筑行业从业人员传统观念。企业内部人员观念的开放有利于新技术的应用，建筑业企业在发展电子商务时，积极开放的思想观念对推动电子商务的发展非常重要。

4）储备专业人才。引进相关技术人才，加强对员工的培训，培养具有建筑业和电子商务知识的复合型人才，弥补建筑业电子商务人才缺乏的不足，同时也是启动电子商务长足发展的原动力。

6 结语

智能建造转型升级背景下，在宏观、中观、微观三个层面力量作用下，建筑业电子商务迅速发展，但总体上发展时间较短，其未来发展的趋势成为关注的焦点。

本文基于《指导意见》，在目前建筑业电子商务发展及研究现状下，积极探索建筑业电子商务发展的动力，对建筑业电子商务发展背景进行 SWOT 分析，并基于产业升级进行了建筑业电子商务发展的路径分析，提出建筑电子商务发展的三个阶段，即《指导意见》所提出的"采购平台"是建筑业电子商务发展的第一个萌芽和雏形阶段，进而进入"全产业链协同"平台阶段，最后进入《指导意见》所提出的"智能建造平台"阶段，即电子商务发展的第三个成熟阶段。论文最后基于目前建筑业电子商务发展所处的阶段，从国家、行业、建筑企业方面提出促进建筑业电子商务发展的对策，以促进建筑产业互联网健康发展，助力建筑业转型升级。

参考文献

[1] 住房和城乡建设部等部门. 关于推动智能建造与建筑工业化协同发展的指导意见（建市〔2020〕60号）[EB/OL]. http：//www. gov. cn/zhengce/zhengceku/2020-07/28/content_5530762. htm.

[2] 辜胜阻，曹冬梅，李睿. 让"互联网＋"行动计划引领新一轮创业浪潮[J]. 科学学研究，2016，034(2)：161-165，278.

[3] 张德群，李忠富. 建筑业电子商务的内容、特点及影响因素分析[J]. 建筑经济，2002(5)：8-10.

[4] 张春林. 建筑行业开展电子商务的难点和思路[J]. 中国建设信息，2012，000(8)：64-65.

[5] 张德群. 电子商务对建筑业的影响分析[J]. 建筑管理现代化，2000(2)：41-42.

[6] 贺亚冬. 电子商务在建筑业的研究与应用[J]. 中国电子商务，2014(12)：1-1.

[7] 李敏. 对于"互联网＋建筑业"商业模式研究[J]. 科技资讯，2017，015(31)：13-14.

[8] 李晨洋. 电子商务在建筑业应用模式探讨[J]. 商业研究，2003，000(18)：172-174.

[9] 邵鹏,胡平.电子商务平台商业模式创新与演变的案例研究[J].科研管理,2016(7):81-88.

[10] 李晓东,杨静.建筑业电子商务框架体系研究[J].土木工程学报,2005,38(5):107-110.

[11] 胡晓明.电子商务在建筑业的研究与应用[J].陕西理工学院学报(自科版),2005(2):94-97.

[12] 杨静.建筑业电子商务框架体系研究[D].哈尔滨:哈尔滨工业大学,2005.

[13] 于颜君.建筑业电子商务平台架构[J].现代物业旬刊,2018,425(6):27.

[14] 王红卫,谢勇,周洪涛,等."互联网+"工程建造平台模式[M].北京:科学出版社,2018.

[15] 罗振华,韩建强.建筑企业发展电子商务的必要性和策略[J].管理观察,2010(22):102-103.

[16] 李晨洋.建筑电子商务项目启动评估可行性研究[J].商业研究,2006(13):195-198.

[17] 李忠富,何丹丹,姜韶华.建筑业电子商务现状SWOT分析与对策研究[J].建筑经济,2017,38(4):76-81.

[18] 巫细波,杨再高.智慧城市理念与未来城市发展[J].城市发展研究,2010(11):56-60,40.

[19] 傅荣校.智慧城市的概念框架与推进路径[J].求索,2019(5):153-163.

建筑电子商务与建筑市场互动发展研究

李　桃　严小丽　吴　静　郭文平

（上海工程技术大学管理学院，上海　201620）

【摘　要】 建筑业转型升级背景下，建筑电子商务与建筑市场相互作用、互动发展。文章首先基于建筑电子商务与建筑市场相关要素构建二者发展互动关系的系统动力学模型，揭示建筑电商平台数量、建筑市场、专业人才规模、资金规模 4 个子系统之间的因果关系和动态反馈机制。然后，基于我国 2013～2018 年建筑业相关统计数据进行仿真；在此基础上设计企业投资发展模式、政府财政支持发展模式、人才培养发展模式三种情景模式，进一步对二者互动关系进行政策模拟。通过比较表明三种模式在建筑电子商务与建筑市场互动发展过程中具有不同的作用方向及领域，且呈现出明显的阶段性特征。最后提出相应对策，以期共同促进建筑电子商务与市场发展良性互动。

【关键词】 建筑业转型；建筑电子商务；互动关系；系统动力学

建筑业生产增加值占 GDP 总量高达 7%，却又是最为传统和落后的产业，当前其正处于转型升级关键时期。2020 年国家十三个部委联合出台的《关于推动智能建造与建筑工业化协同发展的指导意见》提出，"以加快打造建筑产业互联网平台为重点，推进工业互联网平台在建筑领域的融合应用，推进建筑业数字化转型"[1]。建筑业电子商务（简称"建筑电商"）作为典型的互联网平台模式，在融合数字技术、发展平台经济、扩展市场结构、资源配置等方面的作用突出，并为更深层次的建筑业发展储备力量；反之，随着建筑业发展，建筑电商获得在建筑业进一步发展的良好机会和政策保证。同时，建筑电商与建筑市场互动发展吸引众多风险资本进入[2]，促进了对专业人才的需求[3]，并变革建筑市场[4]。即建筑电商与建筑市场具有互动作用，如图 1 所示。研究二者之间的作用效应，探讨如何实现其高效互动，对更好地挖掘电子商务商业价值，及促进建筑业转型升级具有重要意义。

对现有研究梳理发现，学者们已经对建筑电子商务在建筑业实践的必要性、现状进行了相关探索，如建筑业电子商务平台构建[4~6]及商业模式[7]的研究、建筑业电子商务的开展及现状研究[8~10]等，为本文提供了启发，但还鲜有研究关注建筑电商实践与建筑市场发展的相互影响。本文拟根据电子商务及建筑市场动态发展的特性以及相关要素之间作用关系，将建筑电商与建筑市场的互动发展视为一个系统，运用系统动力学方法，建立二者互动关系的系统动力学模型，通过 Vensim 软件进行仿真分析，探究二者互动发展机制，并根据仿真结果进行政策设计与评估。

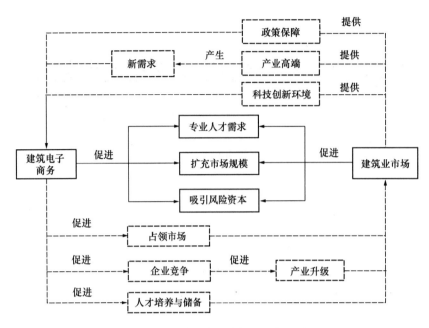

图 1　建筑电商与建筑市场发展作用关系

1　建筑电商与建筑市场互动发展的因果关系分析

1.1　建筑电商与建筑市场互动发展的系统界定

建筑电商与建筑市场互动发展相关重要因素的界定是二者互动关系研究的基础。首先,建筑电商数量变化是二者互动效果最直接的表现;其次,由于建筑电商与建筑市场发展均作用于专业人才、建筑市场、风险投资,同时,四个指标之间存在逻辑关系:①建筑市场扩展离不开人才队伍的建设;②专业人才规模的扩大会提升建筑电子商务创新能力及获利能力,增强企业竞争优势;③建筑电商企业优势能够扩展建筑市场规模;④市场规模的扩张增强建筑电商对各种风险资金的吸引力,实现资金规模扩张,同时需要更多的专业人才;⑤资金的进入吸引大量投资者、专业人才进入电商领域,使建筑电商数量迅速增加。因此,选取建筑电商数量、专业人才、建筑市场、投资资金

规模作为分析要素,反映二者的作用效果,并将建筑电商与建筑市场互动发展关系模型系统界定为建筑市场、建筑电商数量、专业人才、资金规模 4 个子系统,如图 2 所示。

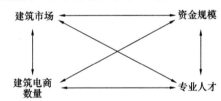

图 2　建筑电商与建筑市场互动
发展关系研究模型

建筑电商与建筑市场互动发展是随时间长期动态变化的,二者互动作为整体具有系统自主性。因此,研究建筑电商与建筑市场互动关系符合系统动力学方法建模要求。建筑电商数量与建筑市场、资金规模、人才发展组成的复合系统具有系统性、复杂性、动态性特征以及多变量、多回路和非线性的反馈回路,而系统动力学方法建模较传统数学模型更能充分描述该系统的非线性结构和动态特征,因此,对于建筑电商与建筑市场互动发展关系的研究具有适用性。

1.2 系统因果反馈关系分析

本文从建筑电商与建筑市场共同作用要素入手：建筑市场规模、资金风险投资、建筑电商数量、专业人才、研究要素之间的运行机制。电子商务和建筑业发展的因果联系如图3所示。

建筑电商与建筑市场互动发展的主要反馈路径包括建筑市场规模回路、专业人才回路、建筑电商数量回路、资金回路。

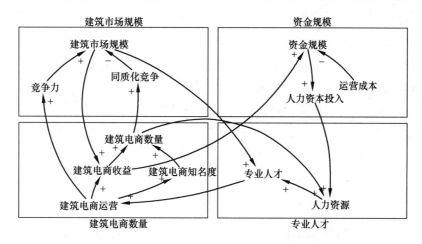

图3 建筑电商与建筑市场互动发展关系的因果关系图

1）建筑市场规模回路

正反馈回路：建筑市场规模→＋建筑电商收益→＋建筑电商数量→＋人力资源→＋专业人才→＋建筑电商运营→＋竞争力→＋建筑产品市场规模（其中正负号分别代表正向与反向反馈）。该反馈回路体现了建筑市场的发展增加建筑电商的收益，从而提升建筑电商的运营效率，提高产品的竞争力，并进一步扩大建筑市场规模，形成良性循环发展。

负反馈回路：建筑市场规模→＋建筑电商收益→＋建筑电商数量→＋同质化竞争→－建筑市场规模。负反馈回路描述了建筑电商数量的快速扩张导致同质化竞争加剧。由于数字化建造趋势以及政策支持，一旦市场存在超额利润，就会吸引大批投资者进入，建筑电商经营模式的复制、扩张导致市场同质化竞争加剧，最终导致建筑市场效率低下。

2）专业人才回路

正反馈回路：专业人才→＋建筑电商运营→＋竞争力→＋建筑市场规模→＋建筑电商收益→＋建筑电商数量→＋人力资源→＋专业人才。该反馈回路体现了专业人才在提高电商运营能力、提升竞争力、扩大市场以及增加电商盈利过程中起到的重要作用。

负反馈回路：专业人才→＋建筑电商运营→＋建筑电商知名度→＋建筑电商数量→＋同质化竞争→－建筑市场规模→＋专业人才。该反馈回路说明专业人才有助于企业知名度的提升。但我国专利保护体系的不健全以及技术的可复制性在给企业带来利润的同时又会因产品高度同质化导致市场竞争加剧，从而对电商收益产生负面影响，进而减少对专业人才的需求，并对专业人才的吸引力降低。

3）建筑电商数量回路

正反馈回路：建筑电商数量→＋人力资源→＋专业人才→＋建筑电商运营→＋建筑电商收益→＋建筑电商数量。该回路中，完善的政策体系为专业人才资源的供给提供有效的政策鼓励及良好的引进条件，电商数量的快速增长必将刺激对建筑电商专业人才的需求。同时，良好的运营效果又会吸引更多投资者加入建筑电商领域。

负反馈回路：建筑电商数量→＋同质化竞争→－建筑市场规模→＋建筑电商收益→＋建筑电商数量。负反馈回路体现了由于市场上同质化产品过多导致过度竞争，降低电商收益水平，最后使电商数量下降。

4）资金回路

正反馈回路：资金规模→＋人力资本投入→＋人力资源→＋专业人才→＋建筑电商运营→＋竞争力→＋建筑市场规模→＋建筑电商收益→＋资金规模。正反馈回路体现了资金的作用机制。在专业人才培养上提供资金扶持，扩充专业人才队伍。同时，推动建筑业发展的财政支持为从事建筑电商的初创者解决资金不足的问题，在一定程度上推动了建筑电商的发展。

负反馈回路：资金规模→＋人力资本投入→＋人力资源→＋专业人才→＋建筑电商运营→＋建筑电商知名度→＋建筑电商数量→＋同质化竞争→－建筑市场规模→＋建筑电商收益→＋资金规模。负反馈回路体现了过度同质化竞争带来的负面效应，缩减电商收益，使部分电商退出市场，降低了对发展资金的需求，同时导致行业对外来资金的吸引力下降。

2 建筑电子商务与建筑市场互动发展关系的系统动力学模型

2.1 电子商务和建筑业发展互动关系的系统流图

本文以建筑市场、建筑电商数量、专业人才、资金规模状态变量为主，根据因果回路图，引入辅助变量及常量，构建建筑电商与建筑市场互动发展关系系统动力学模型，如图4所示。

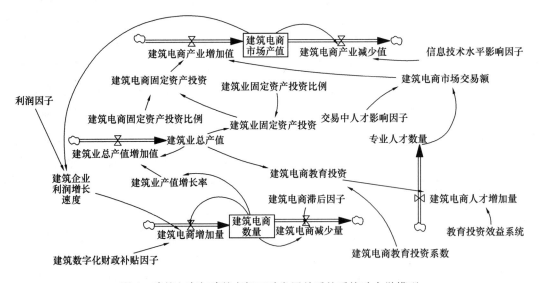

图4 建筑电商与建筑市场互动发展关系的系统动力学模型

2.2 模型参数及相关方程

为了真实展现建筑电商与建筑市场关系，本文运用《中国统计年鉴》中2013～2018年我国建筑业经济发展及建筑电商情况的相关统计

数据，选取"建筑业电子商务网站数量"作为建筑电商数量分析指标的数据来源；选取建筑业总产值、建筑企业利润增长速度作为反映二者互动关系下建筑市场发展状况的指标。设立模拟时间区间为2013～2023年，模拟时间步长

1 年，2018 年之后的时间为所设立模型系统的预测时间，结合数学研究方法和文献资料进行系统性估算，进一步对系统的未来发展进行趋势性分析。对于模型的分析基于以下假设：一是该系统的运行是一个动态、连续的过程；二是模型分析主要考虑与 4 个子系统相关的影响因素，不考虑除此之外的其他因素；三是不考虑延迟性以及时滞性对模型整体运行的影响。建筑电商与建筑市场互动发展关系的系统动力学模型中部分主要模型及参数如表 1 所示。

建筑电商与建筑市场互动发展关系系统动力学建模主要参数及模型设定 表 1

类别	变量名称	类型	模型方程及参数	单位
经济指标	建筑电商市场产值	水平变量	建筑电商产业增加值－建筑电商产业减少值	亿元
	建筑业总产值	水平变量	建筑业总产值＋建筑业总产值增加值	亿元
	建筑电商数量	水平变量	建筑电商增加量－建筑电商减少量	Dmnl
	建筑电商专业人才数量	水平变量	建筑电商人才增加量	Dmnl
辅助指标	建筑业总产值增加值	辅助变量	建筑业总产值×建筑业产值增长率	亿元
	建筑电商教育投资	辅助变量	建筑业总产值×建筑电商教育投资系数	Dmnl
	建筑电商人才增加值	辅助变量	建筑电商教育投资×教育投资效益系数	Dmnl
	建筑电商增加量	辅助变量	建筑企业收益增长速率×建筑电商数量×财政补贴因子	Dmnl
	建筑电商产业增加值	辅助变量	建筑电商固定资产投资＋建筑电商市场交易额	亿元
	建筑业固定资产投资	辅助变量	建筑业总产值×建筑业固定资产投资比例	亿元
常量值	建筑电商固定投资比例	常量	0.036（采用趋势法计算得出）	Dmnl

根据系统动力学模型方程的基本设置规则，将建筑市场、专业人才、建筑电商数量和资金 4 个状态变量分别用相应的辅助量表示，比如在设置衡量电商规模的建筑电商数量时，以建筑电商增加量与建筑电商减少量两个辅助量的差值测算：

设定 $\qquad L = RA \qquad$ (1)

$$\begin{bmatrix} R \\ A \end{bmatrix} = M \begin{bmatrix} L \\ A \end{bmatrix} \qquad (2)$$

式中，M 代表关系矩阵；L 代表状态变量的向量形式，且 $L = [l_1, l_2, l_3, \cdots, l_n]^t$；$A$ 代表辅助变量的向量形式；R 代表速率变量的向量形式，且 $R = [r_1, r_2, r_3, \cdots, r_n]^t$，因此：

$$R = \frac{\partial L}{\partial t} \qquad (3)$$

由于系统的动态变化特性，当状态变量 R 为 $R(t+\Delta t)$，则有：

$$L(t+\Delta t) = L(t) + \Delta R \qquad (4)$$

式中，t 代表时间；Δt 代表时间变化量。

3 系统动力学模型检验及政策模拟

3.1 模型检验

3.1.1 稳定性检验

从本模型中任取一变量，设立不同仿真时间间隔，通过对比该变量在时间间隔发生变化时的运行情况检验模型稳定性。以模型中建筑电商市场产值为检测对象，将仿真时间步长分别设置为 0.25 年、0.5 年和 1 年，其运行结果如图 5 所示。可知，在不同的仿真时间间隔下的 3 条建筑电商市场产值走势图几乎重合，变化趋势保持一致，由此可以判断本文所建立的系统动力学模型是稳定的。

3.1.2 有效性检验

依据指标的重要性及数据的可获取性，选取建筑业总产值为检测对象，通过模拟结果与

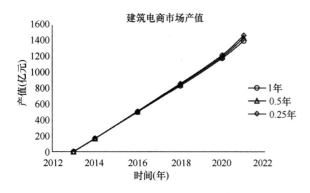

图5　不同仿真时间步长的系统动力学模型模拟的
建筑电商市场产值走势图

实际数据的误差对比验证其有效性，结果如表2所示。选取的建筑总产值历史数据与仿真数据误差值最小为−2.6%，最大为7.9%，误差率均保持在10%的可控范围内，由此可判断本系统动力学模型拟合度较好，能够对实际系统做出分析与预测。

建筑电商与建筑市场互动发展关系的系统动力学模型有效性检验结果　表2

年份	建筑业总产值		
	原始值（亿元）	仿真值（亿元）	误差率（%）
2013	160366	149430	6.8
2014	176713	162730	7.9
2015	180757	177213	2.0
2016	193567	192985	0.3
2017	213944	210160	1.8
2018	235086	228864	−2.6
2019	248446	249233	−0.3

建筑电商与建筑市场互动发展关系的系统动力学模型有效性和稳定性检验均成立，说明本模型能够真实有效地反映二者互动关系的实际系统结构和功能。

3.2　仿真分析

在模型检验成立基础上对建筑电商与建筑市场互动发展关系效果进行仿真模拟分析，结果如图6所示。

由图6（a）可以看出，建筑企业利润增长速度在2013～2023年间稳步增长，2023年末将达到44.9%，从图4中建筑电商教育投资、建筑电商市场交易额两个辅助变量来看，建筑企业利润的增长与建筑电商的发展密切相关，其中电子商务专业人才培养及建筑电商市场发展良好势态起到了关键作用。同时，由图6（b）可以看出，随着我国政府对建筑数字化转型升级的扶持与推进，建筑电商得到快速发展，建筑电商市场产值2013～2023年增长幅度明显，2023年末还将突破2000亿元大关，从图4中建筑固定资产投资和建筑财政补贴因子两个辅助变量来看，建筑电商市场产值的快速增长得益于建筑企业对于电子商务的投资及政府数字化补贴发挥了重要作用。由此可见，建筑电商与建筑市场具有较强的互动作用。

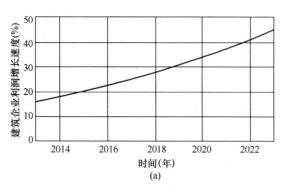
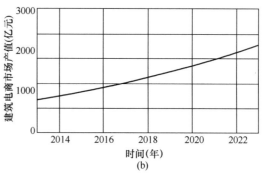

图6　建筑电商与建筑市场互动发展关系的系统动力学模型模拟结果
（a）建筑企业利润增长速度；（b）建筑电商市场市值

3.3　政策模拟

为了比较在未来不同政策情境下建筑电商与建筑市场互动发展关系的变化趋势，根据本系统动力学模型结构特点，从企业角度、政府角度、人才角度对企业投资发展模式、政府财政支持发展模式、人才培养发展模式进行政策模拟，分别选择建筑电商固定资产投资比例、建筑数字化财政补贴因子、建筑电商教育投资系数为调控决策变量，模拟不同决策对二者互动关系的影响。现分别模拟分析如下。

（1）企业投资发展模式。其他参数值保持不变，将建筑电商固定资产投资比例由原来的 0.036 提升到 0.0414，提高 15.0%。

（2）政府财政支持发展模式。其他参数值保持不变，将建筑业数字化发展财政补贴因子由原来的 0.19 提升到 0.2185，提高 15.0%。

（3）人才培养发展模式。其他参数值保持不变，将建筑电商教育投资系数由原来的 0.021 提升到 0.02415，提高 15.0%。

《关于推动智能建造与建筑工业化协同发展的指导意见》提出加强政策支持，在人才培养等方面加大政策支持力度，这与建筑电商提升人才数量和质量的内在发展要求形成交叉，因而选取建筑电商市场产值、专业人才数量展示建筑电商与建筑市场互动发展效果，列出了在不同政策情景下建筑电商市场产值的动态结果，如表 3 所示。

建筑电商与建筑市场互动发展的系统动力学模型在不同仿真策略下的动态模拟结果　　　表 3

年份	建筑电商市场产值（亿元）				专业人才数量			
	初始值	企业投资发展模式	政府财政支持发展模式	人才培养发展模式	初始值	企业投资发展模式	政府财政支持发展模式	人才培养发展模式
2013	800	800	800	800	288.258	288.258	288.258	331.382
2014	904.506	913.398	904.506	908.829	313.804	313.804	313.804	360.874
2015	1015.45	1033.26	1015.45	1024.11	341.732	341.732	341.732	392.992
2016	1133.65	1160.47	1133.65	1146.69	372.147	372.147	372.147	427.969
2017	1259.97	1295.97	1259.97	1277.47	405.268	405.268	405.268	466.058
2018	1395.36	1440.77	1395.36	1417.43	441.337	441.337	441.337	507.537
2019	1540.79	1595.91	1540.79	1567.58	480.616	480.616	480.616	552.708
2020	1697.34	1762.55	1697.34	1729.04	523.392	523.392	523.392	601.901
2021	1866.15	1941.9	1866.15	1902.97	569.985	569.986	570.001	655.483
2022	2048.46	2135.29	2048.48	2090.67	620.799	620.815	620.953	713.928
2023	2245.66	2344.19	2245.77	2293.56	676.783	676.959	678.388	778.395

从模型整体时间段（2013~2023年）来看，三种政策效果均具有阶段性，企业投资发展模式提高建筑电商市场产值的效果最优，而政府财政支持模式效果显现较晚，2022年后才表现出对建筑电商发展的促进作用；人才培养发展模式直接刺激专业人才数量增长的方式，企业投资及政府财政支持发展模式均在

2020年后表现出人才培养作用。以模型仿真过程中2023年的模拟数据为研究对象对比三种发展模式的仿真结果：

（1）企业投资发展模式中将建筑电商固定资产投资比例提高 15.0%，建筑电商市场产值由原来的 2245.66 亿元提升到 2344.19 亿元，增加 98.53 亿元，涨幅约 4.39%，专业

人才数量由原来的 676.783 万人增加到 676.959 万人，增加约 0.0026％。

（2）政府财政支持发展模式中，将建筑业数字化转型财政补贴因子提高 15.0％，建筑电商市场产值由原来的 2245.66 亿元提升到 2245.77 亿元，增加 0.11 亿元，涨幅约 0.005％，专业人才数量由原来的 676.783 万人增加到 678.388 万人，增加约 0.24％。

（3）人才培养发展模式将建筑电商教育投资系数提高 15.0％，建筑电商市场产值由原来的 2245.66 亿元提升到 2293.56 亿元，增加约 2.13％，专业人才数量由原来的 676.783 万人增加到 778.395 万人，增加约 15.01％。

企业投资模式调整建筑电商固定投资水平对建筑电商市场产值贡献明显，而人才培养发展模式调整建筑电商教育投资系数对提高专业人才数量贡献明显，此现象说明，在促进建筑电商与建筑市场互动发展过程中，不同政策的作用方向及领域存在差异。

4　结论

本文运用系统动力学方法，构建了建筑电商与建筑市场互动发展关系的系统动力学模型，并模拟了 2013～2023 年建筑电商与建筑市场相互推进的量化结果，进而通过设置不同政策情景模拟二者互动发展趋势，本文结论及贡献如下：

（1）研究建筑电商与建筑市场互动发展关系系统中各子系统之间的互动关系，运用系统动力学方法建立其互动关系的系统动力学模型，具有可为各种具体、实际的决策模式提供仿真拟合、比较对照、选取合理有效配置资源的优势。

（2）通过模拟 2013～2023 年建筑电商与建筑市场发展情况，设计了企业投资、政府财政支持、人才培养 3 种发展模式，通过仿真分析，明确三种情景模式在建筑电商和建筑市场互动发展过程中具有不同的作用方向及领域，且呈现出明显的阶段性特征。

基于以上因素，为促进建筑电商与建筑市场之间的互动发展，提出以下对策：

（1）建筑企业应注重电子商务应用，主动开展电子商务，加大且适当超前投资，根据企业自身发展规划，合理化投资结构。

（2）根据建筑业发展转型需要，重点围绕人才培养方面激发行业内部活力，促进建筑电商与建筑市场互动发展的深度融合。

（3）政府要激发建筑企业向互联网平台转型的发展活力，逐步加大建筑业财政补贴力度，扩大补贴范围，优化补贴结构，明确补贴重点，使建筑电商与建筑市场互动发展拥有坚实的保障。

（4）要根据建筑电商与建筑市场发展现状，分阶段、有针对性地重点推进相关政策。以促进建筑电商发展为例，在建筑业转型升级初期，应重点提高建筑电商教育投资力度，扩大专业人才规模，确保建筑市场经济发展的创新动力；在成长期，应重点激发企业活力，提高建筑电商固定资产投资水平，加强并完善配套服务体系建设；在成熟期，政府应重点加大建筑业数字化财政补贴的力度，推动建筑电商同建筑市场的高效互动，最终实现建筑业转型升级总体目标。

参考文献

[1] 中华人民共和国住房和城乡建设部．关于推动智能建造与建筑工业化协同发展的指导意见[EB/OL]．http://www.mohurd.gov.cn/zxydt/202007/t20200729_246562.html．

[2] 丁烈云．数字建造导论[M]．北京：中国建筑工业出版社，2019．

[3] 蒋正明，张书凤，李国昊，等．我国科技人才对

经济增长贡献率的实证研究［J］. 统计与决策，2011，000（12）：78-80.

［4］ 王红卫，谢勇，周洪涛，等.“互联网＋”工程建造平台模式［M］. 北京：科学出版社，2018.

［5］ 李晓东，杨静. 建筑业电子商务框架体系研究［J］. 土木工程学报，2005，38（5）：107-110.

［6］ 李忠富，何丹丹，姜韶华. 基于 BIM 和 Linked Data 的建筑业电子商务信息融合框架［J］. 工程管理学报，2017，31（5）：7-12.

［7］ 邵鹏，胡平. 电子商务平台商业模式创新与演变的案例研究［J］. 科研管理，2016（7）：81-88.

［8］ 李忠富，何丹丹，姜韶华. 建筑业电子商务现状 SWOT 分析与对策研究［J］. 建筑经济，2017，38（4）：76-81.

［9］ 李桃，严小丽，吴颖萍.“互联网＋”下建筑电子商务发展的影响因素研究［J］. 工程经济，2020，30（9）：67-71.

［10］ 李晨洋. 建筑电子商务项目启动评估可行性研究［J］. 商业研究，2006（13）：195-198.

工程变更合规管理本体建模与案例推理算法

陈 波[1] 李 迁[1] 张玉彬[2]

(1. 南京大学工程管理学院，江苏南京 210008；
2. 江苏省妇幼保健院基建处，江苏南京 210009)

【摘 要】 由于工程项目的不确定性使得工程变更具有普遍性和突发性，对变更的责任认定和费用处理需满足相关法律规范和合同要求。本文首先将工程项目变更争议司法判决案例作为知识源，通过对近年的法律判决文书以及相关法律规范的研读，利用 Protégé 软件构建了变更的合规案例知识本体，以实现案例知识的共享和重用。其次，在本体的知识管理基础上，运用案例推理算法来获取当前工程变更的相似案例，指导新变更的处理。通过自然语言处理的方法提取案例文本中的隐性特征计算文本相似度。结合时间序列数据挖掘，计算动态时间弯曲距离进行时间序列相似度量化。在此基础上，基于熵值法耦合层次分析法赋权各相似度指标权重。研究结果表明，论文的案例推理算法能在案例库中匹配出较为相似的历史案例供用户进行学习和参考，从而为工程项目变更的合规管理提供支持。

【关键词】 工程变更；合规管理；本体；案例推理；自然语言处理

2018 年，中央出台了《中央企业合规管理指引》和《企业境外经营合规管理指引》，自此拉开了企业合规管理建设的序幕，工程建设领域建立健全合规性管理制度，可以有效防范各种合规陷阱，降低企业法律风险，维护企业合法权益[1]。

案例推理技术是借鉴过去问题的经验来求解新问题[2]，能有效缓解经验缺乏和任务规划不合理所带来的次生危害[3]。同时本体是共享概念模型明确的、形式化的规范说明[4]。结合本体的案例推理可实现知识精准匹配，有利于案例持续更新[5]。本文的工程变更合规案例推理可帮助管理者获取历史相似案例，辅助变更

的合理解决。本研究选用 Protégé 5.5.0 作为本体模型构建的工具，并通过 Python 3.9 进行文本的自然语言处理、相似度计算以及案例推理的实现。

1 工程变更合规管理的知识分析

工程变更的合规性管理是工程项目风险管理的重要组成部分，针对知识源进行分析，变更的合规管理知识来源于以下三大类。由于变更与地方法规、地方计价规则等相关，本文以江苏省为例进行研究。

（1）历史案例变更纠纷的判决文书

数据源为最高法院设立的中国裁判文书网

上的生效裁判文书，具体包括：①诉讼主体信息；②案件重新审理信息；③原被告主张信息；④法院裁定依据；⑤裁定结论。

（2）历史案例的工程项目背景

数据源为"企查查"中的工商信息，住房和城乡建设部"四库一平台"的公开信息，具体包括工程类别、时间、工程规模、争议金额以及涉及司法案件数量等。

（3）各类法条制度

数据源包括法律、司法解释、地方法规以及相关行业标准；通过整理分析，对其中涉及工程变更的规章制度中的各条款进行提取，反映到本体中来。

2　合规视角下工程变更本体的构建

结合 Gruber 提出的本体构建基本原则：明确性、一致性、可扩展性、最小单调可扩展以及最小约束[6]，本文通过学者提出的七步法[7]建立本体。

（1）确定本体的范围

变更合规知识见上文范围。

（2）考虑重用现有本体

已有部分建设工程相关本体，但是由于细

分领域差异导致无法直接复用，因此，重新进行本体设计。

（3）确定本体中的重要概念

通过已有的地方标准，例如《江苏省建设工程费用定额》里的规范化术语作为重要概念。

（4）定义类与类的层次

本体中的类是领域内概念的结构化表述，本文通过自顶而下法，逐层细化建立子类。表1列出了部分变更合规知识类及子类的层次分类。

概念层次表		表 1
层级编码	名称	层次
0	变更合规知识本体	顶层概念
1	工程项目背景	一级概念
1.1	变更信息	二级概念
1.1.1	变更金额	三级概念
1.1.2	变更的施工工序	三级概念
1.1.2.1	土石方工程类	四级概念
……	……	……

由于变更知识和诉讼信息的多样性，本体的规范化分类需不断更新完善，论文篇幅有限，难以完全罗列。以"工程类别"为例说明，其包括建筑工程、市政工程等7子类，7个四级概念可再细分为24子类，24子类还可细分，例如工业建筑可分为生产车间、仓库等，如图1所示。

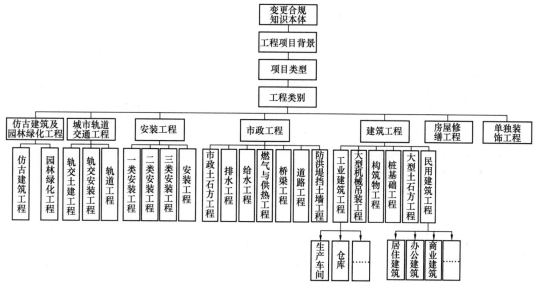

图 1　工程类别层次图

在 Protégé 中构建本体，通过构建父子类、排他、等价关系，逐层建立概念等级结构，并进行类的进一步层次扩展，如图 2 所示。

（5）本体模型的属性设计

本体中存在两种关系属性：数据属性可将概念与数值连接；而对象属性是概念间的连接。属性的定义域是属性的主体，值域是该属性指向的客体；数据属性需要定义属性值的格式，例如变更发生的时间，其值域为"xsd：datetime"。由于本体的类复杂，对应其属性也非常多，因此，以"工程项目背景类"为例诠释属性的创建，在 Protégé 软件中构建相应的属性关系（表2）。

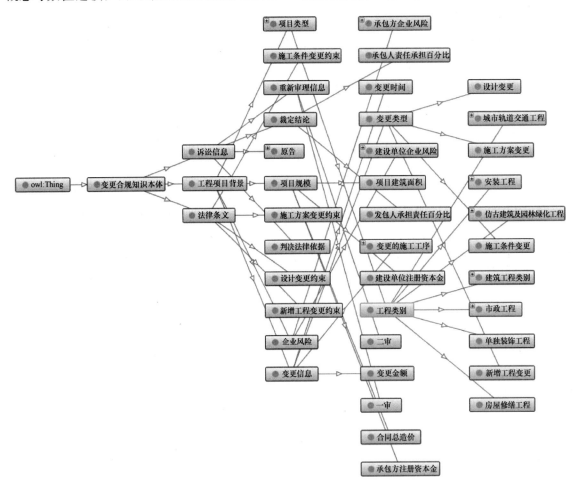

图 2　本体类的 Protégé 图

本体属性表　　　　　　　　　　　　　　　　表 2

属性类别	属性名称	属性含义	定义域	值域
数据属性	lawsuit _ time	描述变更发生的时间	变更类型	xsd：datetime
	employer _ lawsuit	建设单位涉及司法案件数	建设单位涉及司法案件	xsd：int
	has _ area	描述项目建筑面积	项目建筑面积	xsd：float
	has _ capital _ 2	描述承包人注册资本金	项目名称	xsd：float
	……	……	……	……

续表

属性类别	属性名称	属性含义	定义域	值域
对象属性	supported_by	变更合规管理的支持性法律条文	变更类型	法律条文
	has_plaintiff	描述项目原告信息	变更信息	原告
	is_firsttrial	描述变更诉讼为一审	变更信息	一审
	has_process	描述工程变更的工序	工程类别	变更的施工工序
	……	……	……	……

（6）本体模型的实例填充

实例作为具体某个类的实际存在。例如变更相关法律规范类中包含了具体的法条实例，设计变更相关法律条款类，含有实例如《建设工程工程量清单计价规范》第9.3.1条，《江苏省高级人民法院关于审理建设工程施工合同纠纷案件若干问题的意见》第九条等。

（7）本体的评价和修正

使用 Protégé 软件的 Pellet 推理机制对本体进行一致性检验，并结合专家意见完善本体。

3 基于多方法融合的案例推理

3.1 案例推理机制和指标设计

有学者总结了案例推理的经典流程并形成了 R-4 模型[8]；本文采用 R-5 模型[9]，是在原有基础上加入再分配 Repartition，而基于本体进行的案例知识再分配能实现知识重用和精准匹配。本文设计了多方法融合的推理机制，见图 3。基于本体中概念与最近公共父节点的位置关系计算层次相似度，通过动态时间规整计算时间序列相似度；结合自然语言处理，利用三种方法比选计算文本相似度。

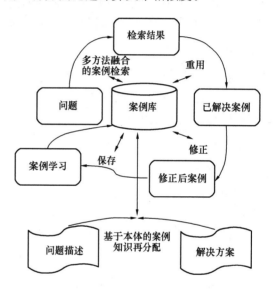

图 3 案例推理系统图

本文针对工程变更合规管理提取了九个检索特征，并将其编号为 f_1、f_2、…、f_9。涉及多种数据类型，对应不同计算方法，具体特征见表 3。

案例检索指标表　　　　表 3

案例特征	编号	数据类型	具体内容
原告	f_1	类别型	发包人、承包人
变更类型	f_2	类别型	设计变更、新增工程变更、施工条件变更、施工方案变更
工程类别	f_3	层次型	建筑工程-民用建筑、市政工程-排水工程等
变更纠纷金额	f_4	数值型	金额具体数值
合同总造价	f_5	数值型	金额具体数值
发包人涉及司法案件数量	f_6	数值型	具体诉讼数量
承包人涉及司法案件数量	f_7	数值型	具体诉讼数量
建设周期内工程价格水平	f_8	时间序列型	固定资产投资价格指数与建设周期的时间序列
判决法律依据、判决结论信息	f_9	文本	隐形特征，通过自然语言处理提取特征计算相似度

3.2 多方法融合的相似度计算

3.2.1 基础特征相似度

（1）类别相似度

通过判断案例间指标是否相同来计算相似度。

$$\text{sim}_1(C_{ij}, C_{0j}) = \begin{cases} 1, C_{ij} = C_{0j} \\ e^{-1}, C_{ij} \neq C_{0j} \end{cases} \quad (1)$$

其中C_{ij}表示第i个历史案例中第j个特征的特征值，C_{0j}表示目标案例的第j个特征的特征值。因后续熵值法权重计算指标不能为0，故取值为e^{-1}。

（2）层次相似度

层次型特征在本体知识结构中按照一定结构层次取值。本文采用WU and Palmer[10]算法来计算概念间相似度，其采用计算概念与其最近公共父节点概念的位置关系计算相似度，是基于语义词典的度量方法。

$$\text{sim}_2(C_{ij}, C_{0j}) = \frac{2H}{H_i + H_0 + 2H} \quad (2)$$

式中，H表示从两案例指标在本体中最近公共父节点到本体根节点的最短路径，H_i、H_0表示i案例、目标案例的层次特征到最近的公共父节点的最短距离。

（3）数值相似度

在计算数值型属性相似度时，使用两个数值的差值除以该属性中最大值与最小值的差值进行计算。

$$\text{sim}_2(C_s^i, C_t^i) = 1 - \frac{|C_s^i - C_t^i|}{\beta - \alpha} \quad (3)$$

其中C_s^i和C_t^i分别表示源案例和目标案例的第i个属性值，α、β分别表示该数值型属性的最小值和最大值。

3.2.2 时间序列相似度

因贸易战、疫情等突发情况，可能导致窝工、原材料价格大幅变动等情况，因此，建设

周期内价格水平是项目背景的必要特征，也是变更纠纷发生的来源。而固定资产投资价格指数可准确地反映建设投资的价格变动趋势和幅度，因此，本文构建了基于固定资产投资价格指数和建设周期的时间序列特征。

在时间序列相似度计算中，动态时间归整（DTW）是一种通过弯曲、拉伸或收缩时间轴来测量两等长或非等长时间序列相似度的方法[11]。该算法通过在两序列构成的矩阵网络上寻找最优路径，用路径累加规整距离来衡量两个时间序列之间的相似性，具体原理见引文[12]。

本文将案例间时间序列的累积距离$\text{dist}(i,j)$求倒数，得到初步指标，并进行最大值归一化处理得到最终相似度指标。

3.2.3 文本相似度

大量知识都蕴含在文字之中，文本特征的加入，可以使得案例推理系统中包含更多非结构化的隐性知识。通过TF-IDF、LSI、LSTM三种方法进行实验，比选文本相似度的计算方法。

（1）TF-IDF算法计算文本相似度

① 分词

使用jieba的第三方库进行分词的操作，并加入了清华大学开源词库的法律领域词典、搜狗细胞词库中的建筑词汇大全作为自定义补充词库。

② 清洗

文本里都有很多无用信息，包括无用的标签、标点符号、停用词。本文采用哈尔滨工业大学停用词表进行停用词清洗。

③ 特征提取

TF-IDF提取特征是基于某一个词在某篇文本中出现的次数较多，并且在其他文本中出现次数少，则认为该词具有较好区分性。

$$W_i = \text{TF} \times \text{IDF} \quad (4)$$

以词频 TF 和逆文档频率 IDF 的乘积作为特征空间坐标系的取值测度。

④ 相似度计算

采用余弦相似度计算，其背后的思想是计算 **A**、**B** 两向量的夹角余弦值。

$$\text{sim}(\boldsymbol{A}, \boldsymbol{B}) = \frac{\boldsymbol{A} \cdot \boldsymbol{B}}{\|\boldsymbol{A}\| \|\boldsymbol{B}\|} = \frac{\sum\limits_{k=1}^{n} A_k \times B_k}{\sqrt{\sum\limits_{k=1}^{n} a_i^2} \times \sqrt{\sum\limits_{k=1}^{n} b_i^2}} \tag{5}$$

（2）潜在语义索引算法

潜在语义索引（LSI）是基于奇异值分解对特征矩阵进行降维表示的方法：

① 同上，计算得到文本的基于 TF－IDF 的向量空间。

② 向量空间的 SVD 分解，将原数据映射到语义空间内。具体算法见引文[13]。

④ 同上，利用余弦相似度，计算奇异值分解后的文本－主题矩阵中的向量计算文本相似度。

（3）Bi-LSTM 算法计算文本相似度

① 原始文本及处理

本文实验选取裁判文书网中案由为建设工程施工合同纠纷，关键词为变更，地域为江苏的 100 例裁判文书进行训练。并同上，对原始文本进行分词和清洗。

② Doc2vec 转换词向量

Doc2vec 包括句向量模型和分布词袋模型，本文利用句向量模型训练向量，其核心思想如图 4 所示。

Doc2vec 工作原理如图 4 所示，每一句话用唯一向量来表示，以矩阵 **D** 的一列来代表。词也用唯一的向量来表示，以矩阵 **W** 的一列来表示。每次从一句话中滑动采样固定长度的词，取其中某词作预测词，其他的作为输入词。输入词对应的词向量和本句话对应的句子

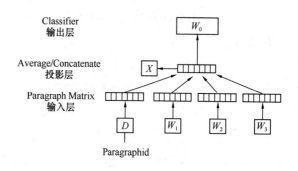

图 4 句向量模型图

向量作为输入层的输入，将本句话的向量和本次采样的词向量平均或者累加构成一个新的向量 **X**，进而使用这个向量 **X** 预测此次窗口内的预测词[14]。训练完了以后，就会得到训练样本中所有的词向量和每句话对应的句子向量。

③ Bi-LSTM 模型提取文本特征

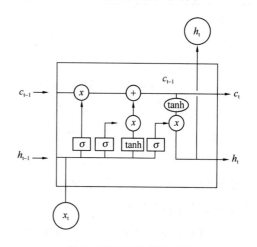

图 5 LSTM 模型原理图

LSTM 是 RNN 的一种变体，通过门来控制传输状态，记住需要长久记忆的，省略不重要的信息（图 5）；而 Bi-LSTM 将正向和反向序列组合以获得输出，可更好地捕捉双向的语义依赖。本文采用网格搜索法调整参数，其中句向量维度 300，窗口大小为 3，每种窗口的个数均为 100，LSTM 隐藏层的单元数取 64，学习速率为 0.01，每个区域的固定长度设置为 30。

④ 同上，用余弦相似度计算提取后的特征相似度。

（4）文本相似度实验结果

对变更纠纷的裁判文书的文本相似性进行人工标注：将历史案例形成多组三元组文书，用（A，B，C）表示一组的三篇文书，采用人工分析方式，保证每组数据中，A 与 B 的相似度大于 A 与 C 的相似度。同时利用算法进行相似度计算并比较，若预测正确，那么该测试数据则准确，否则预测错误。实验通过准确率作为衡量文本相似度度量准确度的指标。

本次实验使用 96 条文本数据，来比较上述 3 种算法的效果，得到表 4 结果。

文本实验结果表　　　表 4

模型	准确率
TF-IDF	62.5%
LSI	59.4%
Bi-LSTM	75.0%

基于 Doc2vec 的 Bi-LSTM 模型实验效果较好，该模型可以通过双向长短程记忆网络有效地过滤文本中的无用信息，以提高准确性。根据上述实验结果，后续案例推理过程中使用 Bi-LSTM 模型来计算文本相似度。

3.2.4　相似度的权重设计

层次分析法是主观赋值方法，利用少数定量信息把决策的思维过程数字化；熵值法是客观赋值方法，是通过指标间的离散程度来确定权重。客观赋权法有着客观优势，但不能反映出决策者对不同指标的重视程度，因此，本次研究采用两者耦合的赋权方法。

（1）层次分析法

通过数名相关领域专家对因素进行 9 分标度法打分，形成判断矩阵。

分析得到特征向量，最大特征根为 8.999，利用最大特征根值计算 CI 值约为 0。九阶判断矩阵随机一致性 RI 值为 1.460，CR 值约为 0 小于 0.1，满足一致性检验，计算所得权重具有一致性。

计算的各项数值见表 5。

层次分析法权重表　　　表 5

项	特征向量	权重值
f_1	0.360	4.000%
f_2	2.520	28.000%
f_3	1.800	20.000%
f_4	1.080	12.001%
f_5	0.720	8.001%
f_6	0.180	1.999%
f_7	0.180	1.999%
f_8	0.720	8.001%
f_9	1.440	15.998%

（2）熵值法

熵值法是依据已构成的判断矩阵，求解评价因子的熵值、信息冗余度，最后求得指标权重，具体计算方法见引文[15]。

（3）组合权重

利用乘数归一法将熵值权重和层次分析法得到的权重相耦合，得到组合权重为：

$$W_j = \frac{w_j^1 w_j^2}{\sum_{j=1}^{m} w_j^1 w_j^2}, (j = 1, 2, 3, \cdots, m) \quad (6)$$

式中，W_j 为综合权重；w_j^1 为层次分析法权重，w_j^2 为熵值法权重。

4　工程实例分析

4.1　实例信息

以"（2017）苏民终 338 号"这一工程实例作为目标案例进行说明：2010 年 11 月 2 日，房投徐州分公司将徐州市新城区 7 号地 7-3 地块二标段的土建安装发包给盐城二建施工，工程招标投标价为 6345.99 万元，工程施工过程中，房投徐州分公司在露台、水电、雨污水、门窗、车库、地砖铺设等方面进行了多项设计变更。而该案争议焦点为：多次因设计变更导致的重大工期延误责任如何认定。

此时若作为管理者，想知道最合理的责任判定方案，可进行案例推理寻找相似案例。选取 5 个真实案例作为历史案例库进行展示，分别用 c_1、c_2、…、c_5 表示，目标案例为 c_0，具体信息见表 6。

案例信息表 表 6

案例	f_1	f_2	f_3	f_4	f_5	f_6	f_7	f_8	f_9
c_0	发包人	设计变更	居住建筑工程	1139707	63459900	51	278	2010s4-2011s4	略
c_1	承包人	新增工程变更	安装工程	5761353	17008034	2	122	2015s3-2016s2	略
c_2	承包人	施工方案变更	居住建筑工程	1040273	8862852	58	1002	2011s2-2012s1	略
c_3	发包人	设计变更	居住建筑工程	41653538	71878284	46	253	2007s3－2010s3	略
c_4	承包人	设计变更	居住建筑工程	24643251	45565967	54	890	2014s1－2015s1	略
c_5	承包人	新增工程变更	市政工程	3326523	29620519	33	125	2010s1-2011s4	略

4.2 案例特征计算

下面分别对目标案例 c_0 与历史案例 c_1、c_2、c_3、c_4、c_5，从九个指标之间的相似度进行计算：

（1）类别型指标

通过案例间指标是否一致确定相似度（表 7）。

类别型相似度指标表 表 7

	c_1	c_2	c_3	c_4	c_5
原告	e^{-1}	e^{-1}	1	e^{-1}	e^{-1}
变更类型	e^{-1}	e^{-1}	1	1	e^{-1}

（2）层次型指标

通过概念与其本体中最近公共父节点概念的位置关系计算相似度（表 8）。

层次型相似度指标表 表 8

	c_1	c_2	c_3	c_4	c_5
工程类别	0.545	0.5	0.5	0.5	0.6

（3）数值型指标

根据案例间指标数值的数量关系计算数值型相似度，结果见表 9。

数值型相似度指标表 表 9

	c_1	c_2	c_3	c_4	c_5
变更纠纷金额	0.886	0.998	0.002	0.421	0.946
合同总造价	0.263	0.134	0.866	0.716	0.463

续表

	c_1	c_2	c_3	c_4	c_5
发包人涉及司法案件数量	0.125	0.875	0.911	0.946	0.679
承包人涉及司法案件数量	0.823	0.177	0.972	0.305	0.826

（4）时间序列指标

时间序列中数值元素为"江苏省固定资产投资价格指数"，时间元素为案例所在的建设周期，时间单元为季度。数据来自 Wind 经济数据库。先计算案例间时间序列的累积距离 $dist(i,j)$，求倒数后进行最大值归一化处理得到最终时间序列相似度 sim_{time}（表 10）。

时间序列相似度指标表 表 10

	c_1	c_2	c_3	c_4	c_5
dist	20.333	6.998	53.823	8.127	20.740
1/dist	0.049	0.142	0.0185	0.123	0.048
sim_{time}	0.344	1	0.130	0.861	0.337

（5）文本相似度——文本特征

通过训练的 Bi-LSTM 模型计算文本相似度，如表 11 所示。

文本相似度指标表 表 11

	c_1	c_2	c_3	c_4	c_5
文本相似度	0.206	0.312	0.413	0.622	0.198

4.3　案例权重计算

根据上述计算，得到目标案例与五个历史案例的九类相似度指标，计算熵值法权重 W_j，见表12。

结合上文中的 AHP 权重，将熵值法和层次分析法权重相耦合，得到组合权重见表13。

熵值法权重计算表　　　　　　　　　　　表 12

	f_1	f_2	f_3	f_4	f_5	f_6	f_7	f_8	f_9
W_j	0.088	0.098	0.002	0.212	0.135	0.102	0.121	0.165	0.077

组合权重表　　　　　　　　　　　表 13

	f_1	f_2	f_3	f_4	f_5	f_6	f_7	f_8	f_9
组合权重	0.036	0.280	0.005	0.261	0.111	0.021	0.025	0.135	0.126

4.4　案例推理结果

将各类相似度赋予组合权重，得到目标案例 c_0 与历史案例的总体相似度水平（表14）。

总体相似度表　　　　表 14

	c_1	c_2	c_3	c_4	c_5
c_0	0.475	0.591	0.528	0.707	0.523

从案例结果可知，第 4 案例与目标案例最为相似。而第 4 案例的裁决书为"2019 苏民终 10 号"，它与目标案例均存在图纸变更，图纸及施工方案未及时下发施工单位，导致严重影响人员组织及周转材料利用，最终延误工期增加成本的问题；且二者设计变更均不涉及对合同内容主体的变更。两案例中，法庭均依靠签订的施工合同和工程联系单等证据文件，判定其变更导致的重大工期延误责任均由发包人全部承担，说明计算结果与实际结果具有较好的一致性。

5　总结

鉴于工程变更无法完全预测和杜绝，当工程变更发生后合规管理需要解决的问题是如何合理处理争议，保证冲突高效解决。工程变更作为突发事项产生的博弈焦点，合理处理难度较大，本文通过结合本体和案例推理的方法，从已有规范、判决文书和案例信息中寻找相似案例提供参考，可在信息不完备情况下实现基于案例的紧急响应，并迅速提供解决方案，减少工程变更法律风险，帮助管理者有理有据地探寻变更处理的"最优解"。

参考文献

[1] 杨蛟龙. 从审计视角浅谈工程建设合同的合规性管理[J]. 质量与市场，2021(9)：31-33.

[2] Esmat Rashedi, Hossein Nezamabadi-pour, Saeid Saryazdi. Long term learning in image retrieval systems using case based reasoning[J]. Engineering Applications of Artificial Intelligence，2014(35)：26-37.

[3] Paul R Kleindorfer, Germaine H Saad. Managing disruption risks in supply chains[J]. Production and Operations Management，2005，14(1).

[4] Thomas R Gruber. A translation approach to portable ontology specifications[J]. Knowledge Acquisition，1993，5(2).

[5] 光晖. 基于本体的企业知识型员工人岗匹配案例推理系统研究[D]. 合肥：合肥工业大学，2020.

[6] Thomas R Gruber. Toward principles for the design of ontologies used for knowledge sharing?

［J］. International Journal of Human-Computer Studies，1995，43(5-6).

［7］ Noy N F，McGuinness D L. Ontology development 101：a guide to creating your first ontology ［R］. USA：Standford University. 2001. SMI Technical Report. SMI-2001-0880.

［8］ Agnar Aamodt，Enric Plaza. Case-based reasoning：foundational issues，methodological variations，and system approaches［J］. AI Communications，1994，7(1).

［9］ Gavin Finnie，Zhaohao Sun. R 5 model for case based reasoning［J］. Knowledge-Based Systems，2003，16(1).

［10］ Wu，Zhibiao，Palmer，Martha. Verb semantics and lexical selection［J］. 1994.

［11］ 曲丽. 时间序列的相似性度量方法研究［D］. 哈尔滨：哈尔滨理工大学，2021.

［12］ 陈海燕，刘晨晖，孙博. 时间序列数据挖掘的相似性度量综述［J］. 控制与决策，2017，32 (1)：1-11.

［13］ 陈攀，杨浩，吕品，等. 基于 LDA 模型的文本相似度研究［J］. 计算机技术与发展，2016，26(4)：82-85，89.

［14］ 杨慎全. 基于 CNN-RNN 网络的图像高层次语义检索［D］. 陕西：西安电子科技大学，2020.

［15］ 周梅华. 可持续消费测度中的熵权法及其实证研究［J］. 系统工程理论与实践，2003(12)：25-31.

制度逻辑视角下重大工程项目供应商网络治理研究——以中交集团供应商管理平台为例

王琳卓　张　超　朱方伟

（大连理工大学经济管理学院，辽宁大连　116024）

【摘　要】 处于不同制度场域中的项目网络参与主体，具有各自独特的价值取向和行为方式，而这些制度逻辑的冲突是导致重大工程项目产生高度的复杂性的深层次原因，会给重大工程项目顺利运行带来巨大的难题，影响项目绩效。在这种实践情景下，本文基于制度逻辑理论和网络治理理论，通过案例研究方法，以中交集团为研究对象，识别出重大工程项目供应商网络治理的筛选进入机制、评价与分级机制、淘汰退出机制和制度拼凑过程，构建了重大工程供应商网络治理的多层次治理体系，从制度逻辑视角丰富了中国情境下重大工程项目的治理理论，同时提出相应的实践建议。

【关键词】 重大工程项目；网络治理；制度逻辑；制度拼凑

1　引言

随着我国的"一带一路"倡议实施以及全面建成小康社会进入决胜阶段，"十四五"期间重大工程项目作为现代化建设的重要支撑，迎来快速增长期。但重大工程的庞大体量，社会专业化分工程度的不断提高，重大工程项目的参与主体数量、技术复杂程度、组织结构管理难度等都显著增加[1]，导致重大工程项目在治理模式上与传统项目存在巨大差异[2]，当前重大工程项目普遍存在工期拖延、成本超支、社会影响重大等问题，如何有效治理重大工程项目受到国内外学界与实践界的广泛关注[3]。

重大工程项目不同于传统项目，是由数量众多的项目参与方，围绕项目目标，形成的临时性合作网络，各参与主体具有典型的异质性特征[4]。基于制度逻辑理论[5]，项目网络中存

在市场逻辑和公益逻辑等多重竞争性制度逻辑，表现为项目网络成员的目标冲突与惯例冲突[6]，而项目网络的临时性特征加剧了制度逻辑冲突带来的负面效应，提高了重大工程的治理难度。现有项目治理研究重点关注双边关系治理和项目中的资源配置情况，主要从契约治理、关系治理等视角出发分析某一组特定治理机制的作用效应，重点关注二元关系中交易成本对项目绩效的影响，缺少对于项目作为一个多种制度逻辑共存的复杂系统的治理模式研究。目前项目治理的相关研究，对于多种治理机制的复合运用和系统组合的相关理论探索较少，不能解释治理机制如何在项目整体层面凝聚多方共识，消解制度逻辑冲突，难以总结港珠澳大桥等中国情境下重大工程项目各参与方齐心协力、攻坚克难背后的制度优势和治理经验[7]，因此，本研究立足于我国重大工程项

目，总结其宝贵经验来系统解决制度逻辑复杂性带来的不利影响，以从制度层面提升重大工程项目绩效。

综上，本文以提升项目组织绩效为治理目标，基于制度逻辑理论与网络治理理论，提出多层治理结构与多种治理机制构成的复合治理框架，并通过探索性案例分析，探究制度逻辑复杂性视角下的重大工程项目网络治理机制的设计和组合，以揭示大型工程项目多层网络治理消解制度逻辑复杂性、提升项目绩效的内在机理。

2 文献综述

2.1 制度逻辑复杂性视角下的重大工程项目

重大项目通常超过 10 亿美元，涉及多个公共或私人利益相关方，开发和建设周期长，对社会和经济发展影响大[8]，同时具有高度的复杂性[9]。重大工程项目的高度复杂性源于地域范围、经济环境、规模、类型等诸多因素的共同影响，从一般项目到重大工程项目治理，在本质上出现了从系统性到复杂性的重大演变趋势[10]，David 和 Baccarnin 将重大工程项目复杂性定义为由许多差异化的部分构成[9]，重大工程项目结合参与者的资源、能力和知识来满足除了项目共同目标之外的参与者的目标，这些目标的不一致性，以及承包商和供应商之间权力不对等、彼此成见以及公共和私营组织之间的目标差异都会使得项目变得复杂[11]，这些问题的核心在于项目参与者之间价值取向和行为方式的差异和冲突。

而这些冲突本质上是项目参与主体的不同制度逻辑间的摩擦，制度逻辑是"社会建构的、历史的物质实践模式、概括、价值、信念和人类生产和再生产他们物质的规则、组织的时间和空间以及为社会提供现实意义[12]"。在

重大工程项目中，至少同时存在两种主要的制度逻辑：一是保证项目符合政府的法律和法规，并契合人民利益的政府逻辑；二是利润最大化的商业逻辑[13]，政府、企业等多种制度逻辑之间的作用关系推动着重大工程项目不断实现社会建构与重构的动态过程，影响着组织中各参与主体的行为模式，并构建出重大工程项目的治理环境[14]。网络组织内部和外部各种权力来源取决于主流的制度逻辑[12]，即项目网络的运行通常需要以挑战某些逻辑和按照某些组织工作的方式完成工作[15]，因而系统地辨识不同时间、空间和情景下重大工程项目的制度逻辑有助于更好地理解各参与主体在项目网络制度场域中的行动策略。

2.2 网络治理理论

网络治理机制影响跨组织项目网络的协调和组织方式，并最终决定项目参与者实现项目目标的承诺和能力[16]，目前学界对网络治理机制的研究多从关系、协调、互动等多个视角展开。关系和协调治理下的网络治理机制在于信任机制和协调机制的构建，认为成员之间的信任关系是网络的运作基础，而构建价值协同、信息共享以及激励和行为的良好协调机制是信任关系建立的必要前提[17]。互动治理下的观点认为任务复杂性促使团队产生互动行为，而个体或团体能够从中获取其他个体或团体资源和知识的机会，同时也能在交互中通过思想交流和资源交换增强信任[18]。Robert 和 Sydow 提出网络治理通过责任、例程、角色、关系四种治理机制来提高灵活性，责任代表基于合同的治理，例程和角色反映了行政控制，而关系则代表了社会治理的形式[19]。

这些治理机制均是从单一视角提出的，但在多主体网络复杂治理情景中，依赖于市场、科层或混合形式的二元关系导向治理方法实现

了"治标"，尚不足以有效应对制度逻辑冲突导致的内在冲突。因此，本文以复杂和网络化的多层供应商网络治理为分析框架[20]，识别网络治理机制及其构建方式，探究其对重大项目的制度逻辑冲突的影响关系[21]。

3　研究设计

3.1　研究思路

上述文献回顾说明，项目网络参与主体具有多元化和异质化的特点，存在制度逻辑复杂性，在不同制度逻辑下的网络成员会因不同的利益倾向和价值观选择，进而表现出行为方式的冲突，如何通过协调和平衡网络参与主体之间的制度逻辑以有效推进项目网络的稳定和发展，有待进一步研究。

中国交通建设集团（以下将简称为"中交集团"）的供应商网络治理主要通过供应商管理平台来实现，该平台容纳了所有参与中交集团重大项目的供应商，形成了供应商的关系网络，网络兼具结构复杂性和制度逻辑混合性，展现出了复杂制度逻辑视角下网络多主体的治理过程。而基于此，本文以中交集团供应商网络为研究对象，来研究供应商管理网络应对复杂制度逻辑的治理机制。

3.2　方法选择

本文重点回答重大工程项目中核心企业如何通过治理结构与治理机制的设计实现制度逻辑拼凑的问题，这一研究属于探索"如何（How）"的过程，尤其适合使用案例研究方法[22]。而且，对同一案例在不同时间阶段做深度分析有利于进行理论探索，能够更好地总结中国重大工程领域的独特经验[22]。因此，本研究采用单一对象的纵贯式阶段性比较分析，以实现对治理结构与治理机制内在设计逻

辑的解剖与重现。

案例研究主要是以案例样本的实践为基础进行理论归纳与提炼，而理论产生主要依赖于案例中构念之间的关系模式以及这些关系所蕴含的逻辑论点[23,24]。基于此，本文将围绕重大工程项目形成的企业网络看作是分析对象，将重大工程项目中核心企业对其他企业的治理结构与治理机制视为分析单元，从制度逻辑视角出发，按照不同阶段的演进分析，在纵贯式研究过程中探索重大工程项目网络中核心企业的治理结构与治理机制的内在设计逻辑。

3.3　案例选择

案例样本的选择需要符合理论抽样[23]，在重大工程项目研究中即为信息导向性选择，即选择具有独特性和重要性的案例[25]。选择独特性案例的目的是为了从特殊案例中捕获突破现有理论的关键信息，而选择重要性案例的目的是为了通过捕获"先进试点"的代表性案例的关键信息，以实现向其他相似案例进行逻辑推演。本文选择曾参与港珠澳大桥、深中通道、大连湾海底隧道的中交集团为研究对象，原因有以下三点：第一，中交集团高度参与港珠澳大桥、大连湾海底隧道以及大连地铁四号线等项目，均为技术复杂、环保要求高、建设标准高的"超级工程"，其中贯穿这三个重大工程项目的合作网络是兼具独特性与重要性的典型案例；第二，中交集团是最早一批启动供应商管理系统的工程类集团公司，其供应商管理经验经历了18年的经验积累，形成了独特的管理体系；第三，中交集团的分级分类的供应商管理模式为重大工程项目网络治理结构与治理机制探索提供了宝贵的实践数据基础。18年来，该治理模式经历了应用试点、丰富机制、全面推广的动态发展过程，形成了丰富的数据支撑。

3.4 数据收集

本研究从多个渠道收集案例数据，其中包括在线资料、人员访谈、文献资料、研究报告、公共媒体资料等，通过数据来源的多样性实现数据层面的"三角验证[26]"，具体包括：①对中交参与项目的人员的半结构化访谈；②对中交供应商人员的半结构化访谈；③对参与政府部门的半结构化访谈；④对行业专家的半结构化访谈。

访谈信息表   表1

访谈人数	受访对象职位	参与项目	访谈时长	访谈日期
2	副总、工区经理	大连湾海底隧道	约2.2h	2020年12月
3	副总、工区经理	大连湾海底隧道	约2h	2021年7月
2	副总、供应商经理	港珠澳大桥项目	约2h	2021年3月
3	供应商经理、总工	大连地铁四号线	约2.5h	2021年6月
3	工区经理、总工	大连地铁四号线	约2.2h	2021年7月
5	总经理、副总、总工	大连地铁四号线	约1.8h	2021年8月

研究团队先后于2012年12月、2021年3月以及2021年6月～8月对中交集团参与项目人员、供应商人员、政府部门以及行业专家等进行了深入的半结构化访谈，每次访谈和讨论的平均时长约为2h（表1）。同时为了保证信息的完整性和准确性，在访谈结束后的当天，研究团队会开展讨论，主要针对在本次调研中所获得的信息进行讨论和核对，同时讨论这次调研过程中信息收集的缺陷与不足，为后续的访谈打下基础，如此迭代，以保证收集到的信息的完整性和针对性。

3.5 数据分析

本文在研究人员对收集的数据进行校正之后开展数据分析，分析过程按照三步走，第一步，在理论文献的基础上，识别出制度逻辑、科层治理、契约治理等等重点概念，将中交集团的供应商管理网络治理模式划分为制度耦合、治理机制和网络结构三部分；第二步，对案例数据进行逐级的概念化提炼与编码，对这三部分的内在逻辑和相互之间的联系进行识别和刻画；第三步，进行理论与数据的往复循环对比，通过不断对比构念间关系与理论探索，最终实现理论涌现，并至结论稳定且不再出现新的发现[27]。在具体的编码过程中，本研究采取多级编码的方式，研究人员首先根据整理的所有资料对案例数据进行初始编码，识别出中交集团供应商网络治理中的关键事件和逻辑，共32项初始编码，初始编码结果如表2所示；其次，通过文献指引，将分别归属于制度逻辑、治理机制、网络结构的关键事件等进行概念化编码，共形成18项二级构念，来表征中交供应商网络治理的维度和特征；最后，对二级构念中的编码结果按照不同的维度进行核心编码，根据相关理论和证据梳理，提出中交集团在多重制度逻辑下的供应商网络治理机制和网络结构，具体编码结果如表2所示。

中交供应商网络治理机制和治理

结构编码   表2

初始编码	关联编码	核心编码
打造创新高效产业链 符合国家产业导向	产业发展	政府逻辑
建造百姓使用的平安百年品质工程 国家行业部门对于产品质量的要求	安全生产	
行业发展的绿色节能要求 环境、健康与安全三项指标	社会公益	

续表

初始编码	关联编码	核心编码
供应商实力强、服务好 质量优先的企业发展理念	实力强	市场逻辑
减少采购成本、利益最大化	降低成本	
良好的合作经历 较好的合作关系	良好的合作 体验	逻辑耦合
供应链整体利益共享 国家逻辑与市场逻辑整合	利益共享	
新供应商准入条件	节点筛选	筛选准入 机制
划清网络界限	边界划定	
标准化准入	门槛标准化	
建立标准化退出机制	标准化退出	淘汰退出 机制
建立黑名单制度	黑名单	
鲶鱼效应，竞争机制 为竞争机制提供合同标准化基础 完善市场化竞争流程	市场化竞争	评价分级 机制
面向供应商动态专项考核 面向自网络及其他主体考核	考核评价	
高级别供应商更易获取资源 优先分配给供应商库中的合作 伙伴	利益分配	
特殊贡献可以调整级别 供应商网络动态优化和流动	供应商网络 动态优化	
网络参与主体 治理参与主体	多方参与 主体	多层治理 结构
多层级结构 层级管控 供应商网络关系	多层级 关系	

4   案例初步分析

4.1   多层网络治理的动因

我国工程建设项目经过多年的发展取得了骄人的成就，但也因政府干预、市场化水平、从业人员整体素质等因素给项目的运行带来棘手的问题，当市场机制、法律手段不能及时有效地防控道德风险与行为偏差时，就需要新的治理手段来激励和约束工程项目参与者的合作关系。中交集团构建的面向工程项目的多层治理网络是我国在基础设施建设大背景下涌现出

的有效实践，主要应对四方面的问题，大规模与定制化的冲突困境、供应商的履约风险、政府干预的不确定性以及工程项目总承包方的治理困境。

4.1.1   规模化与定制化的冲突困境

建筑行业规模化生产与定制化需求的问题根源，是规模化效应的统一性与定制需求的个性化之间的内在矛盾。重大工程项目便面临着这样的矛盾，主要表现为两个方面：其一是工程项目存在一定的情境特殊性，由于地质因素、环境因素，每一个地铁项目都是独特的，因此每个项目都需要一事一议；其二是工程项目追求合理的利润空间，最低价中标的大背景在不断改善，但目前大型项目的合理利润空间还是主要靠原材料的供应，因此，控制材料和人力成本成为保障企业竞争力的关键，物资材料作为成本的主体只有通过规模化的采购才能有效地实现。

4.1.2   供应商的履约风险（市场主导逻辑）

建筑行业末端供应商的受教育程度普遍不够高，存在着较为严重的契约意识不足的问题，"部分小老板的素质并不高，往往会挣钱了按照合同来，赔钱了就不干了"的逆向选择和道德风险情况存在，也说明了合同和法律缺乏有效制约从业者以及维护行业良性发展的效力。

4.1.3   政府干预的不确定性（政府主导逻辑）

工程项目与民生生活息息相关，噪声、环保、拆迁等多种民生需求往往通过民生建议、舆情事件等方式对项目产生压力，在这过程中会产生行政干预与政策波动，施工方往往受制于地方政府的安排，也成为项目网络的包袱，需要项目参与方共同承担。

4.1.4   工程项目总承包方的治理困境（制度逻辑冲突）

总承包企业在工程项目中承担总体协调工作，往往需要协调网络参与成员之间的行为和

利益，而当利益无法实现良好协同时，容易出现部分参与方的机会主义行为，不仅无法凝聚网络合力，甚至会带来严重的内耗，给工程项目生命周期的各个阶段带来成本、施工、环保、社会影响等多方面的风险，严重影响项目绩效。

包括实现大规模定制目标，也包括对冲工程行业存在的履约风险、消减非市场因素给项目带来的不利影响。因此，中交集团通过自身二级子公司，构建了多层次的网络治理结构，通过筛选进入机制、评价与分级机制、淘汰退出机制组合构建了完整的多层网络治理体系，并且在该过程之中实现了制度逻辑的拼凑，如图1所示。

4.2　多层网络治理机制—制度拼凑过程

中交集团进行多层次网络治理的目标，既

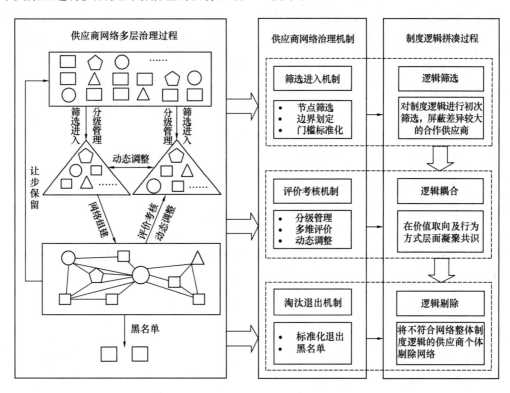

图1　中交集团多层网络治理机制-制度拼凑过程框架图

4.2.1　筛选进入机制

筛选进入机制是中交集团进行多层次网络构建的第一步，筛选进入机制实现的是多层次网络治理过程中的节点筛选、边界划定功能以及标准化门槛效应。节点是指构成中交集团的多层次网络的供应商主体，节点筛选指的是供应商主体的身份确定。对于中交集团而言，成为供应商的首要条件是在软硬件上达到门槛要求，选择"资质齐，实力强，具有较好的服务能力"的供应商作为网络的构成节点，同时兼

顾区域和行业要求的相对指标，在供应商所处行业、业务区域布局、供应商综合竞争优势的基础上，选择契合公司需求的合格供应商。

筛选机制的一个重要功能就是建立统一的标准，将具备一定能力的供应商纳入网络内部，这样可以有效抑制组织间的腐败、利益输送等不道德行为，为多层次网络的持续发展提供了一个良好的环境。而另一个重要功能就是划定边界，只有经过筛选合格的供应商才能加入网络中，能够有效地规避和控制风险，减少

管理成本。

4.2.2　评价与分级机制

评价与分级机制是中交集团实现动态考核和网络良性竞争的重要治理机制。评价功能实现了绩效的量化，以及网络贡献的显性化。中交集团的评价机制主要包括分级管理、多维评价和动态调整三个方面的子机制，在整体上实现了评价机制的有效性、准确性和动态性。

分级管理是网络核心主体通过对供应商进行等级划分和评定，形成不同层次的治理关系，以实现差异化管理的目标。中交集团将供应商分为战略供应商、优质供应商以及合格供应商三个级别，战略供应商是最高级供应商，是在战略层面与中交集团相契合，是合作关系密切的供应商；优质供应商是次高级供应商，是绩效表现优秀的供应商群体，他们能够表现出与中交集团相似的战略目标和行为惯例；而合格供应商是符合供应商入网条件的基础级供应商，经过了基础条件的筛选，是构成中交集团供应商网络的主体成员。

多维评价是在网络治理结构中建立系统评价的主要措施，其建立过程也是网络中多主体建立共识的过程。主要包括供应商自评、体系考核、现场考核与专项考核等四个方面，考核依据战略发展方向、合作意向、采购计划与额度、供应商绩效等因素，覆盖供货、质量、价格、沟通、体系、风险等情况，实现多维立体的考核。

动态调整是指依据多维评价结果和特殊贡献，在不同层次供应商之间形成的内部流动和升降级机制，是保障网络治理优胜劣汰、良性循环的重要机制。

4.2.3　淘汰退出机制

淘汰退出机制是推动网络实现新陈代谢的机制，以维持供应商的良性互动以及供应商网络的健康发展。淘汰机制是通过与不合格的供

应商解除关系，甚至列入黑名单以实现供应商网络关系脱离，主要由两个要素构成，首先是标准化退出机制，即中交集团的各级组织在管理本组织供应商时，实行统一的供应商淘汰退出机制，以维护和建立公开的筛选标准，其次是为了实现网络长期利益，对退出的供应商，在特殊情况下会采用让步保留的方式，保留部分未达到标准的供应商。

最后是黑名单机制，是退出机制的加强版，该机制通过锁定负面供应商，阻断与其的业务联系，来有效降低及规避供应商行为给公司带来的网络风险，通过中交集团作为核心发布统一黑名单，可以有效地促使黑名单企业隔离网络业务。

4.2.4　多层治理结构

中交集团作为庞大网络中的核心主体，主要是通过多个子公司形成的关系网络以实现对于整个网络的治理。在此供应商网络中，中交集团作为核心企业负责网络整体绩效导向、规则制定和战略指引，以管控整个网络的发展方向，中交集团子公司作为网络治理的实施者，在治理过程中按照制度规则，吸收、评价和淘汰供应商，并由相互联系的中交集团子供应商共同形成单极网络。在此单极网络中，既有依靠资源禀赋产生的科层机制，也有依靠签约合同形成的契约机制，还有通过信息与信任构成的关系机制，各种参与主体、多层的网络结构和多样化的治理机制共同构成了复杂的中交集团供应商网络。

4.2.5　制度拼凑过程

政府和供应商的制度逻辑冲突，体现在存在较大差异价值观和行事方式上，政府主要关注项目的社会影响而供应商主要关注经济效益，政府在执行上主要坚持合规原则，而供应商则是结果导向经常忽略过程的合规性。因此，为了减少双方在网络中的制度逻辑冲突，

需要通过中交集团的治理机制实现制度逻辑的耦合，以减少冲突带来的负面影响。

中交集团通过进入筛选、评价与分级以及淘汰退出等综合治理手段，依托多层网络治理结构，在网络层面上凝聚了参与主体的多方共识，并通过各方制度逻辑的有效耦合，实现了制度拼凑。其中，筛选进入机制完成了制度逻辑的初次筛选，屏蔽了制度逻辑差异较大的合作方。评价与分级机制是供应商价值取向及行为方式共识凝聚的体现，实现了网络整体制度逻辑拼凑，淘汰机制是将不符合网络整体制度逻辑的供应商个体剔除网络。制度拼凑过程也是制度逻辑耦合的过程，因非经济因素带来的不利影响也可以通过供应商网络整体实现成本均摊和风险分担。

5 讨论与总结

5.1 理论贡献

本文引入制度逻辑理论来解构供应商网络情景下不同参与方的治理逻辑，首先从制度拼凑的角度来分析大型工程项目通过多层网络治理机制实现制度逻辑耦合的作用过程，剖析了项目情境下多方制度逻辑耦合的治理实现过程，丰富了现有制度逻辑理论在项目领域内的理论解释力。同时，本文还从多层次治理的视角，拓展了现有项目治理理论，现有项目治理理论主要从交易成本和公司治理的视角展开，本文融合这两种理论视角，识别出中交集团多层次的网络治理结构及治理机制，探索并构建了重大工程供应商网络治理的多层次治理体系。

5.2 现实意义

供应商管理网络高度的制度复杂性要求项目总承包企业关注并拼凑网络不同参与方之间的制度逻辑，将网络的整体目标同各项目参与方的目标进行有效结合，凝聚共识形成合力，并充分运用多种治理机制以及多层治理结构，以提升重大工程项目的治理效能。

在实践中，总承包企业可以通过综合信息管理系统的建设，以及进入、分级、退出等筛选机制的运用，构建供应商管理体系，以实现对供应商的全周期管控，提升项目管理过程中对于供应商的有效管控。

5.3 局限性与展望

本文是基于中交集团的案例展开深入研究的，但这种网络治理机制和模式是否能够推广到其他总承包企业并顺利运行，还需进一步采用大样本的数据进行验证。此外，供应商网络运行机制是否还有其他影响因素或机制未被挖掘，以及这些因素或机制对网络的运行造成何种影响，还有待进一步探究。最后，中交的供应商管理网络是经过制度拼凑和耦合形成的，具有较强的中国特色，未来还需对制度逻辑相似的海外国家展开研究，以验证该套治理模式的泛用性和有效性。

参考文献

[1] Senescu R R, Aranda-Mena G, Haymaker J R. Relationships between project complexity and communication[J]. Journal of Management in Engineering, 2013, 29(2): 183-197.

[2] Biesenthal C, Clegg S, Mahalingam A, et al. Applying institutional theories to managing megaprojects[J]. International Journal of Project Management, 2018, 36(1): 43-54.

[3] Qiu Y, Chen H, Sheng Z, et al. Governance of institutional complexity in megaproject organizations[J]. International Journal of Project Management, 2019, 37(3): 425-443.

[4] Sydow J, Braun T. Projects as temporary organi-

zations: An agenda for further theorizing the interorganizational dimension [J]. International Journal of Project Management, 2018, 36(1): 4-11.

[5] Thornton P H, Ocasio W. The institutional logics perspective: a new approach to culture, structure, and process[M]. London: Oxford University Press, 2012.

[6] Ashraf N, Ahmadsimab A, Pinkse J. From animosity to affinity: the interplay of competing logics and interdependence in cross-sector partnerships [J]. Journal of Management Studies, 2017, 54(6): 793-822.

[7] 盛昭瀚, 刘慧敏, 燕雪, 等. 重大工程决策"中国之治"的现代化道路——我国重大工程决策治理70年[J]. 管理世界, 2020, 36(10): 170-203.

[8] Flyvbjerg, Bent. Introduction: the iron law of megaproject management[J]. The Oxford Handbook of Megaproject Management, Oxford University Press, 2017.

[9] David, Baccarnin. The concept of project complexity—a review[J]. International Journal of Project Management, 1996, 14(4): 201-204.

[10] 盛昭瀚, 薛小龙, 安实. 构建中国特色重大工程管理理论体系与话语体系[J]. 管理世界, 2019, 35(4): 2-16+51+195.

[11] Marrewijk A. Digging for change: change and resistance in interorganizational projects in the utilities sector[J]. Project Management Journal, 2018, 49(3): 34-45.

[12] Thornton P H, Ocasio W. Institutional logics and the historical contingency of power in organizations: executive succession in the higher education publishing industry [J]. American Journal of Sociology, 1999, 105(3): 801-843.

[13] Qiu Y, Chen H, Sheng Z et al. Governance of institutional complexity in megaproject organizations[J]. International Journal of Project Man-

agement, 2019, 37(3): 425-443.

[14] 李迁, 武雨欣, 胡毅, 等. 多重制度逻辑视角下重大工程组织模式多样性分析[J]. 科学决策, 2019(11): 49-65.

[15] Biesenthal C, Clegg S, Mahalingam A, et al. Applying institutional theories to managing megaprojects[J]. International Journal of Project Management, 2018, 36(1): 43-54.

[16] Kujala J, Aaltonen K, Gotcheva N, et al. Dimensions of governance in interorganizational project networks[J]. International Journal of Managing Projects in Business, 2021, 14(3): 625-651.

[17] 鄞益奋. 网络治理: 公共管理的新框架[J]. 公共管理学报, 2007(1): 89-96+126.

[18] 彭正银. 网络治理理论探析[J]. 中国软科学, 2002(3): 51-55.

[19] DeFillippi R, Sydow J. Project networks: governance choices and paradoxical tensions [J]. Project Management Journal, 2016, 47(5): 6-17.

[20] Ruuska I, Ahola T, Artto K, et al. A new governance approach for multi-firm projects: lessons from olkiluoto 3 and flamanville 3 nuclear power plant projects[J]. International Journal of Project Management, 2011, 29(6): 647-660.

[21] Stigler, George J. The organization of industry [M]. University of Chicago Press Economics Books, University of Chicago Press, 1983.

[22] Yin R K. Case study research design and methods (5th ed.). Thousand Oaks, CA: Sage, 2014.

[23] Eisenhardt K, Melissa E Graebner. Theory building from cases: opportunities and challenges [J]. Academy of Management Journal, 2007(50): 25-32.

[24] 毛基业, 苏芳. 案例研究的理论贡献——中国企业管理案例与质性研究论坛(2015)综述[J].

管理世界，2016(2)：128-132.

[25] Flyvbjerg B. Five Misunderstandings About Case-Study Research［J］. Qualitative Inquiry，2006，12(2)：219-245.

[26] Glaser B，et al. The discovery of grounded the-ory：strategies for qualitative Research［J］. Nursing Research，1968(17)：364.

[27] Eisenhardt K M. Building theories from case study research［J］. Academy of Management Review，1989，14(4)：532-550.

"一带一路"倡议和"粤港澳大湾区"建筑工程管理可持续发展及国际化的机遇

余伊琪

［英国特许土木工程测量师学会（香港分会）理事兼义务司库］

【摘　要】　近年来，愈来愈多的香港和澳门建筑企业及工程咨询顾问开始关注"一带一路"倡议和"粤港澳大湾区"带来的正面影响。本文旨在探讨建筑企业及工程咨询顾问，尤其是内地与港澳三地工程承包企业及工程咨询顾问在参与融入粤港澳大湾区的过程中遇上"一带一路"的持续稳步发展，如何善用这个千载难逢的机会以满足三地建筑可持续发展的融合与共同参与国际建筑市场的机遇，并符合联合国的环保、社会及管治政策 ESG 和 17 项永续发展目标 SDG。

【关键词】　一带一路；粤港澳大湾区；ESG/SDG 可持续发展；国际工程管理；三地融合

1　香港的优势背景

中国香港毗邻内地，位处我国东南岸，内地各大城市瞬间即达，全球近半人口均在五小时航程之内，是国际企业进入内地的门户以及内地企业进军国际市场的跳板。香港服务业优秀完善，由金融服务、专业服务、基建服务、运输服务、信息科技服务以及环保服务等组成，无论是业务国际化、服务深度或专业水平都在亚洲数一数二，能就"一带一路"沿线国家和"粤港澳大湾区"的相关项目提供专业意见、管理及配套支持，特别是在可持续和智慧城市发展多种形式的产学研结合中，把握相关节能和环保技术创新带来的契机。因此，香港汇集独特优势，凭借深厚知识、丰富经验与通达人脉，全面支持国际投资者及项目把握"一带一路"和"粤港澳大湾区"走出去和引进来

的发展机遇。

香港特别行政区致力推动减排，目标于2050 年前达到碳中和。

中国香港作为国际金融中心，在引领市场资金支持本地、内地公私营机构投资到绿色和低排放活动，协助全球达到零排放的进程上，可发挥积极的关键作用。香港特区在过去几份预算案中，大力推动香港绿色和可持续金融的发展，构建香港成为大湾区绿色金融中心。透过支持内地和大湾区内绿色项目投融资及认证，一方面可助力国家实现绿色经济转型的战略目标，另一方面也帮助香港金融业走在国际趋势的前端，进一步多元化发展。

另外，香港在管理大型项目方面经验丰富，傲视全球。香港建筑工程专才不单具备国际认可的技术，更拥有项目管理及顾问的专业知识，管理、策划及发展各种采购模式的项目

成绩卓越，与内地及国际基建公司亦一直有合作经验。国际经验及提供高技术国际专才两方面均名列世界前茅，拥有庞大的专业服务集群，具备广阔的国际网络及关系。

2 "一带一路"倡议和"粤港澳大湾区"的关系

2.1 "一带一路"

"一带一路"倡议现已覆盖全球近三分之二的国家。"一带一路"倡议的初衷是建设贯穿亚欧非的重大基础设施网络，自提出以来经历了漫长的发展进程。随着中国着力寻求新的解决方案，新的国际化发展机遇亦将不断涌现，定义广泛且不断发展变化，但其主要愿景保持不变，即政策沟通、设施联通、贸易畅通、资金融通和民心相通。"一带一路"倡议正不断覆盖全球更多国家，扩大包容性合作，并吸引更多民间资本参与。亦开始专注于发展优质项目，从而提升透明度，降低风险，加大对尽职调查的重视程度。从这一角度出发，国际承包商和工程咨询机构可有效协助客户评估海外拓展机会，以顺应不同的监管要求和文化背景。在项目融资方面，目前为止，大部分"一带一路"项目均由中国企业（尤其是国有企业）提供资金支持并承接。国有企业已经意识到，他们必须增强吸引投资的能力。跨国企业通过多种方式吸引投资，包括债务或股权融资、并购、"建设—运营—移交"（Built-Operate-Transfer，BOT）合同、公私合营（Public—Private—Partnership，PPP），甚至通过工程、采购和建设（Engineering—Procurement—Construction，EPC）合作项目吸引投资。

推进"一带一路"项目融资的参与方越来越多，包括民间资本、多边银行以及外国政府等。对于擅长为大型项目筹集资金的跨国企业而言，这将带来机遇。未来"一带一路"项目将更具全球化和包容性特征，民营企业的参与幅度将大幅提升。有些跨国企业擅长管理基础设施、地产以及合资企业。长远来说，跨国企业可与中国企业展开合作。例如，跨国企业既可以提供优质产品，提供使用该产品所需的相关技术与管理技能。与此同时，对于亚洲和非洲国家来说，除能源、资源和基础设施外，"一带一路"还将进一步延伸至其他领域，并催生大量机遇。中国香港作为国际金融中心，各大银行等金融机构能为内地企业在"一带一路"市场提供银行服务，包括财资管理、资本市场融资、提高人民币在国际上的使用率等。在工程管理方面，专业人士如工料测量师、工程师、法律专业人员、财务专业人员等，亦能用自己的专业知识来协助各项目参与方设计最好的项目采购方案，并进行全过程的管理直至项目完工交付。

在项目风险管理方面，工程咨询机构可运用其专业能力，针对不同机遇开展服务，帮助客户避免投资失误，削减投资成本。在"一带一路"沿线国家有广泛业务网络的咨询机构具有更加突出的优势。他们能够帮助客户应对不同税务和建设监管要求，评估政治和政策风险，分析文化差异所带来的潜在影响，为客户开展海外投资保驾护航，并助其创造最大收益。正如中国国家发展和改革委员会副秘书长周晓飞所说，国际金融机构和专业化服务机构的参与将大力推进"一带一路"建设。参与"一带一路"建设的跨国企业和中国国有企业也将因此得以更有效地应对不同挑战，并挖掘最佳机遇。

2.2 粤港澳大湾区

粤港澳大湾区是我国开放程度最高、经济

活力最强、国际化水平领先的区域之一。尝试推动形成全开放新格局是《粤港澳大湾区发展规划纲要》对建设粤港澳大湾区的明确定位和要求。粤港澳大湾区将对标国际一流标准，深入推进投资便利化、贸易自由化和人员往来高效，携手港澳打造国际一流营商环境。

粤港澳大湾区将充分发挥衔接国内和国际两个市场、聚合两种资源的独特优势，增强联通国内及国际双循环功能，提升全球资源配置能力。广东携手港澳建立统筹内外的贸易、投资、生产、服务网络，把港澳发达的金融体系、经贸网络优势与广东完备的制造体系、市场优势结合起来，共同谋划用好区域全面经济伙伴关系协定（Regional Comprehensive Economic Partnership，RCEP）等重大合作机制，不断深化与东盟等地区务实合作，强化对"一带一路"建设的升级支撑作用。

积极发挥"一个国家、两种制度、三个关税区"的特点优势，创新体制机制，粤港澳大湾区的建设将为中国乃至全球发展区域经济提供难以复制的宝贵经验。内地与港澳联营律师所已发展至15家，97名港澳律师在联营所执业；1600多名香港专业人士通过互认取得内地注册执业资格；68家港澳企业和256名港澳专业人士在自贸试验区备案执业。粤港澳大湾区在人才培养、资格互认、标准制定等方面的成果不断涌现。

秉承着改革开放前沿之地的创新精神，粤港澳大湾区建设中，通过区分不同领域、不同情况，积极探索具体问题的"一事三地""一策三地""一规三地"，逐步打破壁垒屏障，进一步推动粤港澳三地规则衔接、机制对接。

对于营商环境、城市治理等港澳领先、可以复制的，直接对接实现规则"联通"。对于贸易投资自由化、便利化等港澳领先、但难以直接复制的，加快构建接轨的制度体系，实现

规则"贯通"。对于执业资格、社会保障等由于管理方式不同形成的规则差异，则在充分协商基础上实现规则"融通"。

自2019年《粤港澳大湾区发展规划纲要》发布以来，粤港澳大湾区成为世界经济领域的高频词。打破藩篱、疏通堵点，在共同努力下，粤港澳三地在规则衔接、创新合作等多维度取得实质性进展，大湾区正以坚实的脚步向前行进。一直以来，机制规则的差异成为三地融合与发展的主要障碍，而今这些壁垒正被逐渐打破。2019年《粤港澳大湾区发展规划纲要》出台至2021年4月，各级政府已出台逾230部促进大湾区发展的政策文件，"制度之异"正在逐渐变为"制度之利"。更好地融入国家发展大局，关系到香港未来的繁荣和机遇，特别是"十四五"规划的实施和粤港澳大湾区战略的加速推进，从创科到金融、从国际航空枢纽到中外文化艺术交流中心、从国际法律及解决争议服务到区域知识产权贸易，香港都大有发展潜力。

2.3　大湾区的专业服务融合

在国际工程设计咨询公司看来，大湾区需要进一步消除障碍为境内外企业创造公平的竞争环境。在工程建筑领域，包括总体规划、城市改造和基础设施建设，大湾区的11座城市必须进一步提升协调性，以吸引来自内地和其他地区的人才。提升人、货物、服务和信息间的互联性需要依靠大量的基础设施，这不仅指的是物质基础设施，还包括社会基础设施，例如学校、医院和经济适用房。大湾区将会变得更加开放，并将吸引更多人口。但是城市化进程的持续，可能带来包括交通拥堵、空气污染以及可持续能源和清洁水的供给等挑战，而我们需要面对并克服这些挑战。

随着大湾区加速发展，意在与当地企业合

作、分享国际最佳实践的跨国企业将迎来诸多机遇。设计、工程和项目管理咨询公司的渊源可追溯至20世纪70年代的香港，彼时香港正加大对关键基础设施的投资。随后，扩展至深圳和澳门。鉴于大湾区相关政府部门正努力通过改造棕地、规划未来增长、向清洁型工业和技术过渡，以及持续大规模投资基础设施来打造世界一流的湾区，建筑行业在大湾区将大有可为。然而，对于跨国公司而言，成功抓住机遇的关键在于充分利用国际标准和最佳实践，以及强大的本地经验与合作伙伴关系来满足本地需求。伙伴关系是内地企业追求获取知识技能的最佳途径，因为他们希望合作做到最好，这为大多数境外公司提供了机会。内地设计院和承包商正面临产业结构及持续发展的挑战，而两者之间存在缺口。因此，香港的机遇是去扮演整合者的"走出去和引进来"的角色，帮助他们填补下列缺口，将两者融合。

1）人力资源是企业的重要资产，员工往往是生产力的好与坏的关键因素。一所成功的企业首要做到的就是永不忽视员工的努力。除了竞争优势的薪金之外，为员工提供完善的福利如医疗福利、培训与晋升机会等，让员工感受到工作与生活的平衡，减少职员的流失率，这就是对员工的社会责任。

2）企业要能够可持续发展，对内必要跟股东们有交代，例如要能够通过投资者的同意并考虑到投资者的利益情况之下达到目标营业额，保持与投资者的正面互动是经营企业的其中一个重要环节。

3）经济体系的完整性是依靠各行业的不断运作，并非由某一所企业独自经营，保持业界之间的良性竞争也是企业需要做到的。所以业务上的决策需要考虑到宏观经济的层面，避免对大国经济造成不利影响，亦注意不要损害同业或合作者之间的关系。

4）现在环境保护已成为全球的关注问题，所以企业所作的经济活动亦需兼顾能源和环保效益方面，尤其是建筑业对这方面更不容忽视。试想象一所企业能够完成多项基建之余又能够通过ESG（环境、社会及管治）的新模式（图1），把对大自然的损害降至最低水平，甚至对环境保育作出贡献，其公司的品牌必然得到大众的信赖，这些无形资产对企业的营运带来很大的帮助。

图1　ESG新模式

3　建筑行业对国际工程管理的看法与相应的行动

3.1　国际工程管理

一直以来，香港是一个国际化大都市，经济已经充分融合进全球经济，有非常国际化的营商网络和环境，特别是与海上丝绸之路沿线国家的交往历史悠久，了解所在国的法律法规、风土人情、商业及投资规则。香港真正的优势在于，它拥有大量的国际化的建筑专业管理人才，特别是在普通法系下培养的国际经济法人才、投资融资人才、全过程项目管理人才、财务管理人才、通信技术设备管理人才等。可以说，"一带一路"倡议的发展，不缺国家实力，不缺钱，不缺伙伴，可最缺国际化的专业管理人才；而这恰恰是香港的优势。大湾区建设的另一项主要目标，就是对外连接

"一带一路"和国际市场，对内辐射和带动环珠三角和泛珠三角区域发展。澳门则主要建设世界旅游休闲中心，打造中国与葡语国家商贸合作服务平台。

3.2 国际工程管理与走出去海外企业的影响

据亚洲开发银行估计，随着发展中国家在未来数年积极寻求提高生产力，以及数以百万计的人口迁移至城市，亚太区每年对基建投资需求将高达17000亿美元。汇丰预计，单是东盟六大经济体在2030年前的基建投资需求就高达21000亿美元，但开支预算只能负担约9100亿美元。这就是"一带一路"建设带来的巨大机遇。"一带一路"建设的宏大目标势将大大推动基础设施建设，同时带动这些基建项目的融资活动。基建工程和融资服务双管齐下，加上国家积极推动，"一带一路"倡议将如虎添翼，除了造就商机，推动商品流通，亦会刺激人口数以亿计的沿线国家未来的金融活动。假以时日，"一带一路"基建项目创造的涟漪效应，将远大于每个单独项目效益的总和，从而对经济发展产生前所未见的拉动力。

中国的"一带一路"倡议是双向进行的。即使"一带一路"倡议的核心是推动中国的对外投资，但同时也注重鼓励境外跨国公司参与对中国及"一带一路"沿线国家的投资。"一带一路"倡议（及其背后的资金支持）为中外企业"一带一路"沿线的合作开发项目提供了机会，包括双方共同投资建立合资公司或者外方将"一带一路"项目分包给中国公司，或者由中国公司寻找"一带一路"沿线被投资地的当地合作伙伴共同进行相关项目开发。鉴于"一带一路"项目拥有其专项资金拨备，相较于一般的投资机会，外国公司及机构在与中方就"一带一路"相关的项目合作进行谈判时可

能会更加顺利。

香港在"一带一路"中可以抓到的发展机遇主要在"一路"上，即"21世纪海上丝绸之路"，从中国沿海港口经过南海到印度洋，延伸至欧洲，还从中国沿海港口经过南海到南太平洋，发展中国和东南亚、南亚、中东、北非及欧洲各国的经济合作。这些国家和地区无疑涉及大量基础建设，例如高速公路、铁路、大桥等，这些建设将促进中国与邻近国家，乃至欧洲、非洲及中东国家的贸易。香港地理位置优越，从货物贸易转口到金融货币转口，再到文化传递，都是重要的支点，可以与内地和澳门（特别是葡语系国家）发挥很好的作用，可以对三地经济发展注入新的动力。

3.3 向全过程和数字可持续性管理迈进

在国家的"十四五"规划中，值此开局之年、数字化转型之时，能充分发挥"十四五"规划纲要提出的要点：利用机会通过循环经济商业模式应用全过程管理和数字化转型加速实现可持续发展目标。传感器、建筑信息模型（BIM）、数字孪生、跟踪和追溯、材料护照、3D打印和机器人等数据驱动解决方案的部署将开启价值驱动数据分析的时代。这将帮助企业超越提供洞察力，实现基于人工智能（AI）和机器学习（ML）的优化决策，从而支持可持续发展目标并最终实现。

可持续性的好处诸如更大的竞争优势、员工保留率和通过降低能源使用量和减少材料浪费而节省的好处，加起来支持企业投资于更可持续的实践。可持续发展计划不仅对地球有益——它们也有助于实现业务目标。

在设计过程中越早通过数据获得的确定性越多，拥有关于建筑的信息就越多，因此可以做出更好的设计选择。将此视为帮助我们在未来嵌入真正符合循环经济的实践的基础。例

如，通过收集有关所用材料类型及其使用全寿命周期的数据，我们可以防止建筑物过早拆除。随着新技术和创新的出现，建筑业各界响应香港特别行政区相关政策，已经有共同达到2030年实现净零的具体计划。

能源转型计划网络和性能设计计划等促使我们采用更精益和简单的设计方法。这不仅使我们的建筑更具可持续性，而且还为客户创造了更好的价值。减少用于制造的材料的数量，从而降低成本，使我们能够采购更高质量的产品。针对新项目试用制造和组装设计（Design for Manufacturing and Assembly，DFMA）技术，与传统建筑方法相比，它在减少我们的碳排放方面具有优势，因为运输混凝土会大大增加我们的碳排放量。

挑战之一是，现在我们倾向于凭直觉进行设计，利用多年的经验，而不是使用数据来获得直觉。为了实现我们的可持续发展，设计未来建筑需要全新的知识层。例如，通过将碳数据嵌入 BIM 模型和策略中，我们将从直观的环境转移到数据和证据驱动的环境。

3.4 可持续发展的影响

可持续发展是"一带一路"的关键。集合力量，推动环保事业，为"一带一路"沿线国家如缅甸、孟加拉国、印度尼西亚和杜拜提供绿色能源顾问服务。香港可扮演内地与海外国家地区互联互通的角色，并可按各方需要建立合作网络，发挥优势，把绿色方案带到"一带一路"的国家和地区。

建筑业与许多产业有着十分密切的关系。原材料、技术、人员、资金等各种因素之间相互交错式的企业发展与许多其他企业紧密相关。因此，总承包商应与客户成为合作伙伴，为客户提供服务的售前咨询，以增加其选择的准确性和投资效益；与客户建立对话机制，及

时了解需求；增强售后服务，帮助客户了解项目的正确使用方法。企业应找到自身优势所在，利用产业链上下游企业更为高质和廉价的资源，实现资源的更优配置。一些工程承包企业往往形成利益团体，只看到短期利益，这样影响了企业的可持续性发展，造成资源分散、力量薄弱的问题，如果不积极协作、形成优势互补，谋求共同发展，就很有可能被其他企业推倒，最终市场被全部抢占。

要进一步加强香港与内地的全面合作，实现三地发展新跨越。一是共同推进中国企业"走出去"。二是共同推进产业跨境转移。三是共同发展海洋经济。四是共同培养"一带一路"专业人才。"一带一路"这一充满东方智慧的重大倡议为香港把握新机遇、实现新发展创造了广阔前景，也为每一个参与其中的企业提供了国际化发展的新机遇。

4 建筑行业采取的相应行动与改变

国际咨询工程师联合会（FIDIC）发布了SOW2021-Time to ＄Tn-vest 报告，说明全球基础设施每年需要 7 万亿美元来应对气候变化。7 万亿美元相当于当前基础设施投资的两倍，FIDIC 认为，如果没有这样的支出水平，联合国的 17 个可持续发展目标（SDGs）将无法实现。这是一个简单的结论。FIDIC 经济和战略政策主管（Graham Pontin）表示，这当然引发了很多问题，尤其是在投资和维护支出的平衡以及两者之间的比例应该如何倾斜方面。还有需要考虑的COVID-19 大流行，将如何以及何时走出困境。

可持续性的所有要素都会在某个时候进入全过程采购和合同的各个阶段。无论在世界的哪个地方，我们不断地监控和评估这将如何纳入合同。由于我们的合同在全球范围内的使用方式各不相同，因此有些事情在某些地方比其

他地方更有效，重要的是我们对与我们接触的人保持灵活性。我们的合同不是一成不变的，它们会随着市场不断发展。整个COVID-19危机提高了数字合同的前景，这当然是我们需要探索的一个领域。

在2021年及以后的时间里，项目的成本效益和估算将需要重新考虑。这将是COVID-19之后的关键，因为全球政府和企业对大流行的财务影响做出反应。预算限制将导致项目被推迟、取消和替代融资模式。对新的或现有的基础设施项目进行实时协作合同和成本审计的需求从未像现在这样重要。

5 总结

随着国家经济体量持续扩大，并在"一带一路"倡议和国内国际双循环相互促进的新发展格局下，无论是内地或境外建筑企业，新基建的需求势将大增，香港的专业服务有条件及能力，不单是提供境外的"引进来"，更要为内地企业"走出去"打造好基础。

国家在2026～2030年期间逐步减少煤炭消耗，力争于2030年前碳排放达至峰值，并于2060年前实现碳中和的目标。要达到有关目标，有分析预计所需的绿色低碳投资将超过百万亿元人民币。事实上，全球在低碳转型过程中对资金的需求正不断加大，2020年全球绿色债券发行量便达2900亿美元，再创新高。

要实现降低碳排放这个目标，全球各国在能源结构、基础设施以及产业链模式上均须作出巨大转变，亦需要在各种能源及绿色科技上作出庞大投资，这是自工业革命以来最重大的经济转型。在这个变革的过程中，也伴随着巨大的经济机遇。粤港澳三地的建筑企业和工程咨询顾问公司通过大湾区这个平台参与全球碳转型这个大变革过程，使得三地的相关行业无缝融合，并创造了"一带一路"的投资项目可持续发展的支持能力保证。

对接国家的"十四五"规划，香港也要有相应的建筑工程管理可持续发展及国际化筹谋部署，以站上新台阶的角度来厘定新发展方向，这才能牢牢抓紧发展机遇。另外，要做好政策对接、产业对接和市场对接，而更重要的是理念和思路的对接，把创新的科研成果推动适时落地。只有这样，才能在复杂互动的市场里，最大程度动员全社会的力量和资源，推动香港在国家未来的蓝图中有更好的发展。国家的新发展格局和战略为香港带来了新阶段、新空间，我们必须加速推动，"只有做好今天，定会赢得明天"。

参考文献

[1] https：//research. hktdc. com.

[2] FIDIC State of the World Report 2021-Time to $ Tnvest.

[3] 香港特别行政区立法会秘书处数据研究组. 数据便览：粤港澳大湾区概况. 文件编号 FS03/17-18 2018. 02. 23.

[4] Chartered Institution of Civil Engineering Surveyors. The need for digital sustainability. 2021.

[5] SDGs，National Geographic Society.

建筑工业化与建造机器人

Industrialized Construction and Construction Robot

智能建造理论体系框架与研究发展趋势

尤志嘉[1,2]　吴　琛[1,2]　刘紫薇[1]

(1. 福建工程学院土木工程学院，福建福州 350118；
2. 福建省土木工程新技术与信息化重点实验室，福建福州 350118)

【摘　要】 作为一个新兴领域，智能建造技术研发与工程应用目前已受到学术界与产业界的广泛关注，然而其基础理论研究却相对滞后，尚未形成完整的理论体系。本文描述了一套智能建造理论体系框架，将其划分为"基础理论、支撑技术、管理机制、参考架构、工作机理、集成方案、业务场景、运行机制、实施路径、核心目标、评价机制"11 项关键子领域，分别阐释其科学内涵并揭示其内在逻辑联系。在此基础上，展望智能建造领域未来的研究发展趋势，并识别潜在的研究方向，以推动构建智能建造理论知识体系。

【关键词】 智能建造；工业 4.0；智能建造系统；理论体系；研究趋势

作为典型的劳动密集型产业，建筑业正承受着前所未有的压力。建筑施工面临着低效率、高成本、高污染、高能耗等一系列问题，因此迫切需要转型升级，实现高质量、可持续的发展[1]。然而，建筑业的一些特殊性质却成为阻碍其转型升级的桎梏，例如建筑产品的唯一性，建造资源的流动性，施工过程的离散性、复杂性与高不确定性，以及施工环境的恶劣性等[2]，这就决定了建筑业不能完全照搬其他行业的先进理论与技术，必须建立适合自身转型升级的发展模式。

随着工业 4.0 时代的到来，以物联网、大数据、云计算为代表的新一代信息技术与人工智能技术正日益广泛地应用于工程项目建设中[3]，从而衍生出了"智能建造"的概念。实现智能建造模式被广泛地认为是工业 4.0 背景下建筑业转型升级的必由之路。目前关于智能建造技术研发与工程应用的报道方兴未艾，然而其基础理论研究却滞后于工程实践，尚未形成完整的理论体系。

针对上述问题，本文通过构建智能建造理论体系框架，明确该领域的研究范围与内容，揭示智能技术驱动建筑业创新生产组织方式的机理，从而推动构建智能建造理论知识体系。

1　研究背景

1.1　智能建造的概念内涵

随着智能建造日益受到广泛关注，一些学者尝试阐释其概念内涵。表 1 为通过中国知网（CNKI）检索与智能建造相关的研究文献，总结出国内部分学者对于智能建造做出的定义。分析可得，尽管定义的语言表达不尽相同，但不同学者对于智能建造内涵的认知却趋于同质化，可以凝练出以下几项共性要素：①智能建造是一种新型的工程建造模式；②其范围涵盖工程建造全生命周期；③现代信息技术对于提升施工组织管理能力的驱动作用。

智能建造定义总结　　　　表1

序号	作者	定义
1	丁烈云[4]	智能建造，是新信息技术与工程建造融合形成的工程建造创新模式，通过规范化建模、网络化交互、可视化认知、高性能计算以及智能化决策支持，实现数字链驱动下的工程立项策划、规划设计、施工生产、运维服务一体化集成与高效率协同
2	毛志兵[5]	智能建造是在设计和施工建造过程中，采用现代先进技术手段，通过人机交互、感知、决策、执行和反馈，提高品质和效率的工程活动
3	樊启祥等[6]	智能建造是指集成融合传感技术、通信技术、数据技术、建造技术及项目管理等知识，对建造物及其建造活动的安全、质量、环保、进度、成本等内容进行感知、分析和控制的理论、方法、工艺和技术的统称
4	毛超等[7]	智能建造是在信息化、工业化高度融合的基础上，利用新技术对建造过程赋能，推动工程建造活动的生产要素、生产力和生产关系升级，促进建筑数据充分流动，整合决策、设计、生产、施工、运维整个产业链，实现全产业链条的信息集成和业务协同、建设过程能效提升、资源价值最大化的新型生产方式
5	尤志嘉等[8]	智能建造是一种基于智能科学技术的新型建造模式，通过重塑工程建造生命周期的生产组织方式，使建造系统拥有类似人类智能的各种能力并减少对人的依赖，从而达到优化建造过程、提高建筑质量、促进建筑业可持续发展的目的
6	Andrew DeWit[9]	智能建造旨在通过机器人革命来改造建筑业，以达到节约项目成本、提高精度、减少浪费、提高弹性与可持续性的目的

1.2　智能建造理论研究现状

当前，智能建造已由最初的新兴概念迅速发展成为一个热门的研究领域，涉及土木工程、工程管理、计算机科学、人工智能、自动化、机械工程等多门学科，属于典型的交叉学科范畴。然而，相对于产业应用需求的快速上升，智能建造理论研究却相对滞后，主要体现在以下几个方面：

（1）通过现代信息技术驱动施工组织能力提升的机理尚不明确，未来智能建造模式下的业务应用场景尚不清晰。

（2）如何集成各类异构建造资源，实现其协同工作的机理尚不明确；物理建造过程与虚拟建造过程的双向同步与交互作用机制尚未建立。

（3）如何打通建筑产业价值链的业务流程壁垒，消除施工组织内部信息孤岛的方案尚未建立。

（4）指导推广实施智能建造模式的策略、方法与路径尚不清晰，并且缺乏科学的实施成效评价机制。

众所周知，任何一个新兴研究领域若发展成为一门学科，必然要建立在坚实的理论基础之上，形成明确的研究目标与研究内容，以及系统的研究方法[10]。由此可见，智能建造距离形成一门独立学科还有很长的路要走。当前，智能建造领域的研究应着眼于构建基础理论体系，解决建筑业转型升级中所面临的基础共性问题，以实现通过技术系统进步驱动产业创新发展的愿景。

1.3　基于"信息-物理"融合的智能建造系统

目前智能建造业务场景呈现出单一化、碎片化的特征，各类信息技术被分散地应用于解决特定工程问题，尚缺乏一个集成各类技术以提高整体施工组织能力的统一平台。作为工业4.0的核心技术，信息物理系统（Cyber-Physical Systems，CPS）通过集成各类感知、通信、计算、仿真与控制技术，在信息世界中监控物理世界的运行状态，经过分析、模拟与优化之后，再以最优的策略控制物理世界的运行，从而实现两者的深度融合与实时交互[11]。CPS为智能建造模式下各类信息技术的集成化应用提供了全新的视角[2]。

文献［8］提出将基于"信息-物理"融合的智能建造系统作为智能建造概念的实现形式，并分别从功能和技术维度描述了智能建造系统的通用体系结构。本文研究内容建立在文献［8］所提出的智能建造系统体系结构上，以智能建造系统作为智能建造理论研究成果的

技术实现载体。

2 智能建造理论体系框架

本文描述了一种智能建造理论体系框架，以明确该领域的研究目标、范围与内容，并识别潜在的研究方向。如图1所示，该理论体系框架以智能建造系统作为实现智能建造模式的技术载体，将智能建造研究内容划分为11个关键子领域，分别为基础理论、支撑技术、管理机制、参考架构、工作机理、集成方案、业务场景、运行机制、实施路径、核心目标及评价机制。

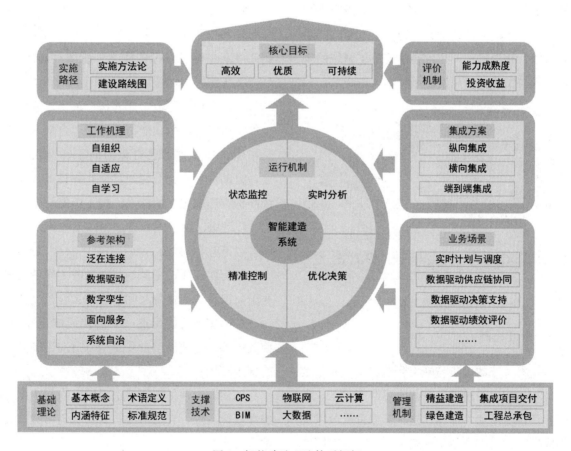

图1 智能建造理论体系框架

3 智能建造理论研究关键子领域

基于上述智能建造理论体系框架，本节分别阐释其各关键子领域的概念内涵与主要研究

内容，并揭示其内在逻辑联系。

3.1 基础理论

阐释智能建造与智能建造系统的基本概念

与内涵特征，定义相关术语，建立智能建造的技术标准与规范体系。

3.2　支撑技术

如图2所示，智能建造系统的支撑技术包括"一项核心技术＋多项智能技术"。一项核心技术即信息物理系统（CPS）技术；多项智能技术包括物联网、大数据与云计算等基础支撑技术，建筑信息模型（BIM）、数字孪生、移动互联网、三维重建、虚拟现实（VR）、增强现实（AR）、面向服务体系架构（SOA）等应用信息技术，以及人工智能与建筑机器人等智能化技术三个部分。该子领域的研究内容包括分析各项支撑技术对于智能建造模式的赋能作用，并揭示不同技术在建筑施工领域的耦合关系与集成化应用发展趋势。

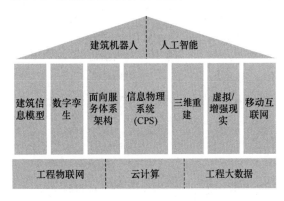

图2　智能建造系统支撑技术体系

3.3　管理机制

技术系统的进步与施工管理机制的变革是一个相辅相成的过程。在智能建造模式下，主要涉及精益建造、绿色建造、工程总承包以及集成项目交付四项新型建造机制。该子领域主要研究如何为各类新型管理机制开发相适应的技术实现方案，以及如何通过现代信息技术驱动施工组织管理能力的提升。

3.4　参考架构

该子领域主要研究如何建立智能建造系统参考技术架构，并揭示其各子系统之间的依赖关系、交互机制与约束条件，为构建面向不同工程类型的智能建造系统提供参考依据。在此基础上，阐明智能建造系统参考架构"泛在连接、数字孪生、数据驱动、面向服务、系统自治"五项基本特征及其内涵[8]，并揭示其技术架构之间与工作机理的耦合规律。

3.5　工作机理

在人工智能领域，所谓"自组织"是指在没有外部指令的条件下，各子系统之间能够协调完成某个任务目标的能力；所谓"自适应"是指系统能够根据环境的变化而不断调整自身的行为，在新的环境中能够保持最优，或者至少保持可以容许的功能；而所谓"自学习"则是指系统通过评估已有行为的正确性或优良度，自动优化自身结构与参数，并且保存在系统结构中形成记忆的能力。揭示智能建造系统的"自组织、自适应、自学习"工作机理，即回答以下几个问题：

如何实现建造资源与建造任务之间的动态匹配，以及建造资源之间的协同工作机制？

如何实现智能建造系统的快速响应能力，使其能够动态地调整自身结构以适应不断变化的施工环境？

如何通过评估智能建造系统已有行为的正确性或优良度，自动修改自身结构与参数，改进自身行为？

3.6　集成方案

如图3所示，该领域主要研究如何以智能建造系统为载体，使人员、机械设备、软件服务及业务流程之间能够互联互通，开发纵向、

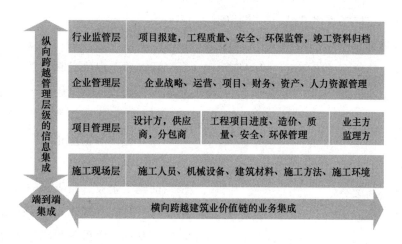

图3 智能建造集成方案

横向以及端到端的集成方案：

（1）纵向集成

纵向集成（Vertical Integration）是指将施工组织内部不同层面（例如，行业监管层、企业管理层、项目管理层，以及施工现场层）的各类建造资源与IT系统集成在一起，建立一个建造资源垂直整合、高效协同的智能建造系统。在这样的体系架构中，各施工组织内部各个业务信息系统之间，以及信息系统与施工现场的物理建造资源之间实现了互联互通，有效地解决了施工组织内部信息孤岛的问题。

（2）横向集成

横向集成（Horizontal Integration）是指整合建筑产业价值链上各利益相关方的业务流程，打破传统企业的边界，将施工企业内部的业务流程向价值链上游的设计方、供应商与分包商，以及下游的业主方、监理方延伸，以智能建造系统为载体实现建筑产业价值链上跨组织的信息共享与资源优化配置，从而建立起各利益相关方的高效协同机制。

（3）端到端集成

纵向集成与横向集成是实现端到端集成的前提和基础。所谓的"端到端"（End-to-End）是指价值链中任意一个业务流程的一端（点）到另外一端（点）都是连贯的，不存在局部流

程或片段流程，即没有间断点。这里的"端"（点）既可以是计算机软件系统，也可以是施工机械、建筑机器人等硬件设备，还可以是供应商、项目经理、现场工人等人员。通过将这些端点连接到智能建造系统，实现各类建造资源的有效整合与业务流程的无缝集成。

3.7 运行机制

所谓运行机制是指影响系统运动的各因素的结构、功能及其相互关系。智能建造系统的运行机制是实现智能建造模式的基础，该子领域的研究内容为如何建立基于"信息-物理"融合的"状态监控、实时分析、优化决策、精准控制"闭环运行机制，并分析智能建造系统技术架构对其运行机制的支撑作用，揭示其工作机理与运行机制之间的交互作用规律。

3.8 业务场景

所谓业务场景是指在特定业务环境中可能发生的一组事件及其相关因素的组合，其内容包括业务事件的时间与空间属性、触发机制与交互过程等。该子领域研究内容为开发智能建造模式下与特定管理机制相适应的业务应用场景，例如，施工过程的实时计划与调度，供应链协同管理，数据驱动的知识获取与管理，以

及数据驱动的施工过程绩效评价等。在此基础上,识别实现特定业务场景所需要的支撑技术,并分析在智能建造系统闭环运行机制下实现该业务场景的工作机理。

3.9 实施路径

智能建造模式的实施路径包括实施方法论与系统建设路线图两部分:

(1)智能建造模式实施方法论

实施方法论是指导实施工作的总体纲要。智能建造模式实施方法论的研究内容包括:协调各利益相关方建立实施团队并划分责任分工,明确实施对象与实施内容,建立实施策略,识别影响实施成功与否的关键因素,评估实施风险并制定应对措施等。

(2)智能建造系统建设路线图

智能建造系统的建设是实施智能建造模式的核心内容之一,两者之间是相辅相成的关系。明确施工企业组织架构、业务处理流程与智能建造系统之间的关系,在此基础上开发智能建造系统建设路线图,包括各个建设阶段的工作目标、主要任务、关键控制节点、需要提交的资料等。

3.10 核心目标

智能建造核心目标可以概括为"高效、优质、可持续"三个方面,其具体内涵如下:

高效:提高施工效率,提高建造资源利用率,缩短项目建设工期,降低建造成本支出。

优质:提升建筑产品质量,降低产品不良率。

可持续:节约建材资源,降低能源消耗,减少污染排放,促进安全施工。

智能建造模式的核心目标将作为评价其实施效果的重要依据。

3.11 评价机制

智能建造评价机制包括能力成熟度评价机制和投资收益评价机制两个方面:

(1)智能建造能力成熟度评价机制

建立定性与定量相结合的评价指标,将能力成熟度由低到高划分为若干个等级,用于评估当前施工企业(组织)的智能建造能力发展水平,或智能建造模式的实施效果。同时,该评价机制也可以反映智能建造模式的发展演进路径。

(2)智能建造投资收益评价机制

针对智能建造模式的核心目标,建立定量化的评价机制,用以衡量施工企业在智能建造领域投入人力、物力、财力等各种资源的综合收益。

4 智能建造研究发展趋势

基于本文所提出的智能建造理论体系框架,可以预见未来智能建造理论的研究方向将沿着以下几个趋势发展:

4.1 技术系统驱动施工管理机制创新

探索通过现代信息技术实现精益建造、绿色建造、集成项目交付等新型建造模式的工作机理;在此基础上建立未来智能建造模式下的各种业务应用场景,并分析其技术实现原理。以智能建造系统为技术实现载体,整合原有的各类信息系统,通过纵向消除施工组织不同管理层级的信息孤岛,横向整合跨越建筑产业价值链的业务流程,实现任意工作流程从发起端到结束端的无缝集成,从而形成扁平化、集约式的施工组织管理模式[12]。

4.2 虚实融合的建造资源协同优化

基于物联网技术同步关联 BIM 设计模

型[13]，形成在建建筑物的实时建造模型，作为信息空间中的数字孪生体[14]。在数字孪生体中建立人、机、料、法、环等各类建造资源要素的虚拟映射，实现物理建造资源与信息资源的深度融合与实时交互。建立对各类建造资源的分布式协同控制机制[15]，使其能以最优的策略动态匹配建造任务，实时响应施工环境的变化，并可通过评估已有行为的优良度改进自身的组织结构。

4.3 数据驱动的施工过程闭环控制

建立在施工组织内部纵向集成的基础上，通过实时采集现场监测数据，获得对物理施工过程与施工环境的状态感知，然后在信息空间中基于数字孪生体进行数据建模与仿真分析，再将经过优化的控制信息发送到施工现场，从而形成数据驱动的施工过程闭环控制机制[16]。

5 结论

本文描述了一套智能建造理论体系框架，将其划分为"基础理论、支撑技术、管理机制、参考架构、工作机理、集成方案、业务场景、运行机制、实施路径、核心目标、评价机制"等11项关键子领域，分别阐释其科学内涵，并揭示其研究内容与目标。分析表明，不同的研究子领域并不是孤立的，而是彼此之间存在一定的内在逻辑联系。可以预见，未来智能建造理论研究将沿着技术系统与管理机制协同创新、虚实融合的建造资源协同优化，以及数据驱动的施工过程闭环控制方向演进。

本文的主要贡献在于通过建立智能建造理论体系框架，确定该领域的研究内容与研究目标，并识别潜在的研究方向，在此基础上展望了未来的研究发展趋势。期待本文研究成果会对学术界有所启示，以推动构建智能建造理论知识体系。

参考文献

[1] 丁烈云. 数字建造导论[M]. 北京：中国建筑工业出版社，2019.

[2] 陈珂，丁烈云. 我国智能建造关键领域技术发展的战略思考[J]. 中国工程科学，2021，23(4)：64-70.

[3] You Z, Feng L. Integration of industry 4.0 related technologies in construction industry：a framework of cyber-physical system[J]. IEEE Access，2020(8)：122908-122922.

[4] 丁烈云. 智能建造推动建筑产业变革[N]. 中国建设报，2019-06-07.

[5] 毛志兵. 智慧建造决定建筑业的未来[J]. 建筑，2019(16)：22-24.

[6] 樊启祥，林鹏，魏鹏程，等. 智能建造闭环控制理论[J]. 清华大学学报（自然科学版），2021，61(7)：660-670.

[7] 毛超，彭窑胭. 智能建造的理论框架与核心逻辑构建[J]. 工程管理学报，2020，34(5)：1-6.

[8] 尤志嘉，郑莲琼，冯凌俊. 智能建造系统基础理论与体系结构[J]. 土木工程与管理学报，2021，38(2)：105-111.

[9] Dewit A, Komatsu. Smart construction, creative destruction, and Japan's robot revolution[J]. The Asia-Pacific Journal，2015，13(5)：2.

[10] 段丹洁. 与时俱进发展新兴学科[N]. 中国社会科学报，2021-03-01(2).

[11] Colombo A W, Karnouskos S, Kaynak O, et al. Industrial cyberphysical systems：a backbone of the fourth industrial revolution[J]. IEEE Industrial Electronics Magazine，2017，11(1)：6-16.

[12] 高小慧，成虎，徐鑫，等. 施工企业项目管理的组织运行模式研究[J]. 项目管理技术，2016，14(6)：11-16.

[13] Tang S, Shelden D R, Eastman C M, et al. A review of Building Information Modeling (BIM) and the Internet of Things (IoT) devices integra-

tion：present status and future trends[J]. Automation in Construction, 2019(101)：127-139.

[14] Tao F，Zhang H，Liu A，et al. Digital twin in industry：state-of-the-art[J]. IEEE Transactions on Industrial Informatics, 2019，15(4)：2405-2415.

[15] 周远强. 分布式系统协调优化控制方法研究[D]. 上海：上海交通大学，2020.

[16] You Z，Wu C. A framework for data-driven informatization of the construction company[J]. Advanced Engineering Informatics，2019(39)：269-277.

基于激光点云的建筑质量自动检测系统

熊　彪[1]　宋金强[2,3]　王　涛[2]　谢宜勋[2]　侯艳彬[1]

（1. 武汉称象科技有限公司，湖北 武汉 430074；
2. 湖北第二师范学院建筑与材料工程学院，湖北 武汉 430205；
3. 武汉验房网啄屋鸟工程顾问有限公司，湖北 武汉 430000）

【摘　要】 建筑质量是关乎人们生命财产安全的重大问题，也是国家重点关注的领域。在施工过程中，建筑几何尺寸是否达到设计指标，是建筑质量检测的重要一环，也是最耗时的一项工作。传统检测方式都是用靠尺、卷尺等一系列的工具，完全依赖手工测量、记录，作业耗时耗力、主观性强、容易出错。这种方式显然不能支撑建筑行业数字化、智能化转型。我们研发了一套基于激光扫描的自动化检测系统，利用三维激光扫描仪采集毫米级精度的点云数据，并通过 4G/5G 网络快速上传到云端，在云端利用人工智能算法自动分析并生成质量报表。自动检测的指标包括平整度、垂直度、阴阳角、方正度、截面尺寸、楼板厚度等。大量实验证明，本系统的测量精度完全满足当前行业规范，测量效率是手工测量的 15 倍以上，极大地提升了测量效率，有效支撑了工程项目的提质增效。

【关键词】 质量检测；激光雷达；实测实量机器人；人工智能；智能制造；BIM

1　引言

这些规范和指导手册有效保证竣工住宅的使用功能和观感质量，促进住宅工程质量整体水平的提高。建筑施工过程中，建筑建造和竣工验收过程需要大量的实测实量工作[1]。实测实量是指应用测量仪器工具，通过现场测试建筑室内空间尺寸（包括长、宽、高等）、丈量而得到的能真实反映产品质量数据的一种方法（图 1）。住房和城乡建设部于 2015 年发布修改版的《混凝土结构工程施工质量验收规范》GB 50204—2015[2]，在充分总结工程实践经验并参考国际先进标准的基础上，对混凝土结

图 1　实测实量
（a）实测实量展示区；（b）手工测量垂直度；
（c）手工测量洞口尺寸

构工程增加了结构位置和尺寸偏差的实体检验规定。各地方政府相关部门也发布了一系列分户验收规范和指导手册[3, 4]，规范包括地面、墙面、顶棚面层、门窗、机电质量等检测规定。

质量检测涉及的指标项繁多，而目前这些测量主要靠手工测量，费时费力，主观性又很强。不同参与方不能充分信任其他方的检测结果，需要反复测量，包括甲方、总包、分包、监理单位以及政府机构等，这又进一步加剧了测量的工作量。当前，进城务工人员数量大幅度减少，且呈老龄化趋势，人力成本也急剧上升[5]。同济大学王广斌团队研究发现，建筑业工人 2014～2020 年减少了 14.45%，预计 2035 年供求缺口将达到 1097 万人，需要通过工业化、智能化等手段提升劳动生产率 45.5% 才能填补此缺口。建筑质量检测这一工序对自动化和智能化也需求迫切。

传统测量采用的是靠尺、塔尺、卷尺、激光投线仪等，完全依靠手工采集、记录，效率低下还容易出错。三维激光扫描仪为当前三维空间数据采集的重要传感器，可以高效、全面、精确地采集现场的点云数据，大量应用于 BIM 建模[6]、桥梁、隧道结构监测[7, 8]等。华中科技大学骆汉宾教授利用深度相机快速采集室内三维模型[9]。考虑到激光扫描仪数据采集的高效、全面，国内外有不少机构将激光扫描仪应用到建筑质量检测中[10, 11]。

针对新加坡政府的需求，南洋理工大学陈义明教授团队研发 QuicaBot 实测实量机器人[12, 13]，利用激光扫描仪、红外相机、超声波传感器以及四轮小车整合一套实测实量机器人，可以自动识别和分析垂直度、平整度、空鼓、裂缝、视觉缺陷等。这一套方案将实测实量需要测量的指标项都考虑到，但传感器精度又不够，不能有效检测需要达到毫米级的测量

项，而且四轮小车在实际应用场景中的移动便携性较差。整套系统研究价值较高，但离中国建筑行业的工程应用还有一段距离。

针对上述方法的局限性，本文提出一种基于激光扫描仪和云计算的高自动质量检测系统。操作员只需要将激光扫描仪搬移到待检测的空间，将激光扫描仪开机，并通过手机 App 连接、发送扫描指令给激光扫描仪。扫描仪数据采集完成后，手机 App 自动将激光扫描仪的数据通过 4G/5G 网络上传到云端。云端利用机器学习和三维建模算法，对点云全自动化处理并生成三维模型和实测实量报告。整个过程操作简单，自动化程度高，数据直接上云，不存在篡改的可能。本系统可以覆盖的实测实量指标包括平整度、垂直度、开间、进深、洞口尺寸等一系列指标。大量实验证明，本系统的实测实量精度达到国家标准，而且测量效率是手工测量的 15 倍以上。本系统的应用推广将很大程度地保障我国建筑质量，也能有效缓解这一工序对大量劳动力的依赖。

2　技术路线

针对当前建筑质量检测耗时耗力，测量点位、读数具有很强主观性的缺点，本文提出一种基于激光扫描自动化检测系统。本系统主要包括四个组成部分（图 2）：①激光扫描仪；②移动端 App；③中央服务器；④点云服务器。整个系统自动化程度高，数据采集和计算客观。

如图 2 所示，移动端 App 与激光扫描仪通过 WiFi 局域网连通后，作业员通过移动端 App 直接操控激光扫描仪，发送指令让其采集数据。移动端 App 自动收集并缓存扫描仪采集的点云数据，通过 4G/5G 广域网自动将点云数据上传到中央服务端。中央处理端将新

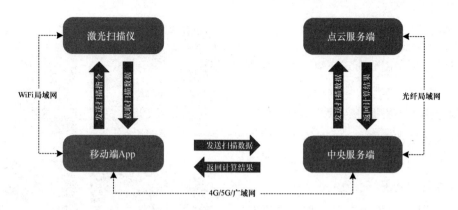

图 2　实测实量机器人系统构架

上传的数据分配给空闲的点云处理端，让其对点云进行解析、分类、建模、模拟靠尺、出分析报告等一系列自动化处理。整个过程中，作业员只需将激光扫描仪搬移到待测的房间，并在移动端 App 操作，一键发送指令收集、处理数据。整个过程自动化程度极高，无须手工数据拷贝、格式转换，也不需要对服务端处理过程进行任何手工干预等。下文将按部件展开介绍本系统。

2.1　激光扫描仪

激光扫描仪是现场数据采集的主要硬件，其性能关乎采集的数据质量。扫描仪性能主要包括扫描精度、量程。此外，考虑到本系统的规模化应用，也需要充分考虑激光扫描仪的扫描速度、质量稳定性以及价格。

本系统支持不同品牌和型号的激光扫描仪，包括法如、天宝、莱卡、Z＋F，以及国产的品牌。但是，综合考虑各项因素和性价比，当前系统客户最常用的是法如 M70，测量精度 10m 处 2mm，有效量程 35m，一个空间测量时间仅需要 1.5min。某些国产组装激光扫描仪，虽然价格便宜，号称精度也能达到 ±1.5mm，但实际精度在 10mm 左右，阴阳角、洞口尺寸等需要高精度点云数据的指标项精度很低，难以达到标准。如果点云质量不高，后续算法难以弥补精度的缺失。

2.2　移动端 App

移动端 App 是作业员操作的主要平台，也是连接扫描仪和中央处理端的枢纽。移动端 App 通过 WiFi 局域网与扫描仪相连接，并利用 WebUI 接口与扫描仪通信，如发送指令、接受反馈信息。发送的指令包括开始扫描、结束扫描、修改扫描参数、开始传输数据、结束传输数据。接受的反馈信息包括扫描的进度、传输数据的进度等。

作业员在 App 中选定待测的项目、楼栋、户号、房间号以及所测阶段，一键发送指令给扫描仪让其开始扫描。等待 1.5min 后，App 接收到扫描仪结束扫描的信息，自动从扫描仪拷贝刚才采集的数据并压缩打包。与此同时，作业员可以将扫描仪搬移到下一个房间并进行扫描。激光扫描仪的扫描和数据拷贝可以同时进行，不耽误数据采集的时间。

在 4G/5G 信号良好的情况下，移动端 App 可以主动将项目元信息及激光点云文件上传到中央服务器，并通知它已经完成了哪些空间的扫描工作。项目元信息包括项目、楼栋、户号、房间号以及所测阶段，这些信息是后续点云分析计算和统计分级报表所必需的。可以上传逐个房间的数据，也可以批量上传。

2.3 中央服务端

本系统的云端采用分布式构架，以提高系统的弹性。云端由一台中央服务端和多台点云处理端两部分组成。根据任务量的不同，可以动态地热插拔地增加或删减点云处理端，以适应波峰的大数据处理量，以及在低谷时期降低整体系统的配置和费用。

中央服务端是云端的中枢机构，负责与多个手机 App 连接通信（1 万～10 万台的规模），并协调多台点云处理端的工作（10～100 台的规模）。点云处理端负责点云的具体解算、分析工作，而点云处理是计算密集型的。单个空间的点云数据，根据计算硬件配置不同，所需时间会有差异。按目前的经验来看，一个空间的点云计算时间通常在 2min 左右。中央处理端可以根据点云处理量，动态启用或者关闭点云处理端，并相应分配点云计算工作。由于中央服务端与点云处理端处于万兆光纤连接的局域网中，数据的传输拷贝时间不是瓶颈。

中央处理端还负责汇总、统计项目的分级分层报表。点云处理端将处理的每个空间的数据上传到中央处理端，中央处理端按照集团公司、地区公司、项目等用户自定义的层级，进行汇总统计，以供移动端 App 调用查看。

2.4 点云处理端

点云处理端负责具体的点云分析、解译以及生成单个房间的报表等工作。如图 3 所示，点云处理端的工作主要包括：（a）点云解析、格式转换；（b）点云分类、识别；（c）全面检测评分；（d）AI 模拟取尺抽检评分，以下详细解释这四个步骤。

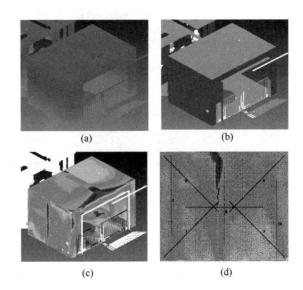

图 3　激光点云自动化处理
（a）点云解析、格式转换；（b）点云分类、识别；
（c）全面检测评分；（d）AI 模拟取尺抽检评分

2.4.1 点云解析、格式转换

本系统支持不同厂商的激光扫描仪，但这些仪器的数据格式不同，需要将原始数据解析成通用的内部格式。点云处理端将移动端 App 上传到中央服务端的原始点云下载到本地磁盘，根据其格式解译成 las 格式。

2.4.2 点云分类、识别

得到点云后，系统首先利用深度学习算法从点云数据中识别关键要素，如墙、吊顶、地板、梁、柱、门、窗等实测实量需要检测的物体。其他物体，包括现场材料、工作人员、护栏等，都作为杂物剔除。

2.4.3 全面检测评分

本系统支持不同厂商的激光扫描仪，但这些仪器的数据格式不同，需要将原始数据解析成通用的内部格式。点云处理端将移动端 App 上传到中央服务端的原始点云下载到本地磁盘，根据其格式解译成 las 格式。在每个墙面或者吊顶上的激光点有上百万个，这些激光点的平均点间距为 1～3mm，全面、准确地记录了墙面的情况。我们通过最小二乘算法，将一个墙面或者吊顶的点，拟合成水平或者垂直的平面。这个拟合的最佳平面为墙面或者板的参考面，然后计算每个点到最佳平面的距离，即得到墙、板上每个位置的凹凸程度。见公式(1)~(3)，S_i 为墙、板平面方程，p_i 为激光点齐次坐标，h_{ij} 为激光点 p_i 到平面 S_j 的距离。根据 h_{ij} 绘制每个点的热力图，即可得到全面检测的三维可视化效果，如图 3（c）所示。

$$S_i = (a_i, b_i, c_i, d_i) \quad (1)$$
$$p_i = (x_i, y_i, z_i, 1) \quad (2)$$
$$h_{ij} = p_i^{\mathrm{T}} \cdot S_j \quad (3)$$

2.4.4 AI 模拟取尺

传统实测实量体系以靠尺、量角器等工具用手工抽样进行测量。为了适配当前实测实量打分体系，我们通过 AI 算法，模拟人工抽样进行测量。不同企业有不同的抽样规则，比如垂直度、平整度，有的公司取 3 尺，有的公司取 5 尺。我们设计了一套可定制的抽样规则，根据不同的公司标准进行配置，如图 4 所示。AI 算法在模拟垂直度、平整度取尺时，通过识别墙面或者吊顶板的 4 个角点的中心点，并根据规则取 3 尺、5 尺，以及按照规则墙面边缘 30cm 或者 50cm 下尺。将 AI 模拟靠尺取值的结果与全面检测的结果绘制到一张图上，即可得到效果展示图，如图 3(d) 所示。

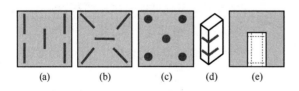

图 4　AI 模拟取尺
(a) 垂直度；(b) 平整度；(c) 吊顶板极差；
(d) 阴阳角；(e) 洞口尺寸

3　实验

本次实验采用 FARO 激光扫描仪 S70，

其量程为 35m，在 10m 处的精度为 2mm。本实验选用某头部地产公司在武汉的 5 个项目，均为超高层住宅楼及商住两用房，涉及四个工程阶段，包括结构、砌筑、抹灰、精装。数据采集过程中，只需要 1 个工作人员移动、操作机器。机器测量共 2745 个房间，计算在房间与楼层之间的移动时间，共耗时 685min，分 21d 完成。日均工作 5.4h，每个房间平均耗时 2.49min（其中扫描时间 1.5min，移动时间 1min）。而人工手工采集数据，一个房间需要一组 2 人 15min 完成测量工作。其中一人负责测量，另一人负责配合及记录数据。平均下来，机器实测实量的效率是人工实测实量的 15 倍以上。

在测量的 2745 个房间中，人工随机抽取复查 110 个房间，人机对比复核数据总计 1535 尺。其中，人工复测垂直度 240 尺、平整度 420 尺、洞口此次 134 尺、房间尺寸 134 尺、方正度 127 尺、阴阳角 162 尺、顶板极差 260 尺。按照机器自动取尺的位置，人工手动测量，并统计差异，具体见表 1~表 4，分别为主体、砌筑、抹灰、精装阶段人、机测量误差对比。总的来说，垂直度、平整度、吊顶板极差、阴阳角平均偏差在 1mm 以内，方差在 1mm 左右。洞口尺寸、房间尺寸的平均偏差在 2~3mm，方差在 3mm 左右。按照现行的国家及行业标准，本系统的实测实量精度能完全满足要求。

主体阶段人、机测量偏差对比　　　表 1

测量项	平均值（mm）	方差	取尺数
垂直度	−0.17	0.96	240
平整度	0.36	0.93	420
洞口尺寸	0.38	4.69	80
吊顶板极差	0.45	1.39	150

砌筑阶段人、机测量偏差对比　　表 2

测量项	均值（mm）	方差	取尺数
洞口尺寸	1.97	2.39	36
房间尺寸	2.74	1.65	40
方正度	0.59	1.39	30
阴阳角	0.29	0.89	42
吊顶板极差	0.52	1.13	35

抹灰阶段人、机测量偏差对比　　表 3

测量项	均值（mm）	方差	取尺数
洞口尺寸	2.15	2.69	40
房间尺寸	2.64	1.93	50
阴阳角	0.29	0.89	55
地板平整度	0.91	2.72	75
吊顶板极差	0.63	1.34	75

精装阶段人、机测量偏差对比　　表 4

测量项	均值（mm）	方差	取尺数
洞口尺寸	2.56	1.48	36
房间尺寸	2.64	1.93	44
阴阳角	0.29	0.89	42
方正度	0.56	1.24	45

4　结论

我们研发了一套基于激光扫描的自动化检测系统，利用三维激光扫描仪采集毫米级精度的点云数据，并通过 4G/5G 网络快速上传到云端，在云端利用人工智能算法自动分析并生成质量报表。

本系统具有以下优点：①指标丰富。系统算法可以自动提取墙面垂直度、平整度、方正度、阴阳角、顶板水平度极差、吊顶板底面平整度和水平度、地面平整度、开间进深、净高、楼板厚度、门窗开洞尺寸等指标。②自动化程度高。设备自动调平，自动扫描、计算、输出实测成果。整个过程大量减少手工作业，

工作效率高，同时杜绝人为干扰，保证数据客观。③双评分机制。系统提供全面检测评分以及 AI 模拟取尺抽检评分。评分规则根据企业标准可以定制。④云管理。所有数据直接在线化，实现云计算、云存储。数据便于追查，可以对历史测量数据进行纵向比较，使检查结果具有不可抵赖性。

本系统的数据流贯穿验收、整改、移交阶段，数据可溯源、可共享、可重复利用，杜绝人为干扰，有效增强工程管理透明度，实现对工程项目进度的远程监控，保证数据准确。所有数据云计算、云存储、云共享，方便管理，可与信息管理平台、BIM 技术结合，赋能工程建设行业的数字化、智能化转型。

参考文献

[1] 黄景忠. 工程质量"实测实量"小组的建立及控制要点. 福建建筑，2014(11)：88-90.

[2] 中华人民共和国国家标准. 混凝土结构工程施工质量验收规范 GB 50204—2015. 北京. 中国建筑工业出版社，2015.

[3] 广东省建设厅实行住宅工程质量分户验收的指导意见. 建筑监督检测与造价，2008，1(9)：7-8.

[4] 北京市住宅工程质量分户验收指导手册. 2015.

[5] 蔡楚瑜，张琦睿，苏星. 我国建筑业劳动力老龄化问题的跨区域分析及应对策略研究. 土木工程与管理学报，2021，38(4)：63-9.

[6] 惠之瑶，张爱琳，王昆，等. 集成 BIM-3D 扫描技术的斗拱建模方法. 土木工程与管理学报，2020，37(2)：151-7.

[7] 孙龚，黄宏伟，薛亚东. 隧道病害检测设备评价方法. 土木工程与管理学报，2018，35(4)：117-22，29.

[8] 吴贤国，刘鹏程，王雷，等. 基于三维激光扫描地铁运营隧道渗漏水监测及预警. 土木工程与管理学报，2020，37(1)：1-7，15.

[9] 雷蕾，朱思雨，骆汉宾. 基于深度相机的实时

室内三维信息采集方法. 土木工程与管理学报,
2021, 38(3): 101-106.

[10] 何涛, 刘震, 李江, 等. 基于三维扫描测量的
住宅产品尺寸检验系统研究. 2020 年工业建筑
学术交流会, 2020: 5.

[11] 李灿, 卫世全, 陈洪根, 等. 三维测量机器人
实测实量智能管理系统. 智能建筑与智慧城
市, 2020(3): 81-83.

[12] Liu L, Yan R-J, Maruvanchery V, et al. Transfer learning on convolutional activation feature as
applied to a building quality assessment robot. International Journal of Advanced Robotic Systems,
2017, 14(3): 1729881417712620.

[13] Yan R J, Kayacan E, Chen I M, et al. QuicaBot:
quality inspection and assessment robot. IEEE
Transactions on Automation Science and Engineering, 2018, 16(2): 506-517.

预制轨道板多批次生产调度优化方法

王雨桐　　王朝静

（北京交通大学经济管理学院，北京 100000）

【摘　要】 高速铁路无砟轨道板预制生产有效提升铁路建设效率、推动铁路工程智能建造转型发展。然而，在实践应用中，轨道板预制生产调度与施工进度计划不协调时常发生，导致预制生产成本高、构件难以按时交付等问题愈发明显。本文以高速铁路无砟轨道板为研究对象，提出了预制轨道板多批次生产调度优化模型，优化配送批次和构件调度方案以实现按时交付及生产成本最少化。与传统生产方法相比，该模型能够降低生产成本，减少延迟、提前交付发生的可能性，推动高速铁路工程智能建造转型发展。

【关键词】 高速铁路；预制轨道板；生产调度优化；多批次生产

交通强国，铁路先行。国家经济发展需要依靠交通，铁路则提供了源源不断的动力。高速铁路具有安全性高、舒适性强、运载量大、低碳环保等特点，是成为当前交通业发展重要趋势的关键。中国铁路的建设目标就是建成具有全球竞争力、世界一流的铁路，实现交通强国战略，推动国民经济高速发展。目前铁路预制生产技术迈向了智能化时代，实施无砟轨道预制技术是提升铁路建设效率、推动铁路工程智能建造转型发展的重要途径。无砟轨道的生产研究具有很高的应用价值和推广意义，合理的生产计划将促进生产质量的提高和效益的提升。

如今，我国高铁的轨道板主要以 CRTSⅢ型为主，轨道板生产工艺由 2017 年前传统"台座法"，发展到规模应用的"流水机组法"[1]。然而国内轨道板生产线大多按照流水节拍 10min 设计，各板厂的生产能力未实现生产线所设计的最优效率。且轨道板的生产需要与轨道工程铺设、人员设施管理紧密结合，

因此不仅需要在生产内部制定适当的生产计划，更需要配合施工节奏合理安排生产批次，以达到低成本、准时制生产的要求。无砟轨道板生产的传统方法已经不能满足我国高速铁路建设的需要，必须寻求高效的生产调度方式，提高工业化生产水平。

Chan 和 Hu（2002）将预制生产流水车间调度作为基本生产模型，应用遗传算法寻找最优的生产顺序[2]。Leu 和 Wang（2004）考虑到预制构件生产车间的起重机和劳动力约束作用，建立以缩短生产完工时间为优化目标的调度模型[3]。Hu（2007）在预制生产调度问题中提出以最小化模具使用率为目标的流水车间排序模型，运用遗传算法求解多目标优化问题[4]。Wang（2017）在生产直接相关的六道工序的基础上加入了模具制造、预制构件存储、运输过程工序，开发了双层模拟—GA 混合预制构件生产模型，采用离散事件仿真优化进度计划[5]。Wang（2019）创建了对预制供

应链管理中的现有研究分类法，提出未来的研究机会，如动态扰动管理、智能预制供应链、参与者之间的协调、同时优化调度和资源分配[6]。Wang（2020）构建了一个基于区块链的预制供应链信息管理框架，扩展了区块链在建筑供应链领域的应用[7]。

然而，以上大多数学者侧重于优化生产顺序、缩短完工时间、管理预制供应链等方面，研究的调度问题主要在传统制造业领域中，对于其他工程领域，特别是土木工程领域中调度问题的研究较少，忽略了预制生产需要与轨道工程铺设、人员设施管理紧密结合，应在生产内部制定适当的生产计划、配合施工节奏合理安排生产批次，达到低成本、准时制、智能化生产要求。

我国在不同程度上存在着预制轨道板生产的计划性较差，进度协调不合理，提前完工或是延迟交付导致资源与时间的浪费等问题。因此，本文构造优化模型，以预制高铁轨道板为例，进行示范研究；以成本最小化、按时交付率最大化为目标，提出预制轨道板多批次生产调度优化模型；以最佳批次数量，降低生产成本和保证按时交付，为推动铁路工程智能建造转型发展的高速铁路无砟轨道板预制生产提供指导建议。

1 高铁无砟轨道板预制生产工艺流程

在高铁建设工程中，无砟轨道板预制生产占据重要地位。其生产是一项复杂的系统工程，具有技术高、批量大等特点。预制生产工艺大致可以分为如下几个流程：（1）模具装配。在生产之前需要进行模具购买准备或模具焊接、边棱打磨光滑等加工操作。该流程属于可中断工序，如果工作在当天正常工作时间内不能完成，操作可以中途停止，并在第二个工作日继续进行。（2）安放预埋件。轨道板生产模具在安放前，必须对模具的平整度、预埋件安装位置等进行严格测试，该流程为可中断工序。（3）浇筑。精确合模完成后，进行混凝土浇筑步骤，同时进行振动平整工作，借助于打磨修光机、铲斗对混凝土表面进行平整，使用抹平工具进行抹平作业。浇筑操作是不可中断工序。（4）固化养护。待浇筑工序完成2h以上，用桁吊将轨道板吊入水池进行固化养护。固化养护分为常温养护和加热养护两类。与浇筑工序类似，预制构件的固化养护操作也属于不可中断工序。（5）模具移除。当达到要求后，对模具进行拆除，用于后期同类轨道板的加工生产。（6）表面处理。进行脱模工序后对预制轨道板表面质量进行检查，修补缺陷并裁剪多余部位。检查表面观感、尺寸规格、预埋件、强度等，满足设计、合同需求。以上两道工序均不可中断，为连续型工序。

进而，将六个环节绘制出完整的高铁无砟轨道板生产流程，如图1所示。

2 预制轨道板多批次生产调度优化模型

本章优化模型以预制高铁轨道板为例，进行示范研究，以成本最小化、按时交付率最大化为目标，提出预制轨道板多批次生产调度优化模型，以最佳批次数量降低生产成本、保证按时交付，达到总成本最小化，如式（1）所示：

$$F(x) = \theta_1 f_1(x) + \theta_2 f_2(x) \quad (1)$$

其中，θ_1、θ_2 表示权重系数；$f_1(x)$ 表示生产成本，包括生产构件的总直接成本、在制品成本、闲置成本和仓储成本；$f_2(x)$ 表示时间成本，即延迟或提前交货的违约罚款成本。目标函数（1）主要包括两大未知变量，生产成本 $f_1(x)$ 和时间成本 $f_2(x)$，可分别由模型进行求解，具体数学表述如2.1～2.3节所示。

2.1 完成时间

n 个待加工的预制构件在 m 台机器上加

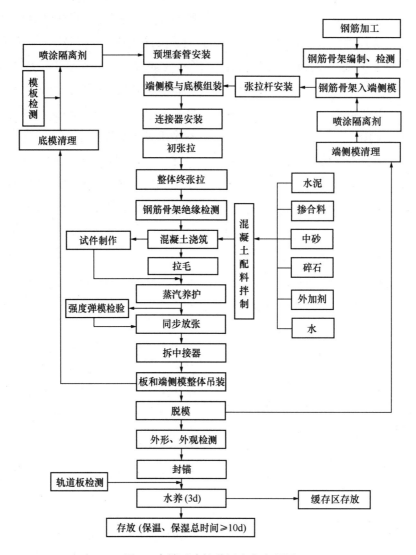

图 1　高铁无砟轨道板生产流程图

工，在轨道板生产调度问题中数学模型表示如下：一个预制构件 J_i 包含 n_k 个加工操作，每个操作的加工时间为 t_{ij}，$F(J_i, k)$ 表示预制构件 J_i 在机器 k 上的加工完成时间，$\{J_1, J_2, \cdots, J_n\}$ 表示一批预制构件的调度。则

$$F(J_1, 1) = t_{J_1 1} \qquad (2)$$

$$F(J_1, k) = F(J_1, k-1) + t_{J_1 k} \quad k = 2, \cdots, m \qquad (3)$$

$$F(J_i, 1) = F(J_{i-1}, 1) + t_{J_i 1} \quad i = 2, \cdots, n \qquad (4)$$

预制轨道板生产分为六个步骤，生产总完成时间等于六道工序的完成时间之和。J_i 在机器 k 上的累积加工完成时间为 J_{i-1}，在机器 k 上累积加工完成时间和 J_i 在机器 $k-1$ 上累积加工完成时间的最大值与 J_i 在机器 k 上加工时间之和：

$$F(J_i, k) = \max[F(J_{i-1}, N_k),$$
$$F(J_i, N_{k-1})] + F_{i, k}$$
$$i = 1, 2, \cdots, n; \ k = 1, 2, \cdots, 6 \qquad (5)$$

其中，$F(J_i, k)$ 表示预制构件 i 在第 k 道工序的完成时间；$F_{i, k}$ 是构件 i 第 k 道工序的实际操作时间区段。第一个加工工件的第一道工序的完成时间可以简化为 $F(J_1, N_1) = F_{1, 1}$。基于以上生产调度理论，构件 i 的总完成时间即第六道工序的完成时间 $F_i = F(J_i, N_6)$。

传统生产调度理论模型假设操作持续不间断进行，但是实际中并不成立，工人工作日的工作时间一般为 8 个小时，非工作时间可再分为允许加班时间和禁止加班时间，生产操作在禁止加班时间是需要中断的。对于预制构件的生产操作，浇筑和固化养护两项工序是不可中断的，其他工序若在允许工作时间内不能完成但可中断的，第二个工作日再继续进行操作。对于轨道板养护工序，无须工人操作，需在混凝土浇筑环节完成后立即进行，工序时间可延长至正常工作时间之外，允许加班。由于养护工序不可中断，如果在正常工作时间和允许加班时间内都无法完成，必须推迟工序开始时间至下一个正常工作日。加工方式分类如下：

（1）连续型工序

如钢筋加工、模具组装、预埋件安装固定等轨道板生产工序，如果在工作日当天的工作时间内无法完成，可顺延至下一工作日。一个预制构件 J_i 包含 n_k 个加工操作，每个操作的加工时间为 t_{ij}，T_w 表示工作时间，T_u 表示非工作时间。其完成时间：

$$F(J_i, k) = \begin{cases} t & （当 t \leqslant 24d + T_w） \\ t + T_u & （当 t > 24d + T_w） \end{cases}$$
$$J_i = 2, 3, \cdots, n; j = 2, 3, \cdots, m \quad (6)$$

$$t = \max\{F(J_{i-1}, k), F(J_i, k-1)\} + t_{ij} \quad (7)$$

$$d = \text{integer}(t/24) \quad (8)$$

（2）非连续型工序

如：混凝土浇筑工序，由于混凝土养护工序无须工人操作，此工序可延长至非工作时间，因此如果在工作时间中无法完成，允许存在加班时间完成工序操作。其完成时间：

$$F(J_i, k) = \begin{cases} t & （当 t \leqslant 24d + T_w） \\ 24(d+1) + t_{ij} & （当 t > 24d + T_w） \end{cases}$$
$$(9)$$

（3）自然工序

无须工作人员操作的工序，如轨道板养护工序，在混凝土浇筑环节完成后立即进行，工序时间可延长至正常工作时间之外，允许加班。但养护工序不可中断，如果在正常工作时间和允许加班时间都无法完成，必须推迟工序开始时间至下一个正常工作日。其完成时间：

$$F(J_i, k) = \begin{cases} t & （当 t \leqslant 24d + T_w） \\ 24(d+1) + t_{ij} & [当 24d + T_w \leqslant \\ & t < 24(d+1)] \\ t & [当 t \geqslant 24(d+1)] \end{cases}$$
$$(10)$$

2.2 时间成本

准时制生产的基本指导思想是通过准时交货来提高企业生产效益，生产工期与铺设工期应尽可能相等，力求达到工业生产准时制生产理念，提前交付与拖后完成均被认为是损失的时间成本。则时间成本，即延迟、提前交货的违约罚款成本：

$$f_2(x) = C_b + C_t \quad (11)$$

生产完成时间晚于交货期的违约费用，即延迟交货罚款成本：

$$C_b = \sum_{i=1}^{n} \beta \cdot \max(0, F_i - D_i) \quad (12)$$

式中，D_i 表示计划交货期；F_i 表示工件 i 生产完成时间；β 表示工件 i 延迟交货的罚款费用系数。

生产完成时间早于交货期的违约费用，即提前交货罚款成本：

$$C_t = \sum_{i=1}^{n} \alpha \cdot \max(0, D_i - F_i) \quad (13)$$

式中，D_i 表示计划交货期；F_i 表示工件 i 生产完成时间；α 表示工件 i 提前交货的罚款费用系数。

2.3 生产成本

预制轨道板生产成本在本文中被公式化，其中包含的三个成本参数，定义如下：闲置成本（C_u）是指与工作人员在批次之间的连续工作中断相关的成本；在制品成本（C_{wip}）是指在长施工周期时间内在制品生产返工的成本；仓储成本（C_s）是指早于交付时间生产完成工件的存储成本。生产成本，包括生产工件的总直接成本、闲置成本、在制品成本和仓储成本，表示如下：

$$f_1(x) = [(1 - P_e) \times C_d] + TC_u + C_{wip} + C_s \tag{14}$$

2.3.1 闲置成本

闲置成本（C_u）是与工作人员在批次之间的连续工作中断相关的成本。假设不会为了避免设备、场地等不可用而导致的更高成本遣散工作人员，则工作人员在生产现场等待。预计不会所有人员都处于闲置状态，则闲置百分比 P_c 被引入，可根据研究灵活更改。

$$C_u = (B - 1) \times P_c \times C_B \times Q_B \\ \times (T_{B,i,1} - T_{B,i,2}) \\ i = 1, \cdots, n \tag{15}$$

$$TC_u = \sum_{u=1}^{m} C_u \quad u = 1, \cdots, m \tag{16}$$

式中，C_u 是批次间的闲置成本；TC_u 是总闲置成本；B 是批次数量；C_B 是批次（B）构件生产的直接成本；Q_B 是分配给批次（B）的工作人员数量；P_c 是批次之间出现闲置的人员数量占比（如 100% 意味着所有的工作人员都遇到空闲）；$T_{B,i,1}$ 是第 B 批次构件 i 的工序开始时刻；$T_{B,i,2}$ 是第 B 批次构件 i 的工序完成时刻。

2.3.2 在制品成本

对于在制品成本，根据经验观察，发现闲置的构件容易损坏。假设某个百分比（P_e）的预制构件闲置不用时，需要返工。根据生产

构件的性质和环境，百分比不同，但施工周期时间越长则越高。考虑到在制品对返工量的影响，引入在制品系数（f_{wip}），该系数是每个预制构件消耗的施工周期时间的函数，用于放大返工效果。

$$f_{wip} = \sqrt{[WIP_{area}]_{PBRS} / [WIP_{area}]_{min}} \tag{17}$$

$$[WIP_{area}]_{PBRS} = \sum_{B=1}^{B} \sum_{i=1}^{N} C_{yi,b} \tag{18}$$

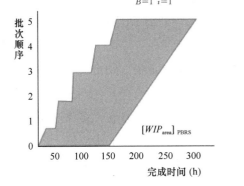

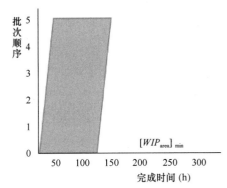

图 2　按施工周期时间计算在制品

式中，$[WIP_{area}]_{PBRS}$ 是全部预制构件生产活动的施工周期时间，$[WIP_{area}]_{min}$ 是当所有生产活动具有相同的最高期望速率时，达到可实现的最小施工周期时间（图 2）。$C_{yi,b}$ 是批次 b 中预制构件 i 的施工周期时间。因此，在制品（C_{wip}）成本计算如下：

$$C_{wip} = P_e \times f_{wip} \times C_d \tag{19}$$

式中，C_d 是总直接成本，f_{wip} 是在制品系数；P_e 是闲置返工百分比。

2.3.3 仓储成本

准时制生产要求加工工件尽量在其预定的

交付时间准时完成，提前完成被认为是损失的仓储成本。则每个工件的仓储成本：

$$C_s = \sum_{i=1}^{n} \gamma \cdot \max(0, D_i - F_i) \quad (20)$$

式中，D_i 表示计划交货期；F_i 表示工件 i 生产完成时间；γ 表示工件 i 提前交货的存储费用系数。

3　遗传算法

本文模型运用遗传算法进行求解，基于案例对模型有效性进行了验证。遗传算法应用求解如下：

（1）编码方案。在流水车间调度问题中，编码方式采用染色体表示工件加工批次，染色体编码方式为整数编码，如：对于 4 批次的流水车间调度问题，第 k 个染色体 $V_k = [1, 2, 3, 4]$，表示相应工件的加工批次顺序为：J_1，J_2，J_3，J_4，设 n 表示待加工的工件总数，批次 J_i 的加工工件为 N_j，此时个体表示为长度等于 $2\sum_{i=1}^{k} J_i N_j$ 的整数串。染色体前半部分表示所有工件加工批次顺序，后半部分表示工件每个批次的加工顺序。

（2）适应度函数。染色体或可行解的优化评价指标，作为预制轨道板多批次生产调度优化模型的目标函数。在本文的遗传算法中，总成本作为适应度函数，如公式（21）所示，旨在以最佳批次数量降低生产成本、保证按时交付，达到总成本最小化。

$$F(x) = \theta_1 f_1(x) + \theta_2 f_2(x) \quad (21)$$

（3）遗传算法参数。种群数量：100；终止条件：200 代；交叉概率：0.75；变异概率：0.01；代沟：0.9。

（4）遗传进化算子。

选择：从当前种群中挑选出相对优良的个体，使它们有机会成为父代繁衍下一代子孙。采用轮盘赌法选择适应度较好的个体，设个体 k 的适应度值为 f_k，个体被选择的概率 p 可通过式（22）计算获得，即适应度高的个体拥有较高的选择概率：

$$p = f_k / \sum_{j=1}^{pop\,size} f_j \quad (22)$$

交叉：种群通过交叉操作产生新的染色体，从而推动整个种群不断进化。从种群中随机选取两个染色体，取出每个染色体的前 $\sum_{i=1}^{k} J_i N_j$ 位，随机选择交叉位置进行交叉。

变异：首先在种群中随机选取一个个体，对选中个体以一定的概率随机改变某一个或某一些基因座上的基因值为其他的等位基因。与生物界类似，遗传算法中变异的发生概率相对较低。

4　案例研究

基于文献，调度模型的部分输入数据，如表 1 所示。数据包括预制构件的种类和数量、构件交付时间、单位交付罚款和存储费用系数（采集数据时间为 2021 年 5 月，采用美元为货币单位，1 美元＝6.4857 元人民币）。

输入数据　　　　　　　　　　　　　　　　　　　　　　　　　　　表 1

批次数（个）	时间成本（美元）	生产成本（美元）				总成本（美元）
		总直接成本	闲置成本	在制品成本	仓储成本	
1	283.4	129076.88	0	74260	2552.1	137976.90
2	264	121023.78	43978	64020.6	972	114378.84
3	263.1	121261.58	46022	63942.6	967.8	129829.32
4	296.5	123810.02	47012	64000	962.1	134360.04

通过算法求得最优方案为 2 个批次，该方案的其他输入数据，如表 2 所示。

输入数据　表 2

批次编号	生产直接成本（美元）	工作人员数量（人）	在制品系数	闲置返工百分比（%）	批次间闲置成本（美元）	闲置占比（%）
1	62341.58	200	1.75	1.24	44858	5
2	58682.20	240	1.69	1.24	44858	5

代入模型公式，可求得各项成本，如表 3 所示。

各项成本　表 3

批次（个）	时间成本（美元）	生产成本（美元）			
		总直接成本	闲置成本	在制品成本	仓储成本
2	264	121023.78	43978	64020.6	972

在此基础上，以最优方案为例，其总成本可由公式（23）进行求解：

$$F(x) = \theta_1 f_1(x) + \theta_2 f_2(x) \qquad (23)$$
$$= \theta_1 \{[(1-P_e) \times C_d]$$
$$+ TC_u + C_{wip} + C_s\} + \theta_2 (C_b + C_t)$$
$$= 0.5 \times \{[(1-1.24\%) \times 121023.78]$$
$$+ 43978 + 64020.6 + 972\} + 0.5 \times 264$$
$$= 114378.84（美元）$$

其中 θ_1、θ_2 表示权重系数，本章均按 0.5 计算。由于企业对按时交付与生产总成本两种目标的关注度不同，其各自的目标权重也并不相同，可以按照实际进行调整。同理，可以求得各可行批次方案的各项成本，如表 4、图 3 所示。

各方案成本对比　表 4

预制构件种类编号	构件数量（个）	交付时间（h）	E&T 单位罚款 [美元/(units·h)]		存储费用系数 γ [美元/(units·h)]
			延迟 β	提前 α	
1	3	164	1.47	0.29	1.34
2	4	140	1.47	0.29	1.27
3	4	164	1.47	0.29	0.34
4	2	160	1.47	0.29	0.39

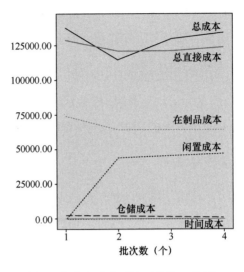

图 3　各方案成本对比（单位：美元）

由上各方案成本对比可得，最优方案即 2 个批次生产的总成本 114378.84 美元低于其他 3 个可行方案的总成本，分别为 137976.90 美元、129829.32 美元和 134360.04 美元。可见，最优批次方案得到了最小的总成本，模型有效地实现了生产成本最少化和预制构件的按时交付。

5　结论

本文以高速铁路轨道板为研究对象，提出预制轨道板多批次生产调度优化模型，对轨道板生产批次进行合理优化，结合遗传算法，以实现预制轨道板按时交付及生产成本最少化。预制智能化生产是发展的必然趋势，对轨道板生产有着积极意义。未来将从预制轨道板生产调度与生产施工协同优化视角深化研究，为高速铁路工程建设提供理论依据，推动铁路工程智能建造转型发展。

参考文献

[1] 李秋全，曹新刚. CRTS Ⅲ型轨道板预制场规划与设计[J]. 工程质量，2020(4)：37-41.

[2] W T Chan, H Hu. Constraint programming approach to precast production scheduling [J].

Journal of Construction Engineering and Management, 2002(128): 513-521.

[3] Wang Z, Hu H, Gong J. Framework for modeling operational uncertainty to optimize offsite. production scheduling of precast components[J]. Automation in Construction, 2004: 86.

[4] Hao Hu. A study of resource planning for precast production. [J]. Architectural Science Review, 2007, 50(2): 104-114.

[5] Wang Z, Hu H. Improved precast production scheduling model considering the whole supply Chain[J]. Journal of Computing in Civil Engineering, 2017, 31(4): 91-97.

[6] Zhaojing Wang, Hao Hu, Jie Gong, et al. Precast supply chain management in off-site construction: A critical literature review[J]. Journal of Cleaner Production, 2019 (232): 1204-1217.

[7] Zhaojing Wang, Tengyu Wang, Hao Hu, et al. Blockchain-based framework for improving supply chain traceability and information sharing in precast construction [J]. Automation in Construction, 2020: 111.

BIM 技术在某大跨连续刚构桥中的应用

王银刚[1,2]　杨　栋[3]　戴晓燕[1,2]　胡　凯[1,2]

(1. 湖北第二师范学院建筑与材料工程学院，武汉 430205；

2. 湖北第二师范学院 湖北省 BIM 智慧建造国际科技合作基地，武汉 430205；

3. 中铁十一局集团有限公司第五工程公司，重庆 400037)

【摘　要】 为研究桥梁设计信息在 BIM 建模软件中的表达方式及探索 BIM 技术在桥梁工程项目中的应用点，本文从 BIM 信息共享和传递的角度，在对某主跨 180m 连续刚构桥精细化 BIM 建模的基础上，对其做了五项具体的 BIM 应用：设计核查、施工模拟、工程量统计、碰撞检测和箱梁 0 号块水化热分析。研究表明，图纸中的非几何信息可以在构件属性中完整表达；BIM 技术的应用应以信息为核心，对 BIM 模型的应用就是提取模型信息并进行信息再加工的过程；主流施工动画模拟软件与 Revit 的互操作性较好，但有限元软件 Midas/FEA 与 Revit 的互操作性有待加强。

【关键词】 BIM；参数化建模；信息共享；互操作性；施工模拟；水化热分析

BIM 技术是未来建筑业发展的大趋势。近 10 年来，越来越多的工程项目应用 BIM 技术，建筑业正在经历一场史无前例的彻底变革。目前，在桥梁工程领域，对 BIM 技术的应用主要在设计阶段，偏重于桥梁几何信息创建及利用，如方案比选[1]、参数化建模[2~4]、力学分析[5]等，这些应用为桥梁专业的 BIM 正向设计提供了参考，但也存在着"重几何模型、轻信息表达及利用"的不足。BIM 技术尤其注重建筑信息在工程建设全寿命周期的共享和传递[6]。BIM 模型中除了包含拟建实体的几何信息外，还包含了图纸中大量的非几何信息，它们共同构成了 BIM 模型强大的信息资源。因此，对 BIM 技术的研究，不仅要关注几何模型的创建，还应关注其非几何信息的表达及利用。

本文基于 Autodesk Revit 平台，对沿溪沟特大桥做了 BIM 的深化应用，应用中重点关注了模型信息的表达与传递，力求体现将 BIM 模型作为建筑信息的载体，对 BIM 模型的利用就是提取这个载体中相应的信息并进行再加工这一核心思想。本文的应用思路可供设计企业和施工企业参考。

1　工程概况

沿溪沟特大桥位于重庆市黔江区，为跨越深沟、沿溪河及国道 G319 而设。本桥平面位于分离式路基段，桥梁分左、右两幅桥，桥梁全长 797m。

全桥平面位于直线上，左线纵断面位于 $R=25000m$ 的竖曲线上，右线纵断面位于 $R=35000m$ 的竖曲线上。桥面为单向 2% 横坡。

每幅桥桥面宽度 12.75m。单幅桥面组成为：0.5m(防撞墙)＋11.5m(行车道)＋0.5m(防撞墙)＝11.25m。

桥梁上部结构孔跨布置为：3×40m 预应力混凝土（后张）连续 T 梁＋（96＋180＋180＋96）m 预应力混凝土连续刚构桥＋3×40m 预应力混凝土（后张）连续 T 梁。主桥上部结构采用变截面预应力混凝土悬浇箱梁，单箱单室截面，箱宽 7.0m，两侧翼缘悬臂宽 2.875m。箱梁端部梁高 4m，墩顶根部梁高

11.6m，为主跨跨度的 1/15.52，箱高以 1.5 次抛物线变化。

主桥桥墩采用钢筋混凝土矩形双薄壁墩，主墩下部采用整体箱形截面。主墩顺桥向宽度 10m，薄壁厚 2m；横桥向宽度 7m；薄壁部分高度 56m。5、6 号墩身横向两侧按 1：80 坡度放坡至承台顶面。大桥主墩均位于岸上，最大墩高 140.6m。沿溪沟特大桥桥型布置如图 1 所示。

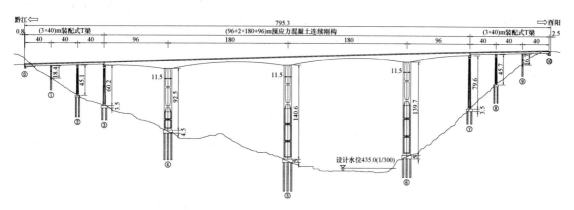

图 1　沿溪沟特大桥立面布置图（单位：m）

2　BIM 实施计划

2.1　项目难点分析

（1）模型构件拆分需与实际工序进行协调。建模前应充分理解项目特点及设计意图，结合施工方案对模型构件进行合理拆分，尽可能避免后期 BIM 应用时对模型的二次加工。

（2）模型制作及定位控制难度大。一是本项目构件种类多且构造复杂，构件族的制作难度大。上部结构包含 T 梁、现浇箱梁；下部结构包含柱式墩、等截面及变截面空心墩、桩柱台、U 台、普通盖梁、L 形盖梁等多种类型。二是桥梁构件定位时需考虑路线平、纵、横的影响，将构件族在空间线位中定位的难度较大。

（3）项目精度要求高。项目的模型精度要求为 LOD350 及以上，该精度下的模型即可用于模型单元的加工和安装[7]，即尺寸和定位需要严格以实际为准，此模型可以作为施工的依据。

（4）参数化建模复杂。"参数化"是 BIM 技术的优势之一，采用参数化建模不仅可以减少同一类构件的建模工作量，也能适应后续图纸变更时快速修改模型并统计工程数量的要求。本项目高墩、箱梁构造复杂，要对同一类构件通过不同的参数反映其几何特征，建模之初就应仔细分析各类构件的构造特点，制定详细的参数化策略。

（5）设计语义丰富，信息集成度要求高。为便于 BIM 模型应用，在 BIM 模型中应对图纸中的设计信息进行集成，这些信息包括上部

结构 T 梁的布梁信息、桩位坐标、墩台标高、现浇梁节段控制标高等非几何信息。

2.2　BIM 应用点解析

沿溪沟特大桥 BIM 的实施从施工阶段介入。首先根据设计图纸建立桥梁 BIM 模型，然后根据施工单位项目部的需求做具体应用。本文将 BIM 模型中的信息划分为几何信息与非几何信息，对沿溪沟特大桥 BIM 技术的应用可认为是从模型中提取所需的信息，在 BIM 软件或其他软件（系统）中做进一步的信息加工。BIM 模型在各系统间的信息共享和传递是 BIM 技术的精髓。在 BIM 模型信息的基础上，本文做了图纸核查、施工模拟、工程量统计、施工碰撞检测及有限元精算五个方面的应用。从 BIM 模型信息共享和传递的角度，可将沿溪沟特大桥 BIM 技术的应用用图 2 来描述。

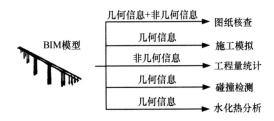

图 2　沿溪沟特大桥 BIM 技术应用概览

3　BIM 模型创建

3.1　几何建模

本桥用 Revit 创建 BIM 模型。几何模型的创建充分利用了 Revit 参数化建模工具，遵循由构件族到项目的建模步骤。本项目对同一类构件均采用了参数化建模，大大减少了建模工作量，图 3 列举了主桥变截面空心墩的参数化实例。而对于构件族在项目中的定位和拼装，则充分利用了 Revit 软件内置的 Dynamo

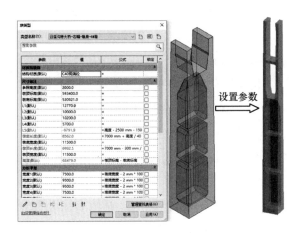

图 3　主墩参数化实例

可视化编程插件。由于 Dynamo 提供了丰富的 Revit 应用程序接口（API），可使用户通过编程来提高建模效率，在建模中被广泛应用[8~10]。本项目应用 Dynamo 放置构件的具体思路是：利用路线设计软件输出路线设计线的逐桩坐标、高程等基础数据，将其整理为 CSV 数据格式或 Excel 文件格式，在 Dynamo 中通过提取的路线桩号信息和设置构件参数（位置、高程、角度等）可实现构件的精准放置。图 4 为沿溪沟特大桥全桥 BIM 模型。

图 4　沿溪沟特大桥全桥 BIM 模型

3.2　钢筋建模

大跨连续刚构桥通常采用三向预应力体系，预应力筋与普通钢筋密集交错，钢筋布置非常复杂。本项目重点考虑了箱梁三向预应力钢筋模型，图 5 展示了竖向和横向预应力钢筋族。而对于普通钢筋，一是由于 Revit 对复杂

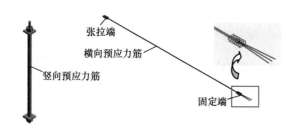

图 5　预应力钢筋族

结构钢筋建模的功能有限，导致建立全桥普通钢筋模型的效率低下；二是建立全桥钢筋模型将占用系统大量内存而降低软件运行性能；三是在实际施工过程中，桥梁普通钢筋的放样常因预应力钢筋干扰而挪动；因此建立全桥普通钢筋模型的意义不大，本项目仅根据施工单位的具体需求有针对性地建立局部普通钢筋模型。本桥对主桥承台、高墩及箱梁 0 号块（图 6）建立了普通钢筋模型。

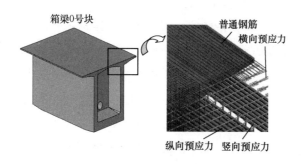

图 6　箱梁 0 号块钢筋模型展示

3.3　非几何信息表达

　　BIM 模型中除了完整建立上述几何模型外，还应将非几何信息集成到模型中，这是 BIM 模型相比于其他三维设计软件的优势所在。在 Revit 中，非几何信息一般添加到构件的属性栏中。这些信息可以预先在 Excel 中编辑，然后通过在 Dynamo 中编程批量添加到模型中。图 7 展示了将引桥上部结构 T 梁的布梁信息添加到构件属性中的实例，这些信息也可类似设计图纸那样以表格形式给出。这种将

箱梁的非几何信息添加到构件的属性中的方法为桥梁的正向设计提供了一个思路。

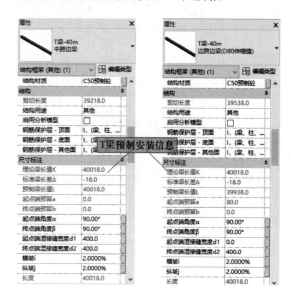

图 7　T 梁布梁信息在 BIM 模型中的表达

4　BIM 技术应用

4.1　设计核查

　　尽管本项目使用二维图纸逆向翻模，但这个建模过程能核查出绝大部分设计上的"差、错、漏、碰"。翻模就是将图纸的二维语义转换成三维模型的过程，可看作是设计图纸在 BIM 软件中的虚拟实现。一方面，建模人员需要在精读设计图纸、理解设计意图的基础上才能建立参数化的 BIM 模型，这个过程本身就是一个设计核查的过程。另一方面，利用 BIM 模型中的基础信息，通过在软件中编译公式可以自动生成其他设计信息，将该信息与设计图纸比对，也可对设计图纸进行核查。因此，在现阶段，利用 BIM 软件对图纸进行翻模仍是核查设计缺陷的有效手段。

　　图 8 展示了 BIM 模型中自动计算的箱梁控制点标高与设计图纸中列出的控制点标高的对比核查的一个例子。通过对箱梁参数赋值，

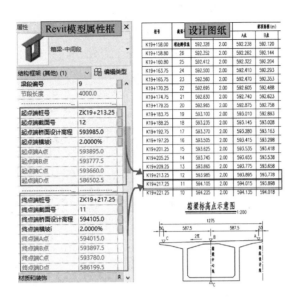

图8　上部结构箱梁非几何信息对比

BIM 软件可自动计算与图纸一一对应的箱梁放样控制点高程，若该数据与图纸一致，则说明图纸给出的标高无误。

此外，利用 BIM 模型中的几何信息，可以将 BIM 模型导入 Navisworks 软件对原设计进行碰撞检测。图9展示了一个引桥墩梁固结处的冲突实例：本桥左、右线2号墩和8号墩为墩梁固结，设计采用 T 梁梁底钢板与墩顶预埋钢板焊接，梁底距盖梁顶仅2cm，施工时盖梁顶应做成与 T 梁一致的纵坡，否则会影响 T 梁的放置。

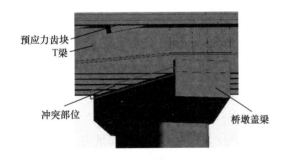

图9　T梁与桥墩盖梁的碰撞

通过设计核查，本项目共检测出图纸重大错误（里程、标高数据有误）5项，重大错误（图纸构造前后不对应）8处，一般错误若干。于施工前应用 BIM 技术对设计图纸进行校核，

有效减少了施工阶段中的变更风险，是 BIM 技术最常用的应用点。

4.2　施工模拟

施工模拟主要体现在施工动画的制作上，其成果不仅可用于施工方案汇报、评审，也可用于可视化交底。施工方案模拟可以根据项目进度计划制作构件生长动画（4D 施工模拟），这类动画 BIM 建模人员常可以胜任。于事前演练施工程序，可以提前发现施工问题，避免现场的错误与停工，能有效节省时间及人力成本，以及提升施工安全、质量。常用的施工模拟软件有 Navisworks，Fuzor，Lumion 等。在这些软件中，Navisworks 可以读取由 Revit 导出的 NWC 格式文件，Fuzor 可以作为一款插件嵌入 Revit 工具栏，Lumion 也有专用的格式转换插件使之能读取 Revit 模型。而对于高品质动画的制作，则需要在 3D Max 中进行渲染方能达到出众的效果，3D Max 可以读取由 Revit 导出的 CAD 格式文件。因此，得益于施工模拟软件与 BIM 软件之间良好的互操作性，利用 BIM 技术进行动画制作不需要另行建立 3D 模型。模型信息在不同系统之间的共享和传递充分发挥了 BIM 技术的价值。本项目的施工方案的模拟应用了 Lumion 和 3D Max 等多个软件（图10）。

图10　桥位动画渲染

本项目施工方案模拟的主要内容（图11）

有：主桥承台采用分段分层连续浇筑法施工工艺，主墩墩身采用液压爬模施工，上部结构箱梁采用挂篮对称悬臂浇筑。

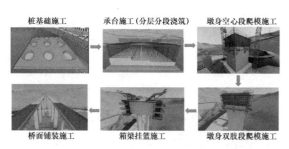

图 11　主要施工方案模拟

4.3　工程量统计

工程量统计是编制工程预算的基础工作，具有工作量大、费时、繁琐、要求严谨等特点，约占全部工程预算编制工作量的 50%～70%。对工程量统计的精确度和快慢程度将直接影响工程预算编制的质量与进度。用 BIM 软件自动计算工程数量是一种行之有效的方法。特别是对于复杂桥梁构件工程数量的统计，传统手算普遍存在易错、效率低等缺点。而应用 BIM 技术自动进行工程量统计的方法不仅有利于加快概预算速度、减轻概预算人员的工作量、提高概预算质量，而且对于增强审核及审定透明度都具有非常重要的意义[11]。利用 Revit 软件中明细表功能，添加相应字段可自动生成工程量（图 12）。本项目的实践也证明，用 Revit 软件生成的工程数量有效地更正了二维算量中的错误数据，项目各参建方对此高度认可。

4.4　施工阶段碰撞检测

大跨连续刚构桥上部结构施工过程中，常需要在箱梁中的适当位置布置浇筑孔。悬浇箱梁大多采用三向预应力体系，浇筑管道的位置与预应力钢筋存在相冲突的风险。因此，需要

<左幅桥箱梁工程量统计>		
A	B	C
梁段编号	节段长度	节段混凝土方量
22	4680	95.61 m³
21	2000	23.49 m³
20	4500	53.29 m³
19	4500	54.53 m³
18	4500	56.38 m³
17	4500	58.67 m³
16	4500	64.13 m³
15	4500	70.22 m³
14	4500	73.85 m³
13	4000	68.93 m³
12	4000	72.22 m³
11	4000	75.69 m³
10	4000	79.32 m³
9	4000	83.13 m³
8	4000	91.14 m³
7	4000	99.67 m³
6	4000	104.39 m³
5	3500	95.34 m³
4	3500	99.18 m³
3	3500	103.12 m³
2	3500	107.16 m³
1	3500	111.30 m³
0	16000	823.59 m³

图 12　左幅箱梁混凝土数量明细表

提前规划好浇筑孔的具体位置。根据施工单位初步的浇筑孔布置方案，本项目对浇筑管道的布置进行了碰撞检测。

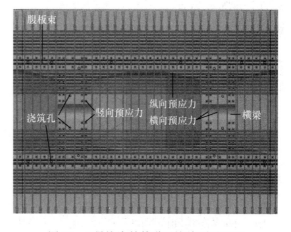

图 13　0 号块浇筑管道碰撞检测平面图

如图 13 所示，对浇筑孔管道的碰撞检测结果表明，孔位基本避开了三向预应力管道，但由于浇筑管道的直径为 15cm，腹板处的纵向预应力管道净距仅 9.3cm，因此，腹板处的浇筑管道建议减小直径或适当增大箱梁腹板束的横向间距。

4.5 箱梁0号块水化热数值分析

大跨连续刚构桥0号块的混凝土总方量大，而且都采用高强度等级的混凝土，因此，在浇筑混凝土的施工过程中会产生大量的水化热而导致较高的温度应力，不利于结构的抗裂。沿溪沟特大桥0号块梁高11.6m，横隔板厚2m，腹板厚1.5m，底板厚1.5m，需按大体积混凝土编制专项施工方案[12]。本项目利用Midas/FEA软件对0号块浇筑过程中的水化热进行了分析。

Revit模型可为Midas/FEA软件提供分析所需的0号块的几何信息。本文采用了如图14所示的数据交互过程。实践发现，几何信息导入FEA后，出现了数据丢失及精度降低现象（如面被分割及过人孔处的圆弧线由多条短直线代替等），需要手工对模型进行修复。因此，Midas/FEA软件与Revit模型的互操作性还有待提高，而不同专业软件之间缺乏互操作性通常是限制BIM应用充分发挥潜力的问题所在[13]。

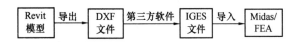

图14　Revit与Midas/FEA数据交互流程

箱梁0号块分两阶段浇筑，第一次浇筑到横隔墙过人孔顶面以上1m位置，浇筑高度5.2m；第二次浇筑至箱梁顶面，浇筑高度6.4m；两次浇筑间隔时间5～7d。由于0号块为对称结构，本文仅建立1/2实体有限元模型。经计算，水化热最大部位在横隔墙中面（两浇筑阶段的最大水化热时刻温度场如图15所示）。通过改变水化热分析边界条件、拆模时间及两阶段浇筑间隔时间，对0号块施工方案提出了优化建议：一是底模及箱梁内模采用木模板，延缓混凝土表面热扩散，从而降低浇

筑体的里表温差；二是延缓拆模时间，使浇筑体缓慢降温；三是两阶段浇筑间隔时间按7d控制，使先浇混凝土有充分的散热时间。

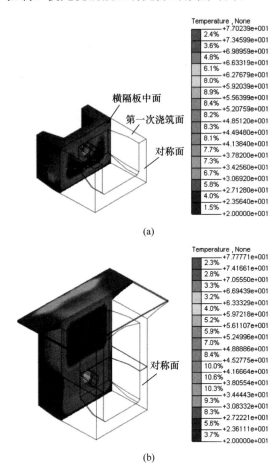

图15　最大水化热时刻温度云图
(a) 第一阶段（144h）；(b) 第二阶段（280h）

5　结论

（1）BIM技术建议从规划阶段介入，从规划阶段至运营阶段逐层深化，这样才能最大限度地发挥BIM价值。

（2）Revit软件建立BIM普通钢筋模型的效率不高。鉴于普通钢筋在施工时经常会有微调，现阶段建议只根据具体需求，在局部复杂的部位建立钢筋模型，不建议创建全桥钢筋模型。

（3）基于现有的BIM平台，对桥梁工程

实现 BIM 正向设计还有很长的路要走，BIM 正向设计除了创建几何模型外，还应着重非几何信息的表达。建议在构件属性中添加设计信息，这些信息应方便其他系统（通过软件自身功能或编程）提取，这是 BIM 信息共享的核心要求。

（4）BIM 信息在不同专业软件之间应用时，需要考虑不同系统之间的互操作性。良好的互操作性可以充分发挥 BIM 技术的潜力。实践发现，主流施工模拟软件 Navisworks、Fuzor、Lumion 等与 Revit 的互操作性较好，但结构分析软件 Midas/FEA 与 Revit 之间的互操作性还有待提高。

（5）作为 BIM 模型的应用方，施工单位应加大 BIM 人才的培养，应养成 BIM 思维，能善于基于 BIM 模型中已有的信息发掘应用点，为工程施工创造价值。

参考文献

［1］ 黄成岑，李洋溢，袁通. BIM 技术在桥隧相接部位方案设计中的应用［J］. 中外公路，2020，40（5）：236-239.

［2］ 李红豫，李恒，吴悦，等. 基于 BIM 的东洲湘江大桥参数化设计应用研究［J］. 公路，2020，65（11）：173-178.

［3］ 傅战工，郭衡，张锐，等. BIM 技术在常泰长江大桥主航道桥设计阶段的应用［J］. 桥梁建设，2020，50（5）：90-95.

［4］ 杜一丛，王亮. 基于 BIM 参数化在桥梁工程设计阶段应用初探［J］. 建筑结构，2019，49（S2）：972-978.

［5］ 陈素华，丁建明，姜严旭，等. 莲花桥 BIM 技术应用及异形索塔力学性能研究［J］. 重庆交通大学学报（自然科学版），2020，39（10）：43-48.

［6］ Eastman C，Teicholz P，Sacks R，et al. BIM handbook：a guide to building information modeling for owners，managers，designers，engineers，and contractors［M］. John Wiley & Sons Inc.，2008.

［7］ 中华人民共和国住房和城乡建设部. 建筑信息模型施工应用标准 GB/T 51235—2017［S］. 北京：中国建筑工业出版社，2017.

［8］ 仇朝珍，贺波，葛胜锦. Dynamo 在桥梁 BIM 建模中的应用［J］. 中外公路，2019，39（5）：179-182.

［9］ 吴生海，刘陕南，刘永哓，等. 基于 Dynamo 可视化编程建模的 BIM 技术应用与分析［J］. 工业建筑，2018，48（2）：35-38.

［10］ Tang F，Ma T，Zhang J，et al. Integrating three-dimensional road design and pavement structure analysis based on BIM［J］. Automation in Construction，2020（113）：103152.

［11］ 周冀伟，郭婧娟. BIM 技术在工程量统计中的应用研究［J］. 施工技术，2017，46（S2）：1233-1235.

［12］ 中华人民共和国住房和城乡建设部. 大体积混凝土施工标准 GB 50496—2018［S］. 北京：中国建筑工业出版社，2018.

［13］ Venugopal M，Eastman C M，Sacks R，et al. Semantics of model views for information exchanges using the industry foundation class schema［J］. Advanced Engineering Informatics，2012，26（2）：411-428.

基于 BIM 与 IoT 技术的装配式建筑施工阶段智能管理研究

苗泽惠　李明慧

（吉林建筑大学经济与管理学院，吉林长春 130118）

【摘　要】　随着建筑工业化飞速发展，信息技术已经成为建筑工业化非常重要的工具手段，以 BIM 技术、互联网、物联网、云计算及大数据等为代表并快速发展的信息技术应运而生。本文通过 IoT（物联网）等技术将传感器、机器、人、物连接起来，实现施工过程的智能收集、定位、监控、处理、分析和管理以采集施工现场数据信息，并与 BIM 模型进行数据交互，从而实现智能化管理，由此可降低工程成本，提升建筑质量，促进我国建筑行业的发展，为建筑智能化提供发展的前进动力。

【关键词】　BIM；IoT；装配式；施工阶段；智能管理

随着我国建筑业可持续发展理念逐渐深入，人们对美好生活需要不断提升，建筑智能化、绿色化、工业化成为建筑行业新的发展趋势。2018 年以来，随着我国新型基础设施建设深入推进，工程建设的质量、效率、安全、成本受到越来越多的重视。

在我国"十四五"规划背景下，建筑工程项目的智能化管理是实现建筑可持续发展的根本。通过将智能化技术与工程建设相融合，可实现施工智能化、科学化。目前，BIM 和 IoT 技术在装配式建筑智能化和工业化上的应用最为普遍，二者的集成应用，实质上是建筑全生命周期信息的融合[1]。在装配式建筑中，施工阶段的管理是极为重要的环节，本文将 BIM 与 IoT 技术结合，建立智能化管理平台，并将其应用到装配式建筑施工阶段的管理中，可达到工程项目降本增效的成果，有助于拓展装配式建筑的研究领域，并为装配式建筑的智能

化发展带来强劲动力。

1　装配式建筑施工阶段智能管理的研究现状

近年来，装配式建筑施工阶段智能化管理发展较为迅速，应用最为普遍的是 BIM 技术与 IoT 技术，国内外学者对此也进行了一系列研究，如：采用可视化技术对结构进行细化，并完成了现场施工模拟，进而在 BIM 平台上对数据信息进行整合，实现对项目建设全过程的有效管控[2]；采用 BIM 技术与无线射频识别集成应用，实现了资源实时跟踪控制，在施工过程中对成本与进度进行了精确控制[3]；应用 BIM 和物联网技术在施工阶段进行价值分析，指出企业信息化需要 BIM 技术和物联网在施工阶段的应用，使项目精益管理[4]；国内学者提出了 BIM＋物联网的智能建造综合管理体系，实现数字化建造，对原有

的传统建造模式进行数字化升级，构建全新的无人化施工系统，最终实现全智能化的建造方式[5]。

在以往的装配式建筑智能管理中，BIM技术与物联网技术大多是各司其职，负责两个不同的阶段。对于将BIM技术与IoT技术有机结合，进行装配式建筑施工阶段智能管理的研究较少。因此，在我国目前装配式建筑发展的基础上，结合BIM技术与IoT技术，对装配式建筑施工阶段进行智能化管理研究是十分有意义的。

2 BIM技术在装配式建筑施工阶段管理分析

2.1 BIM技术概述

BIM技术通过建立虚拟的三维建筑模型，将所得信息整合到三维模型数据库中，可帮助实现建筑信息的集成，也可及时、准确地对相关数据进行共享和传递，应用BIM技术可达到工程项目降本增效的效果，是实现建筑智能化管理的基础。

2.2 BIM技术在装配式建筑施工过程中的应用分析

（1）模拟施工现场管理

应用BIM技术对装配式建筑施工现场进行管理，应在施工准备前应用BIM软件对施工现场的布置情况、施工机械的施工过程、主要材料的现场布置及塔式起重机的吊运范围进行模拟，可减少二次搬运，提升整体装配率。同时，应用BIM技术可预先对施工过程中构件进场、运输、存储等环节进行模拟，根据模拟情况，优化施工方案，提升整体装配的质量。

（2）施工成本管控

在工程项目施工中常出现造价超出预算的现象，利用BIM数据库，可真实地观测到造价所需的信息，并利用这些信息，对各构件进行工程量计算，准确预测工程成本。利用BIM技术与智能化制造的结合，可对各构件进行智能化加工，降低构件的误差，节约施工成本。而在BIM三维模型的基础上加上进度、成本两方面信息，可形成BIM5D建筑信息模型。通过BIM5D模型模拟施工，可查看施工进度及资金的投入是否合理，并及时优化工期，调整资金投入，有效地对成本进行管控。

3 IoT技术在装配式建筑施工阶段管理分析

3.1 IoT技术概述

物联网（Internet of Things，IoT）技术在建筑施工中的应用大多是通过传感器、射频识别等技术，实时采集施工中所需的构件信息，通过各种可能的网络连接，实现对构件的标记、识别与管理。目前，物联网技术中的射频识别（Radio Frequency Identification，RFID）是建筑项目中运用最广泛的技术。其核心技术是在需要识别的对象中植入芯片，通过网络将芯片与识别对象相关联，可赋予对象智能化。

3.2 IoT技术在装配式建筑施工过程中的应用分析

3.2.1 施工各阶段构件管理

（1）制作阶段

在构件制作阶段，将所有构配件的信息录入RFID芯片中，对各构配件进行编号，以便施工人员查找相关构件信息，避免工程延误。

（2）运输阶段

在构件的运输阶段，将 RFID 芯片植入运输车辆中，可及时对运输车辆的行驶路线进行规划，并推荐最短、最省时的路径，以降低运输成本。

（3）吊装阶段

在构件吊装阶段，通过将已录入的构配件信息反馈到施工机械操作人员手中的读写器中，根据读写器信息，按顺序精准吊装，避免错误搭接的同时节省时间。

3.2.2　现场材料管理

应用 RFID 技术对施工材料运输、进场、储存等进行标记，并对其进行实时监控与跟踪，保证材料充足。

3.2.3　机械设备管理

通过 IoT 技术对大型设备施工过程进行监控、录制、反馈，最大限度地降低施工阶段事故发生的概率。

3.2.4　施工安全管理

通过 IoT 技术对施工环境、构配件搭接等因素进行监控，将各类检测到的信息集中反馈到中心控制平台，以便对各因素进行统一管理。

3.2.5　施工人员管理

将 RFID 芯片用于识别施工人员的身份信息及位置信息，在发现施工人员有违规操作等行为时，及时将信息传入管理人员手中，以采取相应措施，避免出现施工事故。

4　BIM 和 IoT 技术在装配式建筑施工阶段集成管理分析

BIM 属于上层信息存储、交叉和管理者，IoT 属于下层信息监测、收集和信息传输者。通过将 BIM 和 IoT 中的 RFID 技术相结合，建立一个智能化管理平台，可对装配式建筑施工阶段进行更精准管理控制。BIM 与 IoT 技术的集成应用，可对装配式建筑智能化管理的发展产生推动作用。

4.1　BIM 和 IoT 技术在装配式建筑施工过程中的集成管理

4.1.1　构件的进度管理

（1）构件制作、运输、进场阶段

在构件的制作、运输及进场阶段，以 BIM 三维模型作为基础，通过 RFID 收集的构件信息，将所得信息输入智能化管理平台进行信息反馈，查看构件制作进度，预测构件是否按时进场，并对实际时间与计划时间进行比对（图 1），及时优化调整，避免出现场地占用等问题。

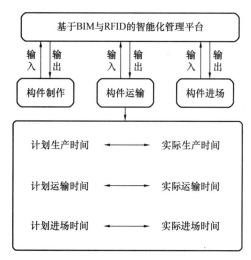

图 1　构件制作、运输、进场
阶段智能化管理

（2）构件储存、吊装阶段

在构件储存及吊装阶段，多以 RFID 技术为主导，并结合 BIM 的三维信息模型对构件进行实时监测，查看已植入 RFID 芯片的构件信息，通过智能化管理平台将所得信息输出给吊装中心及存储中心，吊装控制中心在接收到指令时对构件进行吊装，存储中心则对构件进行监控，预留出充足的存储空间[11]（图 2）。如此，加快施工进度的同时减少可能出现的问题，确保了工程的顺利进行。

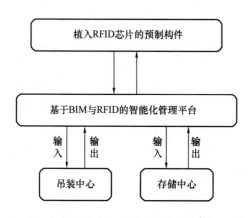

图2　构件储存、吊装阶段智能化管理

4.1.2　构件的质量管理

基于 BIM 与 IoT 技术的构件质量管理，是在施工准备阶段对各构配件进行碰撞检查，并在智能管理平台上对质量控制点进行设置，以此对构件质量进行严格把控。在现场吊装阶段，通过可视化模型及智能管理平台录入的信息对构件的搭接情况、材料的堆放情况等进行控制。保证了构件的质量，施工才能顺利有序地进行。

4.1.3　构件的成本管理

（1）制造阶段

使用智能化管理平台可快速获取装配式建筑各构件的数据信息，以提前准备制造构件的材料，加快生产速度，提高效率，降低构件制造的成本。

（2）运输阶段

在构件的运输阶段，基于 BIM 与 IoT 技术的智能化管理平台可对构件的运输路线进行监控并对构件从运输到现场堆放、储存等各个重要环节进行实时跟踪汇报，以便及时调整优化运输方案，进一步提高构件的运输效率，在一定程度上降低构件运输阶段的成本。

（3）吊装阶段

利用 BIM 与 IoT 技术对装配式建筑施工过程中的构件关键节点搭接情况进行虚拟预演，以便在真正施工过程中能够提高搭接准确度、节约时间。然后，利用 BIM 三维模型在前期进行碰撞试验，并通过智能化管理平台检查出问题所在关键部位，对此及时进行调整，减少施工中的返工情况，提高了施工现场的生产效率，有利于提高施工效率，在一定程度上降低成本。

4.2　BIM 和 IoT 技术在装配式建筑施工过程中的实际应用

上海浦江某大型 PC 保障房项目将 BIM 与 IoT 技术集成应用[13]。在此期间，项目运用 BIM 技术将施工进度加入三维模型中形成 BIM 4D 可视化模型，并结合射频识别技术对构件的编码规则进行研究、制定、定位；指导现场施工；控制构件搭接的准确性；及时定位构件的运输、进场路线，将项目施工阶段信息录入数据库中，实时跟踪记录。由此可见，将 BIM 与 RFID 技术应用到装配式建筑施工阶段，可实现预制构件生产、安装过程全面智能化管理，提高了施工管理效率的同时缩短了施工工期。

4.3　BIM 和 IoT 技术在装配式建筑施工过程管理中的不足

（1）相关标准不完善

对于 BIM 及 IoT 等高新技术，国外起步较早，发展较快，已经形成了一系列标准体系。而我国，信息化技术起步晚，导致许多地区并没有相应的标准可供参考。

（2）社会应用度较低

我国智能化管理仍处于初级阶段，对于 BIM 与 IoT 等先进技术行业认可度较低。虽然国家大力推崇，但许多企业、业主对智能化管理不甚了解，且不愿冒险使用。这些因素导致了我国智能化管理的应用度普遍低于其他发达国家。

4.4 BIM 和 IoT 技术在装配式建筑施工过程管理中的发展路径

（1）出台相关标准

当下，制定智能化管理标准是我国政府和整个建筑行业共同努力的目标。同时，应格外注重在装配式建筑施工阶段的管理，只有制定符合我国国情的标准体系，才能满足目前我国行业发展的需求。

（2）加强人才培养

目前，我国装配式建筑的智能化管理仍落后于许多发达国家，国内建筑行业急需大批专业人才[14]。在往后的高校培养中，可实行高校与企业联合培养，对人才进行相关专业化培训，实行持证上岗模式。

（3）自主研发智能管理软件

目前，我国的智能化软件主要依赖于国外进口，在一定程度上，国外的技术并不能完全满足我国建筑业的现状，也不能真正实现信息共享，这在很大程度上限制了我国智能化管理的发展，因此，应加大我国自主研发智能化软件的力度。

5 结语

建筑业是我国经济发展的五大支柱型产业之一，发展建筑工程管理智能化、信息化是我国现阶段应大力推崇的。应用 BIM 与 IoT 技术，建立装配式建筑施工阶段智能化管理平台，保证施工质量、效率以及安全的同时降低施工成本的投入，提升了装配式建筑施工阶段的管理水平，为我国装配式建筑的绿色可持续发展添砖加瓦。

参考文献

[1] 中华人民共和国住房和城乡建设部. 住房和城乡建设部关于印发 2016—2020 年建筑业信息化发展纲要的通知[EB/OL]. [2016-08-23]. http：//www. mo-hurd. gov. cn/wjfb/201609/t20160918_228929. htm.

[2] Sungyol Song, Jeongsam Yang, Namhy Uk Kim. Development of a BIM-based structural framework optimization and simulation system for building construction[J]. Computers in Industry, 2012, 63(9): 895-912.

[3] Costin, Aaron, Adibfar, Alireza, et al. Building Information Modeling (BIM) for transportation infrastructure -Literature review, applications, challenges, and recommendations [J]. Automation in construction, 2018, 94(10): 257-281.

[4] 裴卓非. BIM 技术与物联网在施工阶段的应用[J]. 建材技术与应用, 2013(1): 60-62.

[5] 买亚锋, 张琪玮, 沙建奇. 基于 BIM＋物联网的智能建造综合管理系统研究[J]. 建筑经济, 2020, 41(6): 61-64.

[6] 杨德林. 物联网技术在装配式建筑中的应用现状[J]. 建筑工程技术与设计, 2021(5): 1897.

[7] 刘秉岩. 基于人工智能物联网等前沿技术在装配式建筑中的应用[J]. 门窗, 2019(16): 250-251.

[8] 李天华, 袁永博, 张明媛. 装配式建筑全寿命周期管理中 BIM 与 RFID 的应用[J]. 工程管理学报, 2012, 26(3): 28-32.

[9] 杨帆. 基于 BIM 与物联网的装配式建筑设计与施工管理[D]. 西安: 西安科技大学, 2019.

[10] 肖阳. BIM 技术在装配式建筑施工阶段的应用研究[D]. 武汉: 武汉工程大学, 2017.

[11] 呙丹, 杨晓华, 苏本良. 物联网技术在现代建筑行业中的应用[J]. 山西建筑, 2011, 37(26): 255-256.

[12] 张家昌, 马从权, 刘文山. BIM 和 RFID 技术在装配式建筑全寿命周期管理中的应用探讨[J]. 辽宁工业大学学报（社会科学版）, 2015, 17(2): 39-41.

[13] 胡斌, 周磊. 基于 BIM＋物联网技术的装配式

建筑协同管理研究[J]. 甘肃科技，2021，37（13）：101-103.

[14] 西崇峰. 基于 BIM 和 RFID 技术的装配式建筑施工过程管理[J]. 中国建筑金属结构，2021（1）：26-27.

[15] 刘诗楠，刘占省，赵玉红，等. NB-IoT 技术在装配式建筑施工管理中的应用方案[J]. 土木工程与管理学报，2019，36(4)：178-184.

[16] 张洋. 基于 BIM 的建筑工程信息集成与管理研究[D]. 北京：清华大学，2009.

[17] 赵敬忠. 基于智慧建造的工程项目施工成本精细化管理研究[D]. 兰州：兰州理工大学，2019.

医养结合与健康住宅

Integrated Eldercare Services with Medical Care

新冠肺炎疫情背景下智慧城市的韧性建设分析

伍昭亮　郭慧锋　顿彤彤

（南昌航空大学土木建筑学院，江西南昌 330063）

【摘　要】　把增强韧性建设作为智慧城市发展的主要方向，对城市的可持续发展至关重要。为探索提升智慧城市韧性的路径，梳理智慧城市的构成要素和韧性城市的特征，定性分析二者之间的内在联系，并以"新冠肺炎疫情"作为背景，从基础设施韧性、经济韧性、制度韧性、社会韧性四个维度出发，对智慧城市在疫情期间的运营状况进行了综合分析，提出增强智慧城市韧性建设的对策与建议。

【关键词】　智慧城市；城市韧性；新冠肺炎疫情；韧性建设

全球城镇化程度越来越高，但由于城市人口剧增导致城市交通拥堵、住房困难，资源的过度开发引起的生态环境恶化等问题也日益显现，给城市的发展带来了潜在的风险。在人们对美好生活的向往和信息技术融合发展的双重驱动下，"智慧城市"的愿景应运而生，其本质在于充分运用信息技术帮助城市主体科学高效地管理城市，更好地提高城市居民的生活水平。

新冠肺炎疫情的突发，打破了城市的热闹喧嚣，疫情造成的负面影响也引发了我们对城市韧性的深入思考。"韧性城市"的建设也成为我国第十四个五年规划和二〇三五年远景目标中的内容。在新冠肺炎疫情防控过程中，一系列信息技术的应用极大地发挥了智慧化手段的作用，大幅减少了疫情对城市造成的冲击，但也显露出目前智慧城市韧性建设的不足之处。城市韧性提升问题是目前城市可持续发展研究的热点，当城市灾害（洪涝、地震、疾病传播等）发生后，我们可以通过采取应急措施恢复城市原来正常的状态，但是如何在城市灾害发生前清楚地认知城市韧性的短板并进行针对性的改善是更值得思考的问题。

本文综合考虑城市"智慧"与"韧性"之间的联系，以"新冠肺炎疫情"作为背景，从基础设施韧性、经济韧性、制度韧性、社会韧性四个维度出发，分析当前智慧城市韧性建设的状况，以期对提升智慧城市的韧性建设提供理论参考。

1　概念辨析

智慧城市与城市韧性涵盖的内容体系庞杂，随着科学技术的不断进步和各类学科的融合发展，二者的定义也随之扩充、演变，目前为止还未达成统一的共识。二者的内涵都符合以人为本的思想，"智慧"的目的在于提升城市居民的幸福感，"韧性"的目的在于提升城市居民的安全感。

1.1　智慧城市的定义

早在 20 世纪 90 年代，建设智慧城市的设想就在国外逐渐显现[1]。2008 年 IBM（International Business Machines，国际商业机器）公司提出了"智慧地球"的概念[2]，"智慧城市"的愿景也是基于这一概念被正式提出，旨在充分运用信息技术提升城市系统处理各项关键工作的能力，对城市生活的各种需求做出智能化的响应，从而为居民创造更高质量的城市生活。Hollands 等[3]认为智慧城市是通过嵌入信息技术的城市基础设施（包括道路、桥梁、铁路等），优化城市基本功能以提高城市的运转效率。中国工程院院士王家耀[4]指出，智慧城市就是通过信息技术在城市各子系统功能中的嵌入，提高城市的感知和响应能力，让城市更加聪明，本质上是让作为城市主体的人更聪明。通过文献的梳理发现，众多学者对智慧城市定义的解读存在一定的差异，但是都强调了信息技术在建设智慧城市中的重要作用。

1.2 城市韧性的定义

韧性一词最早的本意是"回复到原始状态"[5]。自 20 世纪 90 年代以来，相关学者对韧性的研究逐步由自然生态学向人类生态学的方向延展[6]。而城市是人类生态学中不可或缺的研究对象，这为城市韧性理论的形成奠定了思想基础，也为实现城市的可持续发展提供了新思路。Resilience Aliance（韧性联盟组织）[7]基于生态学的角度认为城市韧性是城市具有一定的抵御外部干扰并保持城市系统正常运转的能力。灾害学学者 Jha 等[8]将城市韧性分为基础设施韧性、制度韧性、经济韧性和社会韧性四部分，其中基础设施韧性很关键，包含了城市基本运行的畅通及其应急反应能力。邵亦文[9]从城市规划学的角度认为，城市韧性是通过对城市组成系统的合理规划，增加城市对抗不确定性风险的能力。总的来说，城市韧性就是城市系统及其居民在各种冲击和压力下保持正常运作，能够积极适应并转向可持续发展的能力。

2 智慧城市的韧性特征分析

2.1 智慧城市的构成要素

无论是智慧城市国际标准化组织制定的《新型智慧城市评价指标》，还是智慧城市委员会制定的《智慧城市指数指标体系（SCIMI）》，都是基于物理结构分类智慧城市的构成要素。总的来说，可分为六个维度：智慧基础设施、智慧治理、智慧民生、智慧产业、智慧人群和智慧环境[10]，如图 1 所示。

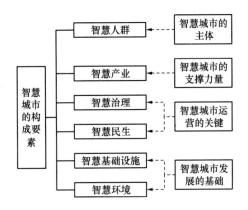

图 1　智慧城市的构成要素

智慧基础设施是现实世界的城市基础设施注入信息网络的智慧化转型，支撑智慧城市的正常运营。智慧治理主要是靠政府自身综合管理水平的提高等。智慧民生是智慧城市建设中的重中之重，与人们的日常生活息息相关，因此，智慧民生主要的建设重点在于交通出行、医疗保障、教育质量、住房等人们最关心的现实问题。智慧产业则对智慧城市的发展提供支撑的力量，它打破传统经济产业的局限性，促进信息技术与传统产业的融合发展，使城市经济的发展更具活力和竞争力。智慧人群是智慧城市的主体，人力资本是智慧城市发展过程中不可或缺的要素，他们参与智慧城市建设的规划，执行具体的建设工作。智慧环境则如同人们生活的载体，优质的环境也会孕育出优质的人才，智慧环境能为智慧城市的健康发展保驾护航。

2.2 韧性城市的特征

美国学者 Ahern[11] 认为韧性城市有五个特征。第一，多功能性：避免因为城市功能单一，使得在城市部分功能受到灾害的冲击时，依然能够为人们提供正常的系统服务。第二，冗余度和模块化特征：如果城市的某一项功能只依靠单一的系统进行提供，在灾害来临时，则更容易出现瘫痪停摆的情况，韧性城市需要有多样的系统提供相同或相似的功能，增加城市系统的冗余性，通过备份或重叠的系统来分散风险，减少灾害冲击带来的损失。第三，多样性：包括生物的多样性、经济和社会的多样性等，当城市受到其他因素干扰时，多样性可以提供不同的对抗干扰的方法途径，增强城市系统的整体韧性。第四，多尺度的网络连通性：如复杂的网络通过重复设置环路，来维持功能连接的正常运转，进而形成韧性。第五，有自我调整能力的规划和设计。此外，国内学者何继新和贾慧[12] 总结了城市社区安全韧性特征，即：复合性、操作性、多样性、恢复性、可变性、共生性。城市社区安全韧性和城市系统韧性的特征基本上能对应韧性城市的特征。通过分析，对于韧性城市的特征，相关学者的论述都具有相同的特点：强调韧性城市具有多样性（城市生态、社会、经济等的多样性）、抗干扰性（在灾害来临时，城市具有一定的应对冲击的抵抗能力）、冗余性（有一定程度的重复和备用部分）的特征。

2.3 智慧城市对应的韧性特征

综合以上分析发现，智慧城市的构成要素

及其实现的具体应用，与韧性城市的特征存在对应的联系，具体如表 1 所示。通过泛智慧城市技术提升城市运转效率，政府智慧治理，公众积极参与，从多方面提升了城市应对不确定性干扰的能力。

智慧城市对应的韧性特征　　表 1

智慧城市的构成要素	应用	韧性特征
智慧基础设施	道路桥梁、城市轻轨等交通设施的智能化建设；5G 基站、新能源充电桩等新基建的建设	多样性、抗干扰性、冗余性
智慧治理	管理者通过利用信息技术提高决策水平；依托"物联网+互联网"创新治理模式，降低城市治理成本	多样性、抗干扰性
智慧民生	智慧医疗的实现，提高现代化的医疗水平；通过交通信息化，改善出行，减少交通事故的发生率	多样性、抗干扰性、冗余性
智慧产业	对投入的资源有效整合，推动产业结构优化升级，使城市经济发展更多元化、更具竞争力	多样性、抗干扰性、冗余性
智慧人群	人是城市运行的主体，人力资本是智慧城市能否建设成功的关键，同时，智慧城市建设过程中也能培养具有创新性与参与性的公民	多样性、抗干扰性
智慧环境	利用信息技术对环境进行监测，通过监测数据分析制定改善环境的有效策略	抗干扰性

3 疫情期间智慧城市的韧性表现

新冠肺炎疫情的出现，使我们能够切身感受到泛智慧城市技术在疫情防控过程中发挥的重要作用，但同时也暴露出智慧城市发展过程中一些不足的地方。经济方面虽然遭受严重损失，但在恢复过程中表现良好，基础设施、制度、社会方面都显示出因数据融合不足而引发的一些问题，如图 2 所示。

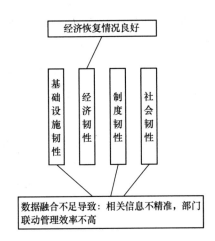

图 2　疫情期间智慧城市的韧性表现

3.1　基础设施韧性

在疫情期间，武汉"雷神山"和"火神山"医院的建造是一个"智慧"应对的方案，运用一系列泛智慧城市技术，如 5G、AI、云计算、大数据等研发出智能化运维管理平台，形成医院"智慧大脑"，实现了"零接触"运维。类似于体育馆之类的基础设施也被作为方舱医院使用，增加了疫情救济的空间，在一定程度上也缓解了医院在面对大面积突发疫情时病床不足的情况。同时，我国发达的交通基础设施也在疫情严峻期间，为人们日常生活所需的重要物资连续供应提供了有力的支撑。

但是，面对突如其来的疫情，作为为城市居民健康保驾护航的医院首当其冲，一时间医疗物资极度匮乏，医院无法承受与日俱增的确诊患者，而且各地医疗系统的数据信息并不能快速、完整地整合，不利于医疗物资的合理调配。

3.2　经济韧性

经济是维持社会稳步发展的重要保障，受新冠肺炎疫情的影响，2020 年我国经济发展遭受了巨大的打击，几乎所有行业都经历了停工停产的状态。由于我国近年来数字经济、电

子商务等智慧金融的快速发展，推动了产业的优化升级，使得城市经济发展更加多元化，在一定程度上减少了疫情的冲击。2020 年，我国是世界上唯一正增长的经济体，GDP 总量超过 100 万亿元。据互联网数据中心（IDC）估计，2022 年全球 GDP 增长的 65％将来源于数字经济，智慧金融在整个经济发展恢复过程中具有举足轻重的地位。同时，我国行业门类齐全，能够形成完整的供应链[13]。由于防疫物资的紧缺，一些企业能够迅速转型调整设备和生产线，并在短时间内生产符合医用标准的口罩、防护服等物资，在能保证国内需求的情况下还能给国外提供货源，这也体现出我国制造业具有强大的韧性。

3.3　制度韧性

通过智慧治理，使人们的生命财产受疫情的损害降到最低。疫情出现时，依托杭州市智慧城市大脑研发的"健康码"，上线 40 天后服务于全国，"健康码"的使用帮助疫情防控工作者实时、动态地掌握出行人员的信息，从而科学高效地制定疫情防控的措施。由于存在信息壁垒，各地健康码不能同用，不方便人们跨省出行，卫健委规定推进"一码通"融合服务，并且简化对不使用、不会操作智能手机的老年群体出示健康证明的方式。这种依托信息化技术进行不断优化调整的智慧治理充分彰显了制度的韧性。

由于疫情期间线下处理政务受限，管理部门将大量线下办理的业务放到线上，短时间内形成了高频的政务服务事项，政务服务系统没有做好压力测试，出现了系统运作崩溃的现象。加上各部门系统平台数据融合不足、缺乏健全的机制和明确的管理内容，很难形成有效的部门联动治理，造成了线上线下管理衔接不明确的局面[14]。

3.4　社会韧性

智慧人群是智慧城市运行的主体,疫情期间,政府积极地倡导民众的社会责任,社会众多组织、团体自发地为疫情防控贡献力量。在信息技术的背景下,民众能够通过社交媒体实时了解我国疫情的动态实况。城市社区的民众也自觉遵守防疫规定,并由社区内部组织或者志愿者为居民配送日常所需的生活物资。目前国内疫情防控可能呈常态化,民众出行之处都有数字化的信息记录,并由相关工作人员管理。我国人口数量全世界最多,人口密度大,疫情防控难度大,但是我国的防疫效果是最好的,这离不开社会民众的团结。

但是在疫情监测过程中,由于各地智慧化手段应用程度不一,且各地数据系统呈孤立现象,跨地区协同合作不足,出现了一些信息盲点,导致防疫机关无法及时掌握相关人员的流动情况,防疫体系容易出现漏洞;在这种与个人隐私信息高度相连的疫情防控形势下,个人有关隐私信息的安全也可能遭受到一定的挑战。

4　增强智慧城市韧性建设的建议

综合以上分析可以总结出,城市的“智慧”与“韧性”存在较强的内在联系,可以融合发展。智慧城市的建设内容体系庞杂,此次新冠肺炎疫情也反映出目前智慧城市的建设还存在一些问题,一系列泛智慧城市技术的嵌入也不能完全能给城市的发展带来良性的影响,智慧城市的建设需要把增强韧性作为主要的发展方向。

4.1　加强智慧城市建设顶层设计

目前我国智慧城市试点数量多,建设投资规模大,但是存在建设水平参差不齐的现象。

要把加强智慧城市韧性作为主要发展方向的理念纳入智慧城市的规划当中,构建城市“智慧”与“韧性”融合发展的框架。完善智慧城市建设的整体设计规划,并且根据各地区的资源配置情况,制定符合地区实际情况的智慧城市建设总体计划,避免盲目建设造成资源浪费。进一步细分建设过程中相关部门的责任和工作内容,完善部门政务平台的工作机制,加强各部门的联动工作能力,提高智慧城市的建设效率。新冠肺炎疫情防控可能呈常态化,有必要把对防控疫情有明显作用的代表性指标合理地纳入智慧城市的评价体系中,提高评价的科学性与实效性,帮助建设者加深对智慧城市建设内容的理解,在实际建设过程中做到及时纠偏。

4.2　加快新型基础设施建设

新型基础设施建设包括 5G 基站建设、数据中心、城市轨道交通、新能源汽车充电桩等诸多领域[15]。从此次疫情中,我们能深刻感知到基础设施的重要性,要加强各地“城市大脑”的建设,通过建立数据中心,打破信息壁垒,实现对数据信息的高度共享互联,加强跨部门、跨地域的协同合作能力,减少疫情下政府的治理成本,提高制度韧性。加快新型交通基础设施的建设,缓解城市交通拥堵的同时,提高冗余性。以信息技术作为支撑,加快推进其他新型基础设施的建设,如智慧医疗、智慧电网等,丰富智慧城市基础设施的多样性,提高智慧城市在疫情下维持正常运转的能力。

4.3　加强信息安全保障

智慧城市以信息技术为支撑,在运营时涉及城市居民、企业单位等各类信息数据的收集、传输、处理等环节。尤其在疫情防控状态下,身份信息、行程轨迹等重要个人信息需要被高频地查询、出示等。要建立健全信息安全法律法规，加强非法利用信息行为的惩戒力度，加强相关部门的执法工作；规范信息数据管理者的职责，对重要信息数据库进行安全测评、落实等级保护制度；加强对城市居民的信息安全教育，提高民众的防范意识，建立良好的信息安全环境，提高智慧城市建设的社会韧性。

参考文献

[1] 刘洪民，刘炜炜. 智慧城市建设理论与实践研究综述[J]. 浙江科技学院学报，2020，32(2)：89-95.

[2] IBM商业价值研究院. 智慧地球[M]. 北京：东方出版社，2009.

[3] Robert G Hollands. Will the real smart city please stand up?[J]. City, 2008, 12(3): 303-320.

[4] 王家耀. 智慧让城市更美好[J]. 自然杂志，2012，34(3)：139-142.

[5] D E Alexander. Resilience and disaster risk reduction: an etymological journey[J]. Natural Hazards and Earth System Science, 2013, 13(11): 2707-2716.

[6] Holling C S. Resilience and stability of ecological systems[J]. Annual Review of Ecology and Systematics, 1973, 4(4): 1-23.

[7] Resilience Alliance. Urban resilience research prospectus[M]. Canberra, Australia: CSIRO, 2007.

[8] Jha A K, Miner T W, Stanton-Geddes Z. Building urban resilience: principles, tools, and practice[M]. World Bank Publications, 2013.

[9] 邵亦文，徐江. 城市韧性：基于国际文献综述的概念解析[J]. 国际城市规划，2015，30(2)：48-54.

[10] 张飞舟，杨东凯，张弛. 智慧城市及其解决方案[M]. 北京：电子工业出版社，2015：19-24.

[11] Ahern J. From fail-safe to safe-to-fail: sustainability and resilience in the new urban world[J]. Landscape and Urban Planning, 2011, 100(4): 341-343.

[12] 何继新，贾慧. 城市社区安全韧性的内涵特征、学理因由与基本原理[J]. 学习与实践，2018(9)：84-94.

[13] 何波. 新冠肺炎疫情对我国在全球产业链地位的影响及应对[J]. 国际贸易，2020(6)：45-52.

[14] 唐斯斯，张延强，单志广，等. 我国新型智慧城市发展现状、形势与政策建议[J]. 电子政务，2020(4)：70-80.

[15] 李晓华. 面向智慧社会的"新基建"及其政策取向[J]. 改革，2020(5)：34-48.

基于医养结合模式的智能养老家具设计研究

李 叶 张 帆 常 乐

（北京林业大学材料科学与技术学院，北京 100083）

【摘　要】 医养结合是一种将医疗与养老相结合的新型养老模式，其中囊括了医疗、保健、照料、休闲等服务。本研究对医养结合型养老家具的种类及特点进行阐述分析。通过 SWOT 态势分析法对智能养老家具的可行性进行分析，提出医养结合养老模式下的智能养老家具发展战略。最终总结了智能医养家具的安全性、易用性、多样性、情感化设计四项设计原则，并从建设智能控制系统、健康检测系统、信息交互适老化设计三个方面提出医养结合模式下智能养老家具的设计方法，并指导概念生成和设计实践。

【关键词】 医养结合；智能家具；养老家具；家具设计

2020 年人口普查显示，截至 2020 年，我国 60 岁以上人口占比 18.7%，呈持续增长趋势，中国老龄化程度加重。而伴随老龄化的增长，患病的老龄人口也逐渐庞大，其中老年人中患慢性病的比率高达 75%，而失能老人与 80 岁以上老人占到老年人口总数的 30%[1]。这一庞大的老龄群体在生活照料及医疗服务上的需求显著增加。为加快养老行业发展实施健康老龄化，国务院相继发布《关于加快发展养老服务业的若干意见（2013年)》《关于印发全国医疗卫生服务体系规划纲要（2015—2020 年)》《智慧健康养老产业发展行动计划（2017—2020 年)》等多项政策，分别从医养结合、智慧养老等方面提出完善养老保障的建设策略[2,3]。而为实现医疗与养老互通诞生的医养结合型养老模式也对建筑空间和养老设施设计提出了新的要求。

1 医养结合型养老家具的发展现状

1.1 医养结合的概念定义

医养结合是将医疗卫生和养老服务相结合的一种新型养老模式，同时为老人提供医疗诊断、健康查体、保健咨询、日常照料等服务[4]。因此，老人在突发身体疾病时能即时就医，为老人的紧急救助提供强大的医疗保障。日常生活中也能定期进行身体检查和医疗咨询，免去老人从家庭到医院的奔波和医院排队挂号的繁琐。

1.2 医养结合型养老空间实施现状

现有的医养结合型养老空间结构包括养老机构与医疗机构相邻而设，医疗机构与养老机构垂直分布，以及医养社区整体规划。但大部分的机构呈现出以下问题：两者之间相邻而设，空间界限过于分明，动线不流畅；空间局促，且利用率低，功能配置不合理；居家性和娱乐性较差；家具智能化程度较低。目前，医养结合型养老空间旨在提供医疗服务，在用户体验和空间设计上还需进行深入研究。

1.3 医养结合型空间中养老家具的配置及特点

医养结合型空间中的老年人用房一般包括

医疗用房、保健用房、生活照料间、休闲空间四大部分，服务人群涵盖全范围的老人，包括自理、介助、介护老人，以及照护人员和工作人员[5]。医疗参与养老时，各类空间的功能性质不同也呈现出不同的设计特点（表1）。

医养结合型养老空间中家具配置及特点　　表1

空间种类	空间名称	家具配置	功能特点
医疗用房	诊断室、检验化验室	诊桌、诊椅、诊床	• 去医疗化 • 适老设计与无障碍设计 • 智能化提高诊断效率
	就诊等候区	适老座椅、边几	• 预留照护人员等待区域 • 家具体量与分布间隔
	临终关怀室	护理床	• 舒适，温馨
康复保健用房	护理间	护理床、陪护床、床头柜、储物柜	• 功能性，提高照护效率
	康复间	康复器械	• 功能性
生活照料间	老年居室	床、床头柜、衣柜、电视柜、座椅、书桌、门厅柜、换鞋凳、储物柜	• 居家性，细节适老 • 不同老人不同配置
	洗浴间	助浴座椅、助浴扶手、洗漱台	• 安全性，防滑设计，危险报警
	护士站	前台、工作椅、问诊椅	• 智能导诊 • 位置显眼，方便寻找
公共空间	活动室	活动桌椅、储物柜、休闲沙发	• 智能化，增加娱乐性 • 隔声效果 • 方便移动，自由组合
	餐厅	餐桌椅	• 干净卫生，防油污 • 餐桌高度考虑轮椅老人 • 智能派餐系统、餐盘回收系统
	闲坐区	沙发、茶几、边几、电视柜	• 保证舒适性和社交性

医养结合中医疗空间中的概念设计应呈现去医疗化、智能化、无障碍设计的特征，削弱空间中的医疗感和机械感，缓解患者的就诊压力[6]。对适老医疗家具进行智能化升级可以提高就诊效率。如诊床可以自动调节，帮助老人就诊起身。保健用房的家具更加注重功能性以及减轻护理人员的看护压力，如将折叠床与床头柜结合，提供陪护人

员休息的空间，且能节省空间。设置护理床，方便老人突发疾病就诊。

由于医疗和养老结合，整体空间中要保证良好的通风和采光效果，将医疗空间和日常空间做出明显的区分。老年居室强调居家性，家具的设计上，更加注重细节设计和人性化，如圆角处理、尺度上的适老化等。根据老人不同的类别配置不同的家具，在介助、介护老人的

空间中预留轮椅的位置，考虑轮椅老人的人体工程学，增设养老辅具的配置。医养结合模式下的公共空间突出娱乐性，在举办各类活动时保证良好的隔声效果，避免需静养的老人被打扰。同时也需保证公共空间的卫生健康，避免老人交叉感染疾病。

2 医养结合模式下智能养老家具的可行性分析

2.1 医养结合模式下智能养老家具的 SWOT 分析

由于医养结合模式下的智能解决方案为概念设计阶段的探索。笔者通过对现有的国家政策和形势的梳理及养老家具市场的调研，运用 SWOT 态势分析法，从内部条件的优势（Strengths）、劣势（Weaknesses）以及外部条件的机遇（Opportunities）、挑战（Threats）多维度对其进行可行性分析（表2）[7]。进而结合内部和外部条件提出了基于医养结合模式下智能养老家具的发展战略。智能养老家具在良好的政策与技术支撑下，有着广阔的市场空间，在将来必定在整个家具市场占据一席之地。

医养结合模式下智能养老家具 SWOT 分析矩阵　　　　表 2

	S 优势（Streng-ths） • 产品科技含量高，易占领市场 • 实现医疗与养老互通，提高对老年人的医疗与看护效率 • 实现自动化操作和对老人的远程看护，使用起来更加便捷高效 • 进行健康监测，并实时危险报警，提高室内空间的安全性	W 劣势（Wea-knesses） • 产品单一，智能化程度较低 • 技术突破不足，易存在产品反应不灵敏，使用不够顺畅等问题 • 缺乏数据安全性和隐私性 • 研发及制造等经济成本过高 • 产品过度智能化，功能繁杂，实用性差
O 机会（Opportunities） • 智慧养老与医养结合政策的大力扶持 • 市场迅速增长，社会关注度高 • 高精尖技术完备 • 互联网渗透率高	Opportunity-Stren-gth strategies • 在政策的扶持下联合养老机构、家具企业进行项目落地 • 与技术科研单位进行合作，丰富和优化智能医养家具的功能	Opportunity-Weakness strategies • 做好基础用户研究，明确用户需求后，再进行技术搭载 • 制定智能家具产业的相关标准
T 挑战（Threats） • 技术更新过快，产品容易过时 • 行业竞争大，智能家具及养老家具行业发展尚未成熟，行业缺乏统一标准 • 产品价格过于昂贵，接受度低 • 线上和线下销售渠道不合理	Threat-Strength strategies • 打好智能产品前端的基础，并不断更新迭代 • 深入行业的产品研发工作，适应市场，最终降低家具成本	Threat-Weakness strategies • 不断完善智能家具行业的研发体系，降低研发成本 • 加大市场监管，控制产品价格和产品质量

2.2 医养结合模式下智能家具的发展策略

2.2.1 贯彻系统布局，打通数据传输链

医养结合型空间中的智能系统包括养老系统、医疗系统与社群系统。养老系统中包含健康体征监测、人身安全监护及紧急报警、自助体检。医疗系统中包括智能导诊、护理呼叫、远程医疗、远程探视。社群系统包括门禁体系、物业服务、支付系统[8]。医养结合模式的系统运行，必须解决产品之间的兼容性问题，

做好系统性布局，保证医疗与养老的数据储存及传输。

2.2.2 突破技术壁垒，避免过于机械化

智能医养家具在现有的高精尖技术加持下，需突破技术与家具环境以及家具生产工艺相融合的瓶颈[9]。在面向老人的使用场景下，应降低家具的操作难度，使用一键式操控、感应控制等。保证产品操作的顺滑和安全，在机械结构自动变化时保证静音效果，避免家具过于机械化。

2.2.3 深入用户研究，挖掘老人实际需求

在医养结合空间中，不同空间对智能家具的功能需求不同。医疗空间旨在为老人提供医疗服务时提高工作效率，休闲空间智能家具为老人提供娱乐性，居室空间保障老人的安全性和便利性。因此，智能医养家居因避免过度智能化，深入挖掘不同空间、不同用户群体的不同使用需求。

3 医养结合模式下智能养老家具设计原则

智能养老家具设计中产品前端的问题解决是非常重要的，必须充分考虑老年人的生理、心理、行为上的需求，在医养结合这种特定的新型养老模式下，做好基础用户研究和理论策略探讨。因此，针对医养结合模式以及智能养老的特殊性，笔者提出了安全性、易用性、多样性、情感化设计四项家具设计原则。

3.1 安全性

智能家具的安全性对于老人而言是第一重要的。由于人老后肌肉骨骼能力退化，导致动作变慢，迅速反应能力降低，容易受伤[10]。再加上医养结合空间的使用者大部分为病患，因此，对安全性的需求更大。智能家具的安全性包括使用安全、造型安全以及

材料安全。首先针对老人的智能家具在操作上保证安全，如避免出现失灵的状况，并增加老人误用时的紧急制动功能，以及针对拿取高处物品设计的家具，在自动变化时确保结构不会松动摇晃等。在造型上保证安全，包括边角的圆角处理、孔隙处避免夹手，及结构的稳定耐用。在材料上的安全，包括使用健康无毒材料，严格把控材料表面有毒有害物质限量，在医疗空间与公共空间使用抗菌抑菌材料避免疾病的传播。

3.2 易用性

针对老年人的智能家具还应保证其易用性。由于老年人的记忆力衰退，学习能力下降，对于智能产品的接受程度降低。智能家具的目的是更方便老人的生活，而非给老人带来更大的电子产品焦虑。因此，应设置为简便操作的一键式操作、红外感应控制等。如在柜类家具的下端设置感应灯带，起到老人起夜照明功能。同时应保证家具中结构连接件的顺滑，避免需用很大力量的开合。

3.3 多样性

医养结合型养老空间是一个综合型养老空间，因为其具备医疗及养老的综合功能，因此，家具也应该具备多功能性，包括功能多样性及服务对象多样化。医养结合型空间中的使用对象包括自理老人、介助老人、介护老人，根据不同的用户特征配置不同的家具。同时空间中还包括老人的家属、照料人员、医疗工作者等不同群体，空间中的家具功能需进行合理配置。如针对介助或介护老人的居住空间需预留轮椅位置，增设智能辅具的配置，家具设计中需根据轮椅老人的人体工程学进行设计。空间内也需预留出看护人员的空间，如配置折叠式床。

3.4 情感化设计

在医养结合空间中，应避免将太重的医疗氛围带入养老空间中，以缓解老年人患病的焦躁感。医疗空间中强调干净卫生，多使用白色作为主色调，针对老人的疗养空间，可以增加暖色调，构建更温暖的氛围。而在日常居室空间中，应尽量布置成居家的环境，避免智能家具带来过于科技的效果。也可将家具与智能机器人相结合，起到陪伴老人的功能。颜色上尽量选择暖色调。公共空间中注重老人们的休闲娱乐，设计上就能更加具有趣味性，合理配置一些高科技的家具增强老人的互动性。在细节设计上，智能家具中显示屏上的字可以适当加大，方便识别，并增加语音播报的功能供视障老人使用。

4 医养结合模式下智能养老家具设计

4.1 智能控制系统

基于老年人记忆力衰退的生理特征，可利用生物密码进行数据储存，如人脸识别、指纹识别等。运用人脸识别系统实现智能门锁、医疗就诊、餐饮服务、娱乐支付等功能的互联互通，避免老人遗忘丢失钥匙、医疗卡、手机等。如飞利浦 3D 人脸识别智能锁，有效辨别图片、影像，保证安全性。同时也可以将数据与手机、手环、实体卡同步，保证生物识别故障时也能使用。

在家具的控制上可以使用红外感应控制、语音识别控制等，避免老人寻找遥控器，降低家具的使用难度。如针对轮椅老人使用衣柜的情况，可以通过语音或触屏的方式对挂衣杆进行拉下和升起的调节。结合家具进行的照明设计中，可以在衣柜、走道、卫生间等地方设置感应灯带，避免老人起夜摸黑发生危险。

4.2 健康检测系统

利用家具检测老人的健康数据，可以实时获取老人的生理信息，从而帮助老人进行医疗诊断。通过智能床、座椅检测老年人身体指数，并上传至医生端，将医嘱返回至老人及其家属[11]。智能床垫、智能枕头能进行睡眠质量检测。建筑新风系统结合室内智能显示屏能实时检测室内环境的空气质量、温度、湿度等信息，进行室内环境的自调节。如艾特智能魔镜，能够接入家庭健康管理系统检测体压、体温等，并进行不正常提醒。

对于室内防跌倒及紧急报警功能，在房间的床头，卫生间坐厕、淋浴，机构内老人活动场所均应设置报警装置，包括紧急拉绳、声控报警、智能手环等设备。

4.3 信息交互适老化

对于老年人使用的交互界面遵从简洁的设计原则，以大字体、大按钮、一键式操作等功能为主，避免功能的繁杂，同时可以增加语音控制与播报的功能为视障老人服务。与此同时，由于老年人的听力弱化，语音播报的音量不宜太低，报警音量应达到 60dB 以上，或比环境声高 15dB 以上[12]。在色彩设计方面，由于老年人辨色能力减弱，警示颜色应设置饱和度高的红色，且空间里中高彩度的原色更能调动老年人的积极性[13]。

5 总结

医养结合型养老模式为老年生活提供了强大的医疗保障，能够在一定程度上缓解社会的养老压力。将智能家具引入医养结合型养老机构中，能够显著提高"医"与"养"的效率，构建一个高效、安全、健康、舒适的医养空间。本文探讨了医养空间中的家具现状，并从发展策略、设计原则、设计方法上提出医养空间中智能家具的理论指导。随着研究的深入，医养结合模式下智能养老家具的这些概念设计也将落地，并服务于无数老年人。

参考文献

[1] 欧阳雪梅. 中国大健康产业如何塑造未来医养模式[J]. 人民论坛，2020(28)：71-73.

[2] 李尊雨，程超，武兴斌. 健康中国背景下医养结合机构建筑设计理念[J]. 中国医院建筑与装备，2021，22(7)：33-36.

[3] 周橙旻，吴智慧，罗欣，等. 康养家具行业发展现状与趋势[J]. 家具，2020，41(2)：1-6.

[4] 张雪莹，黄双，孙欣然，等. "互联网＋"医养结合结构分析：基于结构功能主义理论[J]. 中国卫生事业管理，2021，38(4)：277-279.

[5] 靳茹. 基于医养结合背景下的养老类建筑设计策略研究[J]. 工程建设与设计，2021(15)：28-30.

[6] 李雪莲，吴佳敏. 现代医疗空间家具设计策略分析研究[J]. 家具与室内装饰，2020(2)：36-38.

[7] 陈嘉珍，吴智慧. 我国智能家居发展的 PEST-SWOT 分析[J]. 家具，2020，41(6)：53-57.

[8] 梅淋. 医养结合养老云服务平台的设计与实现[J]. 现代信息科技，2019，3(21)：19-21.

[9] 熊先青，李荣荣，白洪涛. 中国智能家具产业现状与发展趋势[J]. 林业工程学报，2021，6(1)：21-28.

[10] 李叶，张帆，常乐. 老幼复合型空间及家具设计的新材料应用探究[J]. 家具与室内装饰，2020(12)：14-16.

[11] 李江晓. 智能化"适老家具"的交互设计展望[J]. 艺术品鉴，2020(5)：44-45.

[12] 汪利民，陈万恩，陆喆. 医养结合养老机构的智慧医养系统设计[J]. 智能建筑，2020(8)：46-50.

[13] 张金勇，刘齐，李美，等. 老年医院的色彩设计研究[J]. 包装工程，2020，41（2）：109-113.

建筑工人心理韧性对安全行为的影响：
安全氛围的中介作用

刘红勇　周　旭　赵　航　王思佳

（西南石油大学土木工程与测绘学院，四川成都 610500）

【摘　要】 本研究为阐明建筑工人心理韧性对安全行为的影响路径，基于社会认知、积极组织行为、自我决定等相关理论，构建了心理韧性各子维度（情绪调节、现场应对、坚韧、自强、乐观）与安全行为之间的关系，并探讨了安全氛围的中介作用；根据 538 份来自我国建筑工人的有效问卷，利用结构方程模型（SEM）建立三者之间的关系模型，并实证检验该模型。研究结果表明：①心理韧性的情绪调节、现场应对、坚韧、自强四个维度正向影响安全行为；②心理韧性的乐观维度与安全行为呈显著负相关关系；③心理韧性的坚韧维度与安全氛围无显著关系；④安全氛围在心理韧性（情绪调节、现场应对、自强、乐观）与安全行为之间起着部分中介的作用。

【关键词】 心理韧性；建筑工人；安全行为；安全氛围；中介效应

建筑业一直属于高危行业，是最危险的行业之一[1]。根据《关于 2019 年房屋市政工程生产安全事故情况的通报》显示，2019 年全国共发生房屋市政工程生产安全事故 773 起，死亡 904 人，比 2018 年分别上升 5.31% 和 7.62%，平均每天死亡人数超过 2 人。由此可见，我国建筑业安全事故数量不断刷新峰值，安全生产形势严峻。虽然学界已经研究了许多影响建筑事故的因素，但行为因素没有得到充分讨论，并且有研究显示超过 80% 的事故是由工人行为因素造成的[2]。因此，研究建筑工人安全行为的影响因素，并有针对性地对这些因素进行前瞻性干预和管理，是减少安全事故发生的重要途径。

建筑工人面临着高风险和高精神压力的工作环境，根据人因分析与分类系统理论，传统的安全管理，例如：为工人提供规范的标准、成熟的施工技术、先进的设备等"硬科学"的投入，对安全行为的改善所起到的边际效应越来越小，已经无法有效降低安全事故的发生，分析其中的原因发现可能是忽视了"软科学"的支持，如工友之间的安全提醒、安全知识共享、心理干预等[3]。因此，不少学者开展了跨学科的研究，比如将心理学和行为学等领域的研究与安全管理领域相结合[4]，研究重心从施工工具、施工方法等"可观察变量"扩展到个体认知、组织氛围等"潜变量"。在这些"潜变量"中，作为影响工人行为的个体认知因素，心理韧性引起了学界广泛关注。心理韧性作为积极心理资本的核心维度之一，属于积极组织行为学的研究范畴，其有助于个体在组织中获得身份认知，增加对组织的认同，从而影响个体的行为，但是，鲜有研究检验积极组织行为对建筑工人态度和行为的影响。自我决定理论中的因果定向理论也提到，心理韧性强的个体能更好地完成工作，也能更快地在单位中获得归属感，从而对个体行为产生积极影响。鉴于此，从心理韧性的视角对建筑工人安全行为进行探讨既必要又有意义。此外，有学者认为[5]，在心理韧性与安全行为之间可能存在中介变量的作用。社会认知理论也指出，个体行为不仅受心理感知的影响，还受到环境作用。作为影响个体心理状态的重要环境因素，安全氛围日益被关注。基于上述理论和研究，笔者提出，安全氛围在心理韧性和安全行为之间可能存在中介作用。

综上所述，本研究运用积极组织行为理论、社会认知理论、自我决定理论来阐述心理韧性对安全行为的作用机制，包括安全氛围在其中的中介作用。基于已有研究并结合建筑工人特点将心理韧性划分为情绪调节、现场应对、坚韧、自强、乐观5个维度。相比其他学者的研究，本研究提出了"心理韧性—安全氛围—安全行为"的作用机制，将有助于安全管理人员更好地从建筑工人的心理角度去思考如何提高施工的安全性能，为建筑企业安全管理实践与决策提供新思路。

1 理论基础与研究假设

1.1 心理韧性与安全行为

安全行为是指操作人员在完成任务过程中为保障安全而做出的现实反映[6]。目前学界对安全行为影响因素的研究主要从个体特质、安全氛围、环境、群体沟通等视角展开[3]。其中，个体特质因素的关注重心在个体认知和心理因素。已有研究发现人格特质在很大程度上能够影响行为的相关变量[7,8]，并与外部环境

因素对个体的行为产生共同影响。因此，可以推测，个体特质可能对安全行为造成影响。自我决定理论中的因果定向理论也提到：个体心理状态的差异对其潜在行为动机的形成具有深远的影响。因果定向理论认为，自主定向的个体展示出很强的韧性，即在困境中，其能表现出超强的复原能力并超越自己，最终战胜困难取得成功。因此，心理韧性强的个体在面临外部刺激时更有可能做出有利于安全的行为。除此之外，社会认知理论也认为，个体行为不仅与环境有关，而且受到其心理状况的影响。Changquan[5]等认为坚韧与安全行为呈正相关关系；Wang[9]等研究结果表明：心理韧性作为心理资本的一个维度，对于员工安全行为具有显著的正面影响；高静[10]研究发现建筑工人消极情绪与安全绩效具有显著负向作用；Strutton[11]研究发现乐观主义者更易控制自身的行为。据此，可以进一步推测，工人心理韧性作为一种积极的个体资源对安全行为造成影响。

综上所述，本研究认为，心理韧性作为一种独特且积极的心理因素，能够促使工人以更积极的心态面对工作，并投入更多的精力，从而更容易实现与安全有关的行为。因此，提出以下假设：

H1：心理韧性正向影响安全行为。进而，情绪调节（H1-1）、现场应对（H1-2）、坚韧（H1-3）、自强（H1-4）、乐观（H1-5）正向影响建筑工人安全行为。

1.2 心理韧性与安全氛围

心理韧性反映了个体在面对逆境、挫折、失败等一系列消极事件时，展示出复原并超越自己最终战胜困难的能力[12]。社会认知理论中的三元交互决定论认为，环境、行为、个体身心机能三者之间相互联系，每二者之间都具

有双向的互动和决定关系，个体可以通过自身的人体特征如性格、心理状况等身心机能激活不同的环境反应，使其满足人基本的心理需求和生存目的。当工人表现出很强的心理韧性时，意味着工人在面临困境时，会通过自我调节以适应环境并激活周围的环境，即营造一种安全的氛围，从而对工人参与和遵守安全行为的意愿产生积极影响。虽然很少有学者直接研究心理韧性与安全氛围的关系，但自我决定理论的发展已经证实，坚韧、自强等心理韧性的重要因素能有效地改善员工所处环境的安全氛围。Harland[13]等认为员工的坚韧性与领导人格魅力、号召力以及人文关怀等组织氛围的因素呈显著正向影响；Taylor[14]等认为只有当组织中的成员具备韧性时，该组织才会具有组织韧性，进而影响工作环境与安全氛围。因此，提出以下研究假设：

H2 心理韧性正向影响安全氛围。进而，情绪调节（H2-1）、现场应对（H2-2）、坚韧（H2-3）、自强（H2-4）、乐观（H2-5）正向影响安全氛围。

1.3 安全氛围与安全行为

安全氛围是指工人对他们所在单位安全状态的综合感知[15]，属于感知的安全环境因素。良好的安全氛围及其结果会改善工人在工作中的心理状态，使其在面对工作逆境时更能从容应对，形成良性循环。社会认知理论认为个体行为与周围环境和其心理状态的特点密切相关，并且会随着这些因素的改变而变化，其行为存在不稳定性。Zohar[15]等认为安全行为受安全氛围的影响，构建一个和谐的安全氛围，对提升工人的安全行为水平有促进作用；成家磊[16]等研究发现班组级安全氛围对安全行为有显著正向影响。因此，提出以下研究假设：

H3 安全氛围正向影响建筑工人安全行为。

1.4 安全氛围的中介效应

社会认知理论中的三元交互决定论认为，环境、个体行为、个人身心机能三者之间互为因果，每二者之间都具有双向的互动和决定关系，一方面，个体的信念、动机、心理等身心机能可以控制并指导其行为，行为及其结果反过来又会影响行为主体的心理状态与情绪反应。另一方面，个体可以通过自己的人体特征如性格、社会角色等因素引起不同的环境反应，使之适合人的心理需求和达到生存的目的。环境作为人与行为之间的中介，是改善人与行为之间关系的手段。因此，可以推测，安全氛围作为重要的环境因素，在个体身心机能与行为之间起着某种作用。已有研究结果表明，工人心理与安全行为具有直接或间接的关系，并且发现安全动机和安全知识在其中扮演中介的角色。Rahlin[17]等研究发现安全氛围在心理因素和个体安全绩效之间有显著的中介作用。由于安全知识、安全动机等是测量安全氛围的重要因素，而安全行为是安全绩效的重要体现。综上所述，安全氛围既受到心理韧性的作用，又对安全行为产生作用，因此，可以进一步推测，安全氛围在心理韧性和安全行为之间可能存在中介作用。故可作如下假设：

H4 安全氛围在建筑工人心理韧性与安全行为之间的作用是中介，进而，安全氛围在情绪调节（H4-1）与安全行为、现场应对（H4-2）与安全行为、坚韧（H4-3）与安全行为、自强（H4-4）与安全行为、乐观（H4-5）与安全行为之间具有中介作用。

基于上述研究假设与相关理论，构建以心理韧性的5个维度为自变量，安全氛围为中介变量，安全行为为应变量的概念模型。如图1所示。

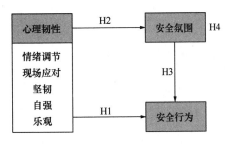

图1　概念模型

者还与本领域的专家对测量题项的措辞、语义和表达反复进行讨论，以符合现场建筑工人的语言习惯。在形成正式量表之前，还对四川某建筑工地的3位建筑工人进行了采访，以评估量表内容的可理解性，根据工人的反馈，又对一些题项进行了修改，以简化语言，使其易于理解，形成初始量表，量表题共包含39个题项。在初始量表设计完成后，以纸质问卷的形式发放给成都周边建筑工地的建筑工人进行小范围的预调研。共发放102份，其中返回有效问卷83份。通过因子分析、信度分析，将量表剩余题目重新排号，得到正式的量表，共35个题项（表1）。

2　数据收集与量表设计

2.1　样本数据

首先，本研究在初始量表设计过程中，参考了国内外相关研究的成熟量表。基于此，笔

题项来源　　　　　　　　　　　　　　　　　　　　　　　　表1

变量	因子	题项及编号	题项来源
心理韧性	情绪调节	R1 我不会因失败而灰心丧气	Connor-Davidson[18]、于肖楠[19]、左翼[20]
		R2 在困难面前，我能保持良好的心境	
		R3 我能找到宣泄口，缓解压力	
		R4 我会找朋友倾诉我的烦恼	
		R5 我能处理自己的负面情绪	
	现场应对	O1 遇到突发情况时，我会迅速让自己冷静下来	
		O2 遇到突发情况时，我能冷静思考如何应对	
		O3 遇到突发事故时，我能做出正确的判断	
		O4 遇到突发事故时，我能迅速做出对自己最安全的行为	
		O5 遇到突发事故时，我有足够的经验来应对	
	坚韧	T1 当事情看起来没什么希望时，我也不会放弃	
		T2 我知道去哪里寻求帮助	
		T3 我有强烈的目的感	
		T4 我觉得自己是一个坚定的人	
		T5 我把工作中出现的困难视为一种挑战	
	自强	S1 过去的成功使我有信心迎接挑战	
		S2 由于经历过磨炼，我变得更坚强了	
		S3 我容易从困难或精神上的疾病中恢复过来	
		S4 我会努力工作实现自己的目标	
		S5 我有很强的适应能力	
	乐观	A1 我有着良好的心态去面对突发事故	
		A2 我能够看到事情积极的一面	
		A3 我与家人、朋友或者工友的关系很亲密、稳固	
		A4 当面对逆境时，命运或神灵会帮助我	
		A5 无论发生什么，我都能处理	

续表

变量	因子	题项及编号	题项来源
安全行为		B1 工作中我严格遵守安全操作程序	Neal[21]
		B2 工作中我及时汇报安全工作情况	
		B3 工作前，我会检查仪器设备和施工环境安全情况	
		B4 我积极为工友提供安全支持与帮助	
		B5 我积极提出安全工作建议	
安全氛围		C1 班组长很关注我的心理和身体健康	夏娅雯[22]等
		C2 我经常跟班组其他成员进行工作上的交流	
		C3 我所在项目部对于查处重大事故有相应的规章制度	
		C4 班组长会根据我的行为是否违反操作进行相应的奖励与惩罚	
		C5 发生施工事故后，我都会上报，无论事故大小	

随后，进行正式调研。样本数据来源分为两部分。一部分数据采用现场收集，课题组成员通过社会关系到建筑工地当面发放问卷，主要利用工人休息或吃饭时间指导受访者根据自己真实的想法填写问卷。另一部分资料，即外省市（河南、重庆、广东、上海等地）建筑工地的工人采用网络问卷的方式，以电子邮件的方式发送本问卷地址，请工人网上填写。网络问卷调查开始于2020年12月2日至2021年3月23日，一共发放300份网络问卷，总计回收235份，扣除填答者不是建筑领域相关人员的24份，仅余211份有效问卷。同一时间探访员利用周六、日前往建筑工地收集问卷，一共发放400卷，共回收有效问卷327份，加上网络问卷，有效样本合计538份。

最后，由于本次调查样本同时来自纸质问卷及网络问卷，为避免错误的推论，在数据合并时进行了同质性检验。因为性别、年龄、教育程度、工作性质、工龄为类别变量，所以使用SPSS23进行卡方同质检验，检验结果为，p值均大于0.05，因此不拒绝零假设，即性别、年龄、教育程度、工作性质、工龄在现场访问及网络调查中并无差别，故可以将其合并，共计538份有效问卷。在本次有效调查样本中，受访者性别以男性491人居多，占

91.3%；年龄在26～35岁的最多，占比39.8%；受访者教育程度分析，高中或中专学历的最多，有168人，占比31.2%；受访者以普通工人最多，有283人，占比52.6%，从事专业技术的人员有155人，占28.8%，其余的类别就颇为分散。

2.2　量表设计

心理韧性量表。在Connor-Davidson[18]、于肖楠[19]、左翼[20]等研究的基础上，结合专家访谈与工人反馈形成建筑工人心理韧性初始量表。随后，进行小范围的预调研。运用项目分析与探索性因子分析对量表题项进行分析，最终形成心理韧性的测量量表，包括情绪调节、现场应对、坚韧、自强、乐观五个维度，共25个题项。示例题项如"当事情看起来没什么希望时，我也不会放弃"。

安全氛围量表。基于夏娅雯[22]等学者的研究进行适应性修正，安全氛围的测量包括5个题项。示例题项如"我经常跟班组其他成员进行工作上的交流"。

安全行为量表。采用Neal[21]等学者开发的量表进行测量，并进行修正，共5个题项。示例题项如"工作中我严格遵守安全操作程序"。借鉴关于心理韧性、安全氛围和安全行

为的研究，结合建筑企业与建筑工人职业特点，编制问卷。调查问卷包括4个部分：①人口统计变量；②心理韧性量表；③安全氛围量表；④安全行为量表。其中人口统计学变量包括性别、年龄、教育程度、工作性质、工龄等，共35个题项。量表均采用Likert5级量表，1～5表示从"非常不认同"到"非常认同"的程度。正式问卷主要潜变量、题项整理成表1。

2.3　偏差检验

由于问卷回收时间超过3个月，因此，运用SPSS23对最开始以及最后回收的四分之一有效问卷（各134份）进行响应偏差检验。首先，对性别进行独立样本T检验，随后，对工龄、年龄、工作性质、受教育程度进行单因素方差检验，结果显示都不存在显著差异，说明不存在严重的响应偏差问题。最后，对所有题项采用Harman单因素测试进行共同方法偏差的检验，未旋转的探索性因子分析结果提取出7个特征值大于1的因子，累计解释方差的76.5%，其中，最大因子方差解释率为37.9%（<40%），表明数据不存在严重的共同方法偏差问题。

3　数据结果分析

3.1　量表信度检验

对量表进行可靠性检验，若量表Cronbach's α系数高于0.7，表明量表内部一致性良好。本研究量表的Cronbach's α系数，情绪调节为0.946、现场应对为0.953、坚韧为0.925、自强为0.890、乐观为0.862、心理韧性为0.927、安全氛围为0.902、安全行为系数为0.944，均大于0.85，表示量表内部有很高的稳定性。

3.2　量表效度检验

验证性因子分析（Confirmation Factor Analysis，CFA）是SEM分析的一部分。Jackson等研究者认为应该评估测量模型，当这个测量模型适配度确实能够反应因子时，才进行SEM结构方程模型的评估。本研究对所有因子进行CFA分析，测量模型适配指标如表2所示，除了自强与安全氛围测量模型中的$\chi^2/\mathrm{d}f$稍大于3外，其余均满足模型适配标准。

测量模型适配指标　　　　表2

因子	$\chi^2/\mathrm{d}f$ <3	CFI >0.9	TLI >0.9	IFI >0.9	RMSEA <0.08
情绪调节	2.654	0.997	0.993	0.997	0.056
现场反应	2.823	0.997	0.993	0.997	0.058
坚韧	2.196	0.997	0.994	0.997	0.047
自强	3.304	0.992	0.984	0.992	0.065
乐观	2.684	0.993	0.986	0.993	0.056
心理韧性	2.966	0.952	0.946	0.952	0.061
安全氛围	4.168	0.990	0.980	0.990	0.077
安全行为	2.744	0.996	0.993	0.996	0.057

（1）收敛效度检验

如表3所示，所有因子的负荷量均在0.609～0.931，且显著；组成信度（CR）为0.869～0.954；平均变异数萃取量（AVE）在0.573～0.804，都大于0.5，符合Hair及Fornell and Larcker的标准。故七个因子均具有收敛效度。

（2）区别效度检验

用Fornell and Larcker的方法来评估区别效度，在表4中，对角线数据为开根号后的平均变异数萃取量（AVE），非对角线为因子之间的相关性系数，每个对角线数据都大于其各自的非对角线数据，这表明所有因子与因子之间具有区别效度。

收敛效度表　　　　　　　　　　　　　　　　表3

因子	因素负荷量	组成信度	收敛效度
	Std.	CR	AVE
情绪调节	0.846～0.927	0.946	0.778
现场应对	0.836～0.931	0.954	0.804
坚韧	0.813～0.896	0.925	0.713
自强	0.688～0.820	0.892	0.623
乐观	0.609～0.888	0.869	0.573
安全氛围	0.720～0.850	0.903	0.651
安全行为	0.857～0.905	0.944	0.773

3.3　模型拟合和假设检验

区别效度表　　　　　　　　　　　　　　　　　　　　　　　　表4

因子	AVE	平均值	标准差	情绪调节	现场应对	坚韧	自强	乐观	安全氛围	安全行为
情绪调节	0.778	3.583	1.038	0.882						
现场应对	0.804	3.450	1.081	0.561**	0.897					
坚韧	0.713	3.452	1.028	0.470**	0.554**	0.844				
自强	0.623	3.609	0.958	0.398**	0.351**	0.367**	0.789			
乐观	0.573	3.818	0.868	0.209**	0.175**	0.337**	0.152**	0.757		
安全氛围	0.651	3.146	0.967	0.576**	0.566**	0.558**	0.475**	0.160**	0.807	
安全行为	0.773	2.917	1.022	0.498**	0.445**	0.408**	0.432**	0.306**	0.585**	0.879

注：对角线数据为开根号后的平均变异数萃取量（AVE）；* 表示 $p<0.05$，** 表示 $p<0.01$，*** 表示 $p<0.001$。

（1）初始模型构建及检验

将建筑工人心理韧性的 5 个因子、安全氛围、安全行为分别作为自变量、中介变量、因变量，在 AMOS23 软件平台构建它们的结构方程模型 SEM；并验证模型。模型 1 的标准化路径系数和检验数据见表 5。

（2）初始模型 1 修正

通过表 5 发现，坚韧→安全氛围路径不显著，模型 1 需要修正。删除坚韧→安全氛围路径后检验新模型卡方值由 1207.3 增加到 1208.6，变化很小，表示删除此路径可行。删除此路径后各因子荷载均处于 0.50～0.95，所有测量误差均为正值，且都达到显著性水平，模型拟合度理想；通过对模型 1 的修正，所有路径都通过了验证，模型 1 修正结束，得

模型 1 路径系数及检验结果　表5

路径	标准化系数	标准误差	C.R.	|CR|>2
情绪调节→安全氛围	0.254	0.055	5.008	是
现场应对→安全氛围	0.160	0.052	3.069	是
坚韧→安全氛围	0.059	0.055	1.133	否
自强→安全氛围	0.227	0.058	4.911	是
乐观→安全氛围	0.170	0.050	3.941	是
情绪调节→安全行为	0.191	0.050	4.373	是
现场应对→安全行为	0.165	0.047	3.738	是
坚韧→安全行为	0.240	0.050	5.403	是
自强→安全行为	0.145	0.052	3.663	是
乐观→安全行为	−0.100	0.045	−2.721	是
安全氛围→安全行为	0.278	0.046	6.544	是

到模型2。模型2的路径系数和拟合系数分别如表6、表7、图2所示。由于坚韧→安全氛围路径不显著，故对应的假设H2-3不成立。乐观→安全行为路径呈负相关，对应假设H1-5不成立。

模型2路径系数及检验结果　表6

路径	标准化系数	标准误差	C.R.	|CR|>2
情绪调节→安全氛围	0.263	0.055	5.229	是
现场应对→安全氛围	0.183	0.048	3.799	是
自强→安全氛围	0.236	0.057	5.156	是
乐观→安全氛围	0.184	0.048	4.443	是
情绪调节→安全行为	0.190	0.050	4.357	是
现场应对→安全行为	0.164	0.047	3.713	是
坚韧→安全行为	0.242	0.050	5.444	是
自强→安全行为	0.144	0.052	3.644	是
乐观→安全行为	−0.100	0.045	−2.734	是
安全氛围→安全行为	0.288	0.046	6.566	是

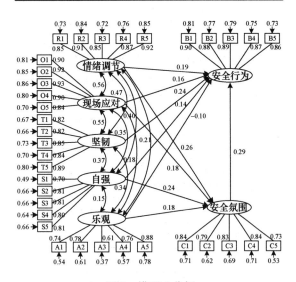

图2　模型2分析

模型2拟合系数表　表7

指标	χ^2/df	RMSEA	NFI	CFI	TLI	PNFI
准则	<3	<0.08	>0.9	>0.9	>0.9	>0.5
数值	2.238	0.048	0.926	0.958	0.953	0.841

（3）中介效应检验[23~25]

运用AMOS23中有偏校正Bootstrap程序检验中介效应的显著性，Hayes[26]建议在估计路径系数时Bootstraping至少重复5000次，本研究在95%信度区间下，执行Bootstraping程序重复5000次，检验结果见表8。

Bootstraping中介效应检验分析　表8

路径	效应值	标准误 SE	95%置信区间	
			下限	上限
间接效应1	0.087	0.024	0.046	0.145
间接效应2	0.056	0.022	0.021	0.108
间接效应4	0.089	0.026	0.047	0.153
间接效应5	0.065	0.018	0.035	0.109
直接效应1	0.218	0.059	0.100	0.336
直接效应2	0.173	0.056	0.062	0.287
直接效应4	0.190	0.059	0.074	0.306
直接效应5	−0.124	0.045	−0.216	−0.039

通过表8发现，第1条中介链：情绪调节→安全氛围→安全行为，第2条中介链：现场应对→安全氛围→安全行为，第4条中介链：自强→安全氛围→安全行为，第5条中介链：乐观→安全氛围→安全行为，4个间接效应中有偏校正信度区间的上、下界均不含0，故情绪调节、现场应对、自强、乐观对安全行为通过安全氛围传递的中介效应是显著的；且直接效应的置信区间也不包含0，则情绪调节、现场应对、自强和乐观对安全行为直接效应显著；综上，安全氛围在情绪调节、现场应对、自强和乐观对建筑工人安全行为之间具有部分中介作用，假设H4-1、H4-2、H4-4、H4-5成立；结合表5可知，坚韧→安全氛围路径不显著，故假设H4-3不成立。

4　结论

本研究基于社会认知、自我决定等相关理论和已有研究，对建筑工人心理韧性是否以及如何影响安全行为进行实证研究。鉴于建筑工人安全行为对建筑企业安全绩效的重要性，以建筑行业一线建筑工人作为研究对象。总体而言，本研究以一种新的视角切入，即建筑工人心理的角度展开研究，结果表明：安全氛围部分中介了心理韧性与安全行为的关系。本研究是对目前安全行为研究的重要补充，丰富了该领域的研究内容，具有一定的理论和实践意义。

4.1 理论意义

（1）心理韧性对安全行为的影响

心理韧性的部分子维度（情绪调节、现场应对、坚韧、自强）对安全行为存在显著的正向影响，有趣的是，乐观对安全行为呈显著负向影响。虽然这与一些研究结论不同[11]，但与 Changquan[5]等的发现一致。有一部分学者将这种现象归因于情景因素，另一部分学者认为可能是由于压力与行为之间的复杂关系导致的[27]。针对这种情况，表明还可以进一步挖掘其他潜变量在其中发挥的作用，例如工作压力，这种变量存在何种效应同样值得研究。

（2）安全氛围的中介作用

研究结果与预期假设大部分一致，即心理韧性大部分的子维度（情绪调节、现场应对、自强、乐观）与安全氛围之间存在显著的正相关关系，安全氛围又进一步正向作用于安全行为。安全氛围在心理韧性（情绪调节、现场应对、自强、乐观）与安全行为之间起着部分中介的作用，虽然乐观对安全行为表现出显著的负向影响，但可以通过安全氛围来降低这种影响。

4.2 实践意义

（1）甄选心理韧性强的工人

建筑企业在甄选工人时，往往通过考察工人与工作的匹配情况，从而判断其是否能够成为单位的潜在工人。然而，由于信息的不对称，且目前在单位中以团队工作的方式很普遍，判断某个工人是否胜任某项工作不只要测试他的工作技能，更重要的是一种潜在能力的对应。心理韧性越强的工人，在处理复杂和棘手的工作时越能从容应对，对环境和工作的适应能力也越强，从而能更好地完成任务。同时，心理韧性作为个体的一种特殊资源，具有稳定且不易变化的特点，因此，可以作为建筑企业筛选工人的一项重要参考指标。

（2）对建筑企业安全管理的启示

通过上文的研究结果表明，心理韧性对安全行为的影响可能是一把双刃剑，心理韧性的子维度（情绪调节、现场应对、坚韧、自强、乐观）对安全行为的不同影响，对建筑工人和安全管理人员都有重要的意义。对建筑工人而言，盲目乐观可能会浪费他们的安全精力，同时也可能会让他们忽视建筑中潜在的危险和困难，维持建筑工人适当的乐观水平更有助于提高安全性能。对安全管理人员而言，可以帮助管理人员从心理学的角度来评估工人工作状态，安全管理人员还可以根据一线建筑工人心理韧性子维度的状况和整体的心理状况来规划心理干预的具体方案，有助于提高安全培训和安全干预的有效性和针对性。

参考文献

[1] Wang J, Zou P X, Li P P. Critical factors and paths influencing construction workers' safety risk tolerances[J]. Accident Analysis & Prevention, 2016(93)：267-279.

[2] Choi B, Ahn S, Lee S. Role of social norms and social identifications in safety behavior of construction workers. I：Theoretical model of safety behavior under social influence[J]. Jour-

nal of Construction Engineering and Management, 2017, 143(5): 4016124.

[3] 何长全, 贾广社, 孙继德. 建筑工人安全行为研究进展与展望[J]. 中国安全生产科学技术, 2018, 14(5): 188-192.

[4] 王丹, 秦云云. 家长式领导对员工安全行为的影响: 心理资本的中介作用和犬儒主义的调节作用[J]. 中国安全科学学报, 2020, 30(8): 25-30.

[5] He C, Jia G, Mccabe B, et al. Impact of psychological capital on construction worker safety behavior: communication competence as a mediator[J]. Journal of Safety Research, 2019(71): 231-241.

[6] Hinze J, Thurman S, Wehle A. Leading indicators of construction safety performance[J]. Safety Science, 2013, 51(1): 23-28.

[7] 杨霞, 李雯. 伦理型领导与员工知识共享行为: 组织信任的中介作用和心理安全的调节效应[J]. 科技进步与对策, 2017, 34(17): 143-147.

[8] 王亦虹, 黄路路, 任晓晨. 变革型领导与建筑工人安全行为——组织公平的中介作用[J]. 土木工程与管理学报, 2017, 34(3): 33-38.

[9] Wang D, Wang X, Xia N. How safety-related stress affects workers' safety behavior: the moderating role of psychological capital[J]. Safety Science, 2018(103): 247-259.

[10] 高静. 建筑工人消极情绪对其安全绩效的影响机理研究[D]. 西安: 西安建筑科技大学, 2018.

[11] Strutton D, Lumpkin J. Relationship between optimism and coping strategies in the work environment[J]. Psychological Reports, 1992, 71(3): 1179-1186.

[12] 吴婷, 张正堂. 积极心态的员工更认同组织吗——匹配视角下心理韧性对员工组织认同的影响[J]. 财贸研究, 2017, 28(4): 101-109.

[13] Harland L, Harrison W, Jones J R, et al. Leadership behaviors and subordinate resilience[J]. Journal of Leadership & Organizational Studies, 2005, 11(2): 2-14.

[14] Taylor C, Dollard M F, Clark A, et al. Psychosocial safety climate as a factor in organisational resilience: implications for worker psychological health, resilience, and engagement[M]. Springer, 2019: 199-228.

[15] Zohar D. Safety climate in industrial organizations: theoretical and applied implications[J]. Journal of Applied Psychology, 1980, 65(1): 96.

[16] 成家磊, 祁神军, 张云波. 组织氛围对建筑工人不安全行为的影响机理及实证研究[J]. 中国安全生产科学技术, 2017, 13(11): 11-16.

[17] Rahlin N A, Majid A H A, Mustafa M. Mediating effect of psychological safety climate in the relationship between psychological factors and individual safety performance in the Malaysian manufacturing small enterprises[J]. International Academic Research Journal of Social Science, 2016, 2(2): 10-23.

[18] Connor K M, Davidson J R J D, Anxiety. Development of a new resilience scale: the Connor-Davidson Resilience Scale (CD-RISC)[J]. 2003, 18(2): 76-82.

[19] Yu X, Zhang J. Factor analysis and psychometric evaluation of the Connor-Davidson Resilience Scale (CD-RISC) with Chinese people[J]. Social Behavior and Personality: an International Journal, 2007, 35(1): 19-30.

[20] 左翼. 警察心理韧性量表的编制[D]. 重庆市: 西南大学, 2013.

[21] Neal A, Griffin M A. A study of the lagged relationships among safety climate, safety motivation, safety behavior, and accidents at the individual and group levels.[J]. Journal of Applied Psychology, 2006, 91(4): 946.

[22] 夏娅雯, 张力, 李鹏程, 等. 核电厂建造施工班组安全氛围测量量表研究[J]. 中国安全科

学学报，2018，28(9)：159-164.

[23] Mackinnon D. Multivariate applications series [M]. 2008.

[24] Baron R M，Kenny D A. The moderator-mediator variable distinction in social psychological research：conceptual，strategic，and statistical considerations.[J]. Journal of Personality and Social Psychology，1986，51(6)：1173.

[25] Sobel M E. Asymptotic confidence intervals for indirect effects in structural equation models [J]. Sociological Methodology，1982（13）：290-312.

[26] Hayes A F. Beyond Baron and Kenny：statistical mediation analysis in the new millennium [J]. Communication Monographs，2009，76（4）：408-420.

[27] Sullivan S E，Bhagat R S. Organizational stress，job satisfaction and job performance：where do we go from here? [J]. Journal of Management，1992，18（2）：353-374.

教学研究

Teaching Research

"建设工程造价管理"课程产教融合实施探索

闫 辉[1] 何晓晴[1] 刘惠艳[2] 张 磊[3] 申琪玉[1] 张海燕[1]

(1. 华南理工大学土木与交通学院，广东广州 510640；

2. 中山大学政府采购与招投标管理中心，广东广州 510275；

3. 广州大学管理学院，广东广州 510006)

【摘 要】 在新一轮科技革命、建筑产业变革以及经济发展的潮流下，现有土木类专业中的部分核心课程已经无法满足新形势下对土木类人才的需求。为加快培养面向国家战略需求和建筑业转型升级需要的新型人才，本文以课程的产教融合为突破口，以华南理工大学"建设工程造价管理"课程为例，对课程的教学改革与企业发展的合作模式进行探索，从前期准备、教学活动、实习实践和后续深化四个环节提出课程产教融合的实施路径。研究成果也可为相关领域其他课程体系的改革提供方向和借鉴。

【关键词】 教学改革；产教融合；"建设工程造价管理"课程

随着工程教育的变革和新工科理念的提出[1]，国内高等教育在不断创新工程教育方式与手段，探索新工科的发展模式[2~4]。而高校与企业实现产教深度融合正是对现有工科专业进行优化和改革的一个突破口[5]。高校与企业实现产教深度融合不仅有利于人才的培育，使校内相关的课程知识与行业需求及前沿保持一致，学生的知识体系及实践能力与行业实际需求接轨，而且还有利于促进产业发展，为产业发展带来人才聚集[6~8]。

在土木类专业中，为加快培养面向国家战略需求和建筑业转型升级需要的新型人才，工程管理和土木工程等专业，也正在向智能建造专业方向迈进。因此，本文拟以华南理工大学土木类专业核心课程体系之一的"建设工程造价管理"课程为例，对课程校企产教融合路径进行探索，一方面期望促进教学与工程实践、产业发展相结合，另一方面期望带动其他课程的改革，最终提升人才培养质量。

1 课程产教融合总体目标及拟解决的关键问题

1.1 课程产教融合总体目标

"建设工程造价管理"课程产教融合的总体目标是：将专业与产业深度融合，依托行业、背靠产业、融入企业，通过建立校企产学研合作关系，努力实现专业核心课程"建设工程造价管理"的校企合作、产教融合、资源互补、共赢发展，与企业共同开展协同育人项目，打造产学研用的一体化教学模式。具体目标有以下三点：

(1) 提高学生对智慧建造应用的认识以及应用 BIM 等技术进行计量与计价的水平；

（2）修订课程的教学大纲，使学生能更好地适应建筑行业数字化与智能化的转型；

（3）通过专家进课堂、学生实践项目的参观学习等活动，增强学生理论联系实践的能力。

1.2 拟解决的关键问题

"建设工程造价管理"课程产教融合，需要解决以下三个关键问题：如何寻找合适的企业，如何落实产教融合，如何保持合作的可持续性。

（1）如何寻找合适的企业

校企合作的目标企业应满足三个条件：引领行业前沿；有意愿；有条件。本研究拟选择广联达科技股份有限公司（以下简称"广联达公司"）作为合作企业，该公司致力于数字建筑最前沿工作，推动智慧建筑发展，与其合作符合新工科建设理念；为开拓市场、储备技术力量、进行科研创新和技术转化，广联达公司有与高校深入合作的迫切需求；广联达公司着力发展数字教育，与学校人才教育息息相关，同时具备强大的资金支持、场地支持和技术支持，有能力推进产教融合项目的落实。双方合作有利于融合高校工程领先优势和企业科技领先优势，开创企业发展和专业发展共赢新局面。

（2）如何落实产教融合

促进学校、企业资源共享和师资融合。要落实与企业达成产学研合作协议，共同研讨修订课程的教学大纲，筛选和企业有关的 BIM＋智慧工地、BIM＋造价管理等实际项目案例及数据，邀请和协调企业的相关专家和讲师进课堂的教学活动，组织学生进入企业实际案例项目进行参观学习。

（3）如何保持合作的可持续性

通过陆续寻求与企业的多点合作、多事项合作，与企业进行长期的产学研合作，才能切实保证产教融合项目落地，推动产教融合的可

持续发展。在项目开展的同时共同寻找持续合作的内容，如通过共同探索企业先进技术与课堂教学深度融合的路径、挖掘教改方式、举办相关比赛，学校为企业发展提供行业领域前沿人才资源和技术支持，保持企业技术领先优势等方式，不断深化双方产教融合的合作，构建校企共同体，实现合作的长期可持续性。

2 课程产教融合实施路径

"建设工程造价管理"课程的产教融合从前期准备、教学活动、实习实践和后续深化四个环节开展，具体实施路径如图 1 所示。

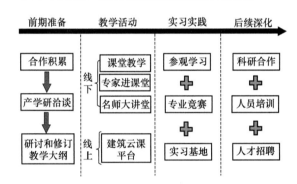

图 1　课程产教融合的实施路径

2.1 前期准备

在实施课程产教融合的前期准备阶段，可初步建立与企业的合作关系，进行前期合作的积累，并通过磋商或洽谈会等方式与企业针对生产、教学以及研发进行深入探讨，最后明确人才培养的目标，共同研讨和修订相对应课程的教学大纲。前期准备阶段的主要环节包括合作积累、产学研洽谈以及教学大纲的研讨和修订三部分内容。

2.1.1 合作积累

在课程产教融合的前期准备过程中，事先奠定好高校与企业间的合作关系，有利于增进高校师生与企业间的熟悉度，促进课程的产教融合。高校与企业前期的合作积累可通过多种方式进

行，包括组织培训、举办竞赛、资源共享等。

组织培训是指企业组织行业内专业人士为校内的学生安排与专业前沿知识、先进技术等相关的培训，同时为课程的任课老师提供课程相关技能培训。例如，2014年10月，华南理工大学工程管理仿真实验室（现BIM中心）购买了广联达公司的造价软件和项目管理沙盘，广联达公司安排了讲师为华南理工大学师生进行了培训；2020年3月，广联达公司通过网络视频为华南理工大学老师进行了为期三天的装配式结构深化设计及造价分析专业培训；2020年6月，广联达公司免费为华南理工大学工程管理近20名师生进行了为期2周的软件开发思维培训。

举办竞赛是指企业通过组织举办行业相关竞赛的方式为高校师生提供专业知识应用和发挥的平台，此外，高校师生在参与竞赛的同时增强了对企业的认知度。例如，2018年6月，华南理工大学师生组队参加了广联达公司举办的"2018'粤价杯'广东省工程造价技能竞赛"；2019年8月，华南理工大学师生组队参加了广联达公司和东南大学共同举办"首届全国大学生智能建造与管理创新竞赛"。

资源共享是指企业与高校之间就专业相关课程、软件、平台等进行信息交流和资源的共享，让企业了解高校人才所学专业知识，同时让高校师生了解行业前沿知识及技术。例如，2019年12月，广联达公司向土木与交通学院BIM中心赠送了价值4万元的造价软件，并进一步商谈产学研全面合作；2020年2月，广联达公司为16级土木工程、17级工程管理近100名师生开通了免费的云造价软件账户；2020年2月，广联达公司开发的"建筑云课"平台免费提供给华南理工大学师生使用。

2.1.2　产学研洽谈

随着高校与企业之间的合作与联系不断加强，在课程实现产教融合的前期准备过程中，高校可以与企业开展进一步的产学研洽谈，进一步确认高校与企业间的合作关系，共同搭建实习实践基地。例如，2020年9月3日，广联达公司以及华南理工大学土木与交通学院工程管理系相关负责人在华南理工大学土木与交通学院举行了全面产学研合作暨校企产教深度融合洽谈会，双方就提升自主创新能力、实践能力、人才竞争力和品牌度，共同推进名师大讲堂、协同育人项目、数字建筑与智能建造学生社团、教学实习基地等具体合作事项进行交谈；对加强数字建筑领域科研合作、项目实践案例平台共享、学生数字化技术与管理思维培养等方面进行了深入交流；并对接下来常态化、可持续性的合作交流机制进行探讨。

同年9月24日，双方在广联达广州分公司进行实习基地签约并授牌，为加强华南理工大学与广联达的校企合作，促进双方的共同发展，广联达公司为华南理工大学提供教学实习基地，以构建优势互补的人才培养共同体。实习活动是高等学校培养合格人才的重要实践性教学环节，实习基地是高等学校教学条件建设之一，良好的实习条件可以提高教育质量，培养大学生的实践能力和创新精神；同时，通过校企合作，有利于共同培养与国家政策吻合、与企业人才紧缺口对应、与院校基础建设相匹配的高素质综合性行业主流人才。

2.1.3　研讨与修订教学大纲

在开展课程的产教融合前期，高校应当与企业结合当前行业发展需求、高校教学理念、专业课程知识、行业前沿知识及技术等因素，共同提出高校人才培养目标。之后，围绕人才培养目标对课程的教学大纲进行研讨与修订。

以"建设工程造价管理"课程为例，围绕培养学生的工程思维以及编程思维的目标，从BIM、物联网、数字化领域对课程的教学大纲进行研讨和修订。

2.2　教学活动

在实施课程产教融合的过程中，需要进一步推进校企融合，企业可以为高校提供丰富的案例、技术、资源、数据、师资、平台资源，高校也应当配合企业开展各项教学活动。教学活动可采用线上和线下相结合的方式开展，实现高校、企业资源共享，建立产学研一体化教学模式。

2.2.1　课程教学资源的支持

加强学生实践能力的培养，解决课程知识与工程需求及行业前沿脱节的问题，促进教学与工程实践、产业发展相结合，离不开优秀的课程教学资源的支持。在产教融合的中期阶段，企业可以为高校提供一些课程教学资源，帮助教师灵活地选用案例教学、项目教学、演示教学等教学方法，激发学生的学习兴趣，从而更好地理解和应用所学知识。

例如，广联达公司为"建设工程造价管理"课程提供了丰富的视频、造价学习 PPT、案例资料等资源，让学生学习知识的同时，增加对工程行业前沿技术的了解。图 2 展示了广联达公司提供的部分视频资源。

图 2　广联达公司提供的部分视频资源

2.2.2　建筑云课平台

随着高校和企业合作的不断加深，企业应当给予高校技术上的支持，如为高校提供自主开发的教学平台或小程序，方便教师采用混合

式、翻转课堂等教学新模式，实现师生课堂内外的同步教学、线上线下的教学互补，从而改善教学模式，提升教学效果。

在"建设工程造价管理"课程产教融合中，广联达公司开发的"建筑云课"平台，免费提供给华南理工大学师生使用。建筑云课平台包括智能考试测评平台和线上云课教学系统。智能考试测评平台支持自主开放式出题、自动测评出分与评价，自带防作弊安全防护，实现无人监考与疫情下考试的需要。线上云课教学系统提供了线上教学、直播、录课等授课方式，教师可以借助系统上传每节课的 PPT、视频，开展混合式教学，也可以发布课后习题，检验学生对课程的掌握程度。系统上拥有大量教学资料，包括一些微课视频和电子资料，支持移动化、碎片化的学习，学生在学习造价课程的同时，可以了解平法识图、钢筋算量等相关知识。图 3 展示了建筑云课平台的部分教学资源。

图 3　建筑云课平台的部分教学资源

2.2.3　专家进课堂

产教融合需要解决学生的实践能力不满足实际需求、课程知识与工程需求及行业前沿脱节等问题，为了培养行业所需高素质人才，高校有必要与企业共同开展协同育人项目，邀请专家进入课堂、组织学生进入实际项目参观等方式，增强学生理论联系实践的能力。

以"建设工程造价管理"课程为例，2021年 6 月 1 日邀请了广联达公司的广东区域造价BG 陈旭文总经理，举办"造价改革形势下的人才需求"讲座，介绍工程造价行业的发展历程和发展趋势。陈经理对造价行业现状和工程造价管理改革进行了分析，陈述了自己对政府出台的两份文件的解读和看法，并探讨了行业和个人如何应对造价改革带来的变化。最后，陈经理分析了造价行业人才需求的趋势，启发学生什么是关键素养，以及如何提高自己。

举办专家进课堂的活动，邀请企业的相关专家和讲师，进入本课程的课堂教学，分享他们的从业经验，有效促进课堂教学与工程实践相结合，同时也可以为学生的未来发展提供一些参考建议，有助于高校的人才培养与行业产业实际需求接轨。

2.3　实习实践

在与企业共同完成一系列课程教学活动，学生掌握一定行业基础知识或专业知识后，可联系课程相关企业进行实习实践，这不仅有利于提升学生的知识应用能力以及动手实践能力，还有利于促进课程教学与产业发展的深度融合。实习实践主要可以通过参观学习、专业竞赛、实习实践等多种方式开展。

2.3.1　参观学习

学校通过前期与企业之间建立的联系，组织课程任课老师及学生进入企业内部，对企业先进知识、生产流程以及实际案例等进行参观和学习；企业可提供企业文化、企业生产工艺、典型案例等材料供师生参考，不仅能够提升校内师生对行业前沿知识、生产工艺以及实际案例的认识，还能提升企业的知名度。

2.3.2　专业竞赛

由企业组织举办行业内专业竞赛，并邀请校内师生积极参与，或企业与校内师生共同组队参加行业内竞赛，有利于促进课程知识与企业实践之间的碰撞，让学生的知识应用能力、实践能力以及企业的生产研发能力、创新能力

得到提升，从而促进课程教学模式的升级，推动企业的产能提升。

2.3.3　实习基地

在课程的基础教学之外，可以组织校内师生前往企业的实习基地进行实习实践，为师生提供将课程所学知识应用到实际实践过程中的平台。实习活动是高等学校培养合格人才的重要实践性教学环节，实习基地是高等学校教学条件建设之一，良好的实习条件可以提高教育质量，培养大学生的实践能力和创新精神。例如，华南理工大学与广联达科技股份有限公司签订教学实习基地协议，广联达公司为参与"建设工程造价管理"课程的师生提供教学实习基地，以构建优势互补的人才培养共同体。

2.4　后续深化

在课程实现产教融合的后期，高校可与企业开展后续深化合作，从科研合作、人员培训以及人才招聘层面进一步深化合作关系，不断深化双方产教融合的合作，构建校企共同体，建立常态化、可持续性合作交流机制，有力推动双方在科研项目合作、技术成果转化以及人才教育培养等方面的进一步深化合作，以实现双赢的局面。

2.4.1　科研合作

高校与企业之间可以通过课题研究等方式开展科研合作，不断加强高校与企业间的联系，从而促进课程的产教融合。例如，2020年8月，广联达公司与华南理工大学开展了"基于数字化技术的项目协作方法研究"课题合作研究，进一步促进了学校与企业之间的科研合作关系。

2.4.2　人员培训

教师综合素质的提升同样是课程体系建设的一部分，后期可针对课程授课老师举办师资

与人员培训，不断提升授课老师的专业知识、技能水平、案例分析等综合能力；同时企业可以鼓励员工共同参与人员培训，从而提升企业内部员工的综合素质。

2.4.3　人才招聘

企业的发展离不开人才的需求，在实现课程产教融合的后期阶段，企业可以通过与高校间的合作关系，在校内进行企业宣传以及人才招聘等工作，为企业的发展提供人才保障，一定程度上解决了学生的就业需求，促进高校学生就业率的提升。

2.5　阶段性实施效果

目前，本课程的产教融合实施已经完成了前期的合作积累、产学研洽谈和教学实习基地的签订等工作，也与广联达公司共同研讨并修订了教学大纲；企业也已完成了向课程提供教学资源、向学校共享了建筑云课平台、选派专家进课堂授课等工作。从目前学生们反馈的教学效果来看，实施效果良好，如学生反馈"讲得挺全面，方向也很新""形式挺好的，内容挺丰富""觉得和新鲜，而且接触过的东西结合实践去讲感觉很受用""结合了一些实际例子来介绍还是挺有趣的""形式新颖，而且学习到了很多课本之外的东西""让我们看到了专业前景，信息化发展也让我们眼前一亮"等。

3　结论

随着建筑行业向数字化、工业化、智能化不断转型升级，"产教融合、校企协作"成为现阶段培养适应社会经济发展和符合行业需求人才的最有效的培养模式。校企产教融合，有利于建立产学研一体化教学模式，推动高校人才培养结构升级转变，培养大批具有较强行业背景知识、实践能力的创新型、应用型和技术技能型人才，推动科学研究与实际工程需求与

前沿相结合。

本文以华南理工大学"建设工程造价管理"课程为例，从前期准备、教学活动、实习实践和后续深化四个方面探讨了课程产教融合的实施路径。前期的主要环节为合作积累、产学研洽谈以及教学大纲的研讨和修订；实施过程中高校可以借助建筑云课平台开展线上教学，举办"专家进课堂"等线下活动，并组织学生进行参观学习、专业竞赛等实习实践活动，采用线上与线下结合的方式进一步推进产教融合；后续高校和企业可以从科研合作、人员培训以及人才招聘层面进一步深化合作关系，促使双方达成共赢。

参考文献

[1] 顾佩华. 新工科与新范式：概念、框架和实施路径. 高等工程教育研究，2017(6)：1-13.

[2] 邓朗妮，赖世锦，廖羚，等. 新工科背景下的工程管理"专创融合"教育模式探索[J]. 当代教育实践与教学研究，2020(2)：195-196.

[3] 覃亚伟，孙峻，余群舟，等. 新工科理念下工程管理专业教学改革研究[J]. 高等建筑教育，2019，28(5)：1-7.

[4] 范圣刚，刘美景. "新工科"背景下土木工程专业建设与改革探讨[J]. 高等建筑教育，2019，28(4)：16-20.

[5] 字应坤. 产教融合开启工业软件新征程[N]. 中国电子报，2020-09-01(2).

[6] 赵中华，温景文，齐庆会，等. 产教协同的土木工程专业实践教育基地建设研究[J]. 高教学刊，2020(26)：58-60+64.

[7] 王慧. 建立基于"合作教育"的土木工程学科专业实践模式[J]. 高教学刊，2020(23)：68-70.

[8] 王婉，李怀健，刘匀. BIM技术在校企联合毕业设计中的应用和实践. 高等建筑教育，2018，27(6)：161-165.

华南理工大学工程管理本科专业特色比较分析

王幼松　闫　辉　黄文炜　曾理菁　吴　凡　张扬冰

（华南理工大学土木与交通学院，广东广州 510640）

【摘　要】　新形势对我国工程管理人才和专业教育提出了更高的要求和新的期许，但目前工程管理专业的培养还存在着一些问题。因此，探索适应未来的专业培养特色与发展方向，是高校的重要任务。研究通过对比华南理工大学、东南大学、大连理工大学和重庆大学四大工学院工程管理专业的相关要素，对华南理工大学工程管理的专业特色进行提炼和提出改进建议，研究结果也可为其他学校工程管理的发展提供参考。

【关键词】　华南理工大学；工程管理；专业特色；对比与改进

自改革开放以来，我国综合国力不断提升，投资与项目规模增大，且愈加复杂，国际化趋势更加明显，新形势对工程管理人才提出了更高的要求，也对工程管理本科专业教育提出了新的期许。

但是工程管理专业还存在着人才宽而不专、师资总体薄弱、学生认知模糊、缺乏国际水准、忽略实践培养等问题[1]，而且单一的技术型人才或者管理人才已不能满足发展需求。要管理好项目并使之效益最大化，我们需要更多的复合型高级管理人才[2,3]。因此，工程管理将是未来建筑业人才培养的重要方向之一。

本文收集华南理工大学、东南大学、大连理工大学和重庆大学四大工学院工程管理专业的相关信息（数据来自各学校及院系官网），通过比较分析，尝试对华南理工大学工程管理本科专业的特色进行提炼，进而提出改进建议，也同时为其他学校工程管理的发展提供参考。

1　四校工程管理历史沿革

1.1　华南理工大学

华南理工大学工程管理的历史最早可以追溯到 1953 年，华南工学院土木工程系成立了建筑施工教研组。2013 年 4 月，工程管理系成立。

目前已拥有完整的工程管理本科、硕士和博士人才培养体系。招收工程管理专业本科，土木工程建造与管理学术学位硕士，土木水利、工程管理专业学位硕士，工程经济管理学术学位博士，土木水利专业学位博士。每年招收本科生约 40 人，硕博研究生约 140 人。

1.2　东南大学

东南大学工程管理专业的历史最早可以追溯到 1921～1928 年间营造门的设立。1983 年招收建筑经济与管理硕士研究生。1986 年招收建筑管理专业本科生。1992 年将工程管理并入土木工程系。1997 年正式成立建设与房地产系。

目前设有工程管理本科专业,管理科学与工程一级学科硕士、博士点和博士后流动站,土木工程建造与管理二级学科硕士、博士点和博士后流动站,以及项目管理、建筑与土木工程、工程管理领域的工程硕士点,形成了工程管理学科本、硕、博完整培养体系。

1.3　大连理工大学

大连理工大学工程管理学科的历史最早可追溯到 20 世纪 90 年代中期,几位教师开始工程管理方面的科研和教学。1999 年在土木工程专业分出工程管理专业方向;2003 年招收工程管理专业本科生;2011 年 11 月,建设管理系成立;2014 年工程管理专业通过了住房和城乡建设部的工程管理专业评估。

1.4　重庆大学

重庆大学工程管理本科专业已有 30 余年的办学历史。1980 年,建筑经济与管理本科专业开始招生。1981 年,建筑管理工程系正式设立。1996 年,国际企业管理本科专业开始招生。1998 年,建筑管理工程、国际企业管理、房地产经营管理三个专业被整合并更名为工程管理专业。2004 年,房地产经营管理专业开始招生。2005 年,工程造价本科专业开始招生。

重庆大学工程管理本科专业 2009 年被教育部批准为国家级特色专业建设点,2011 年获重庆市高等学校"专业综合改革试点"项目立项,2013 年获"重庆市高等学校特色专业"称号。

2　师资力量与科研能力比较

2.1　华南理工大学

华南理工大学工程管理系现有专任教师

14 人,其中教授 2 人、副教授 7 人、讲师/博士后/工程师 5 人、博士生导师 3 人、硕士生导师 11 人。专任教师中有 12 人拥有博士学位,教师博士化率达 86%,主要来自清华大学、哈尔滨工业大学、华南理工大学、同济大学等著名学府。

近 5 年来,工程管理系教师主持国家自然科学基金、省部级、横向及教研等各类项目 50 余项,发表 100 余篇高水平学术和教研论文,出版 10 余部教材和专著;并参与广州大学城建设工程监理、港珠澳大桥珠海口岸工程监理与咨询等多项工程实践活动;获广东省科学技术二等奖 1 项,教育部优秀工程勘察与岩土工程一等奖 1 项,专利 1 项。

2.2　东南大学

东南大学建设与房地产系教师共 24 人,包含教授 6 人、副教授 10 人、讲师 8 人,其中博导 3 人、博士 21 人,博士化率 87%,国际化背景 80%,国内兼职教授 4 人、国际客座教授 3 人。

近年来,建设与房地产系主持国家自然科学基金、省部级、横向及教研等各类项目 60 余项;获国家科技进步二等奖 2 项、中国公路学会科技进步特等奖 1 项,江苏省科技进步奖、江苏省建设科技奖 10 多项;发表高水平 SCI 论文 30 多篇;主编或参与编制国家行业标准规范 10 多部;出版材料和专著 60 多部。

2.3　大连理工大学

大连理工大学建设管理系现有教授 3 名,副教授 5 名,讲师 1 名,主要来自哈尔滨工业大学、大连理工大学、南开大学等著名学府,师资博士后 1 名,交叉学科共享师资 7 名,外聘教授 5 名,国外聘请的海天学者 1 名,师资博士化率 100%。有博士生导师 6 人,硕士生导师 10 人。

近年来，建设管理系教师承担完成了 30 多项研究课题，发表论文 300 余篇，其中 SCI 检索 20 多篇，获得大连市科技进步一等奖 1 项，辽宁省科技进步三等奖 1 项。出版国家级规划教材 2 部，国家级精品资源共享课 1 门。目前，承担多项国家自然科学基金项目、省部级科研项目、大连市建设科技项目和企业委托项目。

2.4　重庆大学

重庆大学工程管理系现有专任教师 32 人，其中教授 8 人，副教授 19 人，讲师及师资博士后、工程师 5 人。专任教师 22 人拥有博士学位，主要来自清华大学、哈尔滨工业大学、上海交通大学等著名学府。

2011～2015 年底，学院共承担 13 个国家级项目，发表 SCI 论文篇 40 余篇，其他高水平论文（北大核心及以上）将近 110 篇，出版教材、专著 20 多部。在科研奖励方面，学院获得五项教学改革计划奖、两项国家奖项和三项省级奖项。（注：其他成果没在公开的数据中查找到）

3　生源与毕业生比较

3.1　华南理工大学

华南理工大学工程管理专业本科每年招生 40 人，其中广东省约招生 10 人，其余省招 30 人，广东省内录取率约为 1‰。2016～2018 年，在广东省的招生情况如表 1 所示。

华南理工大学工程管理专业 　　表 1
广东省内录取情况（2016～2018 年）

年份	最低分	最低排位	考生人数	录取率
2018 年	600	8882	75.8 万人	1.17%
2017 年	582	7805	75.7 万人	1.03%
2016 年	598	7212	73.3 万人	0.98%

据统计，华南理工大学工程管理专业 2014 届到 2017 届的毕业生中，有 32% 的毕业生选择在国内或国外继续升学深造。选择就业的毕业生中，有一半进房地产企业工作，其余进施工单位、设计院、咨询公司等，具体如图 1 所示。

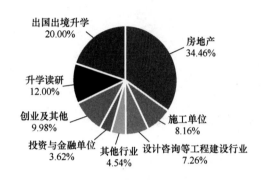

图 1　华南理工大学工程管理专业毕业生
去向分布（2014～2017 届）

3.2　东南大学

东南大学工程管理专业本科每年约招生 60 人，各省招生人数较均匀，在江苏省内录取率约为 0.8%，其他省录取率大部分低于 0.8%。2016～2018 年在江苏省的招生情况如表 2 所示。

东南大学工程管理专业　　表 2
江苏省内录取情况（2016～2018 年）

年份	最低分	最低排位	考生人数	录取率
2018 年	388	2825	33.09 万人	0.85%
2017 年	384	2606	33.01 万人	0.79%
2016 年	393	2279	36.04 万人	0.63%

据统计，东南大学 2012 届工程管理专业毕业生中，近三成选择在国内或者国外继续升学深造。选择就业的毕业生中，接近一半的毕业生选择房地产企业，其余大部分选择施工单位，少部分进设计单位或者其他行业，具体如图 2 所示。

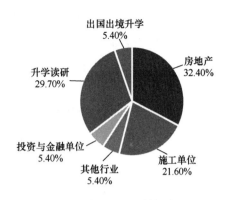

图 2　东南大学工程管理专业
毕业生去向分布（2012届）

3.3　大连理工大学

大连理工大学工程管理专业本科每年招生 30 人，辽宁省招生指标约为 6 人，占 20％，其余各省只有部分省有招生指标，为 2 人。辽宁省内录取率约为 2.0％。2016～2018 年在辽宁省的招生情况如表 3 所示。

大连理工大学工程管理专业辽宁省内　表 3
录取情况（2016～2018 年）

年份	最低分	最低排位	考生人数	录取率
2018 年	637	3845	18.5 万人	2.08％
2017 年	613	4166	20.85 万人	2.00％
2016 年	628	3521	21.83 万人	1.61％

注：网上公开的数据中未能查找到大连理工大学工程管理专业毕业生去向。估计可能跟其余三校相似，大部分学生选择升学深造或者进入房地产企业工作。

3.4　重庆大学

重庆大学工程管理专业本科每年约招生 85 人。本地招 20 人，约占 25％，四川在其余省中占比最大，约 10％，其余各省招生人数较均匀，约为每省招 2 人。在重庆市内录取率约为 2.3％。自 2018 年起，工程管理专业与工程造价、财务管理专业并入管理科学与工程类大类招生，招生总人数不变。2016～2018 年在重庆市的招生情况如表 4 所示。

重庆大学工程管理专业重庆市内　表 4
录取情况（2016～2018 年）

年份	最低分	最低排位	考生人数	录取率
2018 年	620	6377	25.047 万人	2.55％
2017 年	598	5722	24.749 万人	2.31％
2016 年	621	5629	24.889 万人	2.26％

据统计，重庆大学工程管理系 2012～2015 届的毕业生中，约 30.5％选择在国内或国外继续升学深造。选择就业的毕业生中，接近一半的毕业生选择房地产企业，其余大部分选择施工单位，只有少部分进设计单位或者其他行业，具体如图 3 所示。

图 3　重庆大学工程管理专业毕业生
去向分布（2012～2015 届）

4　教学体系比较

4.1　华南理工大学

华南理工大学工程管理专业 2019 年培养方案中，各课程类别模块的学分、学时和占比情况如表 5 所示。

华南理工大学工程管理专业 2019 年　表 5
培养方案各模块学分比例

课程类别	课程要求	学分	学时	学分所占比例
公共基础课	必修	55.5	1096	38.5％
	通识	10.0	160	

续表

课程类别	课程要求	学分	学时	学分所占比例
专业基础课	必修	49.0	784	28.8%
专业领域课	选修	23.5	376	13.8%
集中实践教学环节（周）	必修	32.0	39 周	18.8%
毕业生学分要求		170.0（含必修 136.5，选修 23.5，通识 10.0）		

华南理工大学工程管理专业以土木工程为专业依托，同时增加了管理类、经济类和法律类的选修课程。培养方案以实践教学为核心（13.8%＋18.8%＝32.6%），既可更大限度地进行专业培养（专业课程总占比为 61.5%），也可更好地实现个性化发展（专业课均为选修课程，占比 13.8%）。

4.2　东南大学

东南大学工程管理专业 2016 年培养方案中，各课程类别模块的学分、学时和占比情况如表 6 所示。

东南大学工程管理专业 2016 年　表 6
培养方案各模块学分比例

课程类别	学分	学时	学分所占比例
通识教育基础课	55.5	1116	36.67%
专业相关课程	68.25	1120	45.50%
集中实践教学环节 & 短学期课程	26.75	24＋，课程周数：36	17.83%
毕业生学分要求	150.0	2260＋，课程周数：36	100%

东南大学工程管理专业目前的课程设置比例结构中，通识教育基础课包括思政类、军体类、外语类、计算机类等，占比为 36.67%。专业相关课程包括大类学科基础课、专业主干课及选修课，占比为 45.50%。实践课程类学分为 32.125，包括集中实践教学环节和部分专业相关课程，占比为 21.42%。

4.3　大连理工大学

大连理工大学工程管理专业 2017 年培养方案中，各课程类别模块的学分、学时和占比情况如表 7 所示。

大连理工大学工程管理专业 2017 年　表 7
培养方案各模块学分比例

课程类别	课程要求	学分	学分所占比例
通识与公共基础	必修 选修	49.5 18	38.57%
大类与专业基础课程	必修 选修	36.5 9.5	26.29%
专业与专业方向课程	必修 选修	48.5 5.5	30.86%
创新创业教育与个性发展课程	选修	6	3.43%
第二课堂	必修	1.5	0.86%
毕业生学分要求	必修（选修）	136（39）	100%

大连理工大学工程管理专业目前的课程设置比例结构中，通识与公共基础课占比 38.57%，与其他高校基本一致。专业基础课和专业主干课占比为 57.14%，其中实践环节占比约为 17.14%。培养方案中最具特色的有创新创业教育及个性发展课程，课程比例占比为 3.43%。

4.4　重庆大学

重庆大学工程管理专业 2014 年培养方案中，各课程类别模块的学分、学时和占比情况如表 8 所示。

重庆大学工程管理专业 2014 年　表 8
各模块学分比例

课程类别	课程要求	学分	学分所占比例
通识与人文素质课程	选修	8	4.71%
公共基础课程	必修 选修	37 10	27.65%

续表

课程类别	课程要求	学分	学分所占比例
专业基础课程和专业课程	必修 选修	49.5 19	40.29%
特色专业选修课程组	选修	12	7.06%
集中实践环节	必修	22.5	13.24%
非限制性选修课程	选修	10	5.88%
第二课堂	选修	2	1.18%
毕业生学分要求	必修（选修）	109（61）	100%

重庆大学工程管理专业目前的课程设置比例结构中，通识与公共基础课占比 32.36%，专业课程占比为 60.59%，课程类型偏向于经济和管理两个方面。培养方案中设置了特色专业选修课程组，分为房地产开发与经营管理专业选修课程组（A 组）、国际工程管理专业选修课程组（B 组）和推免研究生专业选修课程组（C 组），对学生进行特定方向的专业培养。

5 结论与建议

在人才市场需求呈现多样化、多层次的今天，传统四大工学院工程管理专业培养目标各有侧重和特色，总结如下：

（1）华南理工大学工程管理专业生源优秀，培养体系具有一定创新性，依托 BIM 实验室，课程融入数字化、信息化、智能化，国际化程度较高。组建特色课程群组，重视学生综合能力的培养，以实践教学为主。主要体现为：①近年的课程加强模拟教学、案例教学，以开放式课堂的形式，强调学习和创新能力、团队合作与组织能力的培养；②采用基于项目的教学模式，增加体验式、探究式、参与式工程管理工作坊实践教学形式，引导学生自己研讨、研究完成特色课程等；③加强实习、及时与用人单位沟通，通过联合培养等渠道，实现与实际工作的无缝对接。

（2）东南大学工程管理专业以培养专业知识能力过硬的学生为目标。其培养方案与注册师制度相衔接，致力于培养国家注册建造师、注册造价工程师、注册监理工程师、注册房地产估价师等专业人才。

（3）大连理工大学工程管理专业重视培养学生的创新创造能力。充分尊重学生的个性化发展，在课堂设置上指引学生根据自己的兴趣爱好或个人所长，选择自己感兴趣或者合适的方向。

（4）重庆大学工程管理专业以大量向市场输送优质人才为目标。尤其重视对学生经济和管理能力的培养，设置特色专业选修课程组，根据学生的选择进行特定方向的专业培养。

虽然四大工学院工程管理专业培养的侧重点稍有差异，但最终目标是一致的，即培养具有管理学、经济学和土木工程技术的基本知识，掌握现代管理理论、方法和手段，能在国内外工程建设领域从事项目决策和全过程管理的复合型高级管理人才。

根据四大工学院的对比结果，针对国内工学院工程管理专业的发展，有如下几点建议：

（1）进一步深化国际化管理人才的培养。引进具有海外经历、实践经验丰富的教师，逐步开展全英、双语教学，增加对学生专业外语交流能力、国际工程管理综合能力、国际视野的培养。聘请外教讲授国际工程口语、交际、管理等课程。

（2）师资力量有待加强。从目前四大工学院及国内主要高校工程管理专业的生师比来看，工程管理专业的师资力量相对薄弱，考虑从国外或港澳台地区引进博士后或更高层次人才。

（3）培养方案与建设领域职业资格考试相结合。考虑到工程管理相关职业发展，可将专业教育与注册造价师、注册建造师、咨询师和估价师等执业资格教育进行有机结合，提高教学的针对性和实践性，为学生未来执业规划和发展奠定理论基础。

（4）增加必修专业课程内容。为培养懂技术、懂管理、懂经济、懂法律的复合型高级管理人才，可适当增加必修专业课程的内容，增加专业知识的学习。

参考文献

[1]　白妤，刘喜琴. 工程管理本科专业建设问题及对策浅析[J]. 中国多媒体与网络教学学报（上旬刊），2019(6)：101-102.

[2]　李辉. 工程管理专业从理论型向应用型转变的改革措施探讨[J]. 中国管理信息化，2018，21(4)：206-207.

[3]　李垠龙. 我国工程管理学的现状与未来发展[J]. 居舍，2018(28)：2.

海外专题

Overseas Specials

How Achievable are the Governments Targets to Reach Net Zero Emissions by 2050?
A View from the UK Commercial Construction Sector

Priyen Halai[1]   Hua Zhong[2]

(1. Linesight, London, EC4A 1LT, United Kingdom;
2. School of Architecture, Design and the Built Environment, Nottingham
Trent University, Nottingham NG1 4FQ,United Kingdom)

【Abstract】 The climate crisis we face today has reached tipping point. Evidence gathered shows that despite international efforts to slow the pace of global warming, the consequential changes to our planet caused by global heating are becoming irreversible. It is evident that the built environment plays a vital role in reducing global emissions and this research explores the situation within the UK construction industry. A literature review was conducted to investigate the rationale behind the 2050 net zero emission targets being set by the UK government, and the current context around the targets.

Primary research was carried out through interviews to obtain the views of key stakeholders within the UK construction industry on how achievable they believe the UK's 2050 net zero target is. The participants were specifically chosen to represent the UK construction industry as a whole in order to form a comprehensive view on the achievability of the target. A case study analysis was then undertaken to analyse the feasibility of moving the UK commercial construction sector towards a net zero future. The findings from the research show that the lack of government legislation and incentives, alongside the increased initial cost of designing and producing a net zero built asset are the largest obstacles facing the industry.

The data collected from the research reveals that despite progress being made within the sector in terms of moving to a net zero future, the 2050 target will be tough to achieve if changes are not made by all key stakeholders within the UK construction industry. This is mainly due to the lack of government legislation and incentives in place to induce clients to build net zero buildings.

This research analyses the current issues facing the UK construction industry in achieving the target and suggests ways in which they can be overcome and what measures can be implemented in order to improve progress within the sector.

【Keywords】 Sustainability; Net Zero; Carbon Reduction Legislation; 2050 UK Target; UK Construction

1 Introduction

Concerns around the climate emergency we face today have been voiced by leading scientists such as the late Stephen Hawking, who stated "We cannot continue to pollute the atmosphere, poison the ocean and exhaust the land. There isn't any more available." (Hawking, 2007) More recently, the United Nations (UN) Secretary-General António Guterres labelled the Working Group I contribution to the Sixth Assessment Report by the Intergovernmental Panel on Climate Change (IPCC) as a "code red for humanity" (Guterres, 2021).

His statement following the report, which addresses the most up-to-date data relating to the climate system and the issue of climate change, is a stark reminder of the dangers facing our planet:

The alarm bells are deafening, and the evidence is irrefutable: greenhouse gas emissions from fossil-fuel burning and deforestation are choking our planet and putting billions of people at immediate risk. Global heating is affecting every region on earth, with many of the changes becoming irreversible (Guterres, 2021).

Currently, Carbon dioxide concentration levels in the earth's atmosphere are the highest on record, at circa 417 parts per million (ppm) compared to circa 392 ppm this time 10 years ago (NASA, 2021). This shows the situation is not improving despite continued efforts globally to slow global warming.

In June 2019, legislation was passed in the UK which legally bound the UK to reduce carbon emissions relative to levels in 1990, by at least 100%. This was a commitment that made the UK the first major economy globally to pass laws to end its contribution to global warming by 2050. This action was a step in the right direction in terms of solving the global climate crisis, however progress needs to improve if the UK is to meet the ambitious target.

According to the UK Green Building Council (UKGBC), the built environment contributes to around 40% of the UK's total carbon footprint, therefore it is clear that the construction industry has a key role to play in achieving the target (UKGBC, 2020). However, the progress indicators show there is a problem within the industry as the Committee on Climate Change (CCC) submitted their "Reducing UK Emissions" progress report

2020 to parliament which stated that despite building emissions falling 14% between 2008 to 2018, the sector will struggle to transition to a net-zero future (CCC, 2020). The report also states that the only one of the five indicators for the buildings sector being met. Therefore, it is clear that there is an issue within the industry. This study could be beneficial as it can form the basis for further research into the area, and obstacles alongside potential solutions will be identified and explored within the dissertation to help guide the industry.

It is timely to research the topic: in November 2020, Savills director of environmental economics, Tom Hill, explained that "The current policy trajectory of where we're headed is not enough. It's quite a long way to get on the net-zero track" (Hill, 2020). With the legislation being fairly new, the issue is very current and research that can help identify the obstacles at an early stage could prove vital in achieving the target in the long run.

There has not been research carried out to date focusing on the UK commercial construction sector and how achievable the net zero 2050 target is within the sector therefore this research aims to plug the gap and form the basis for further research to be carried out in the area.

2 Literature Review

Why was the 2050 target set by the UK government?

The current climate change legislation in the UK has evolved periodically since the es-

tablishment of the IPCC. The IPCC was established in 1988 by the United Nations Environment Programme (UNEP) and the World Meteorological Organisation (WMO) with the aim to provide all 195 governments involved relevant scientific data to inform climate change legislation and policies in order to drive global change (IPCC, 2021).

The UK's first carbon reduction target by 2050 was outlined in the 2003 Energy White Paper, in which a target to reduce carbon emissions by 60% below 1990 levels by 2050 was set (CCC, 2019). This target was set based on a recommendation in the "Energy-the Changing Environment" report by the Royal Commission on Environmental Pollution. This report is key as for the first-time climate change was cemented as a key aspect of government energy policy (Institute for government, 2020). Within the report the recommendation to reduce carbon emission levels in the UK by 60% by 2050 was made based on supporting the proposal that 550 parts per million by volume (ppmw) should be an upper limit on the carbon dioxide concentration in the atmosphere. This upper limit came from the IPCC's second assessment, which is a very reliable and compelling report due to the sheer quantity of expert scientists involved from around the world, and the extensive peer review procedure in place. However, as it was published in 1995, the data and climate models which calculated the upper limit of 550 ppmw for carbon dioxide concentration in the atmosphere, were superseded by a new upper limit of well below 400 ppmw.

This new limit was set as a result of new evidence indicating that carbon dioxide levels needed to be kept well under 400 ppmw in order limit global temperature increases to 2 degrees celsius above pre-industrial levels (House of Commons, 2006).

Subsequently, pressure started to grow on the UK government to improve the 60% target which was deemed to be insufficient. The most influential driver behind the introduction of the Climate Change Act 2008 (CCA) were the non-governmental organisation (NGO), Friends of the Earth (FoE). Their "Big Ask" campaign played a critical role in pushing the government to introduce ambitious legislation, and with around 170000 people backing the campaign, the government improved their target and passed the CCA (Carter and Childs, 2017). The success of the campaign can be deemed to be mainly due to the political support it received across all parties. According to Davies (2018), this was a result of the campaign cleverly leveraging political competition between the main parties, with both wanting to be seen as the drivers behind slowing the rate of climate change. In 2005, arguably one of the most significant political moment of the campaign occurred as David Cameron was elected leader of the Conservative Party (Carter and Childs, 2017). According to Bale (2016) and Carter and Clements (2015), David Cameron's intention was to modernise the party and shift media attention to more positive topics such as the environment, to aid the detoxification of the image of the conservatives. This played perfectly in to the hands of the "Big Ask" campaign as the government were now being pushed from all angles to act effectively on climate change. However, the NGOs pushing for policy change alongside FoE had a tough task in doing so, as according to Carter and Childs (2017), NGOs have limited capacity to influence policy and instead have to rely on "issue attention cycles"——a theory outlined by Downs (1972). The FoE and other NGOs pushing for climate policy change overcame this issue by ensuring the pressure was not cyclical in nature and instead made sure the public remained behind the campaign until tangible change was implanted.

When introduced, the CCA set a legal framework for the UK to cut greenhouse gas emissions to 80% below 1990 levels by 2050 and its introduction saw the UK become the first country to set a legally binding climate change mitigation target (LSE, 2020). The Act also saw the CCC who are an independent committee created to advise the government on climate change and progress being made to reduce emissions (CCC, 2021). Additionally, under the Act, the requirement was set for the government to set five-yearly carbon budgets until 2050.

Following the CCA, the UN announced the adoption of 17 Sustainable Development Goals in 2015, with protecting the planet being one of the key aims (UN, 2020). The Paris Agreement 2016 (PA) was brought in the following year with the aim.

To strengthen the global response to the threat of climate change by keeping a global temperature rise this century well below 2 degrees Celsius above pre-industrial levels and to pursue efforts to limit the temperature increase even further to 1.5 degrees Celsius (UNFCCC, 2020).

The PA was an outcome of global civil society pressuring the governments involved into signing an agreement that is highly ambitious and may prove difficult to meet. According to research by Jacobs (2016), global collaboration between NGOs, academics and businesses continued to apply pressure by building on the scientific communities' data backed reports such as the landmark reports by the IPCC. Following the PA, the CCC (2016) initially recommended in their "UK climate action following the Paris Agreement" report that net zero should not be pursued as a target, and that it should be kept under review as further evidence comes to light. In the key "Net zero, the UK's contribution to stopping global warming" report by the CCC (2019), in which robust new evidence did in fact become available, the CCC made the recommendation that the UK should aim to be net-zero on all greenhouse gases by 2050. This evidence was mainly from the IPCC special report "Global Warming of 1.5°", in which it was concluded that if current targets and pledges were to continue unchanged globally, global warming could reach 3 degrees Celsius by the end of the century which would fall well behind the aim set out in the PA (CCC, 2019).

Despite the UK already being obliged to an 80% reduction in carbon emissions relative to levels in 1990, secondary legislation was passed in June 2019 that extended the target to at least 100% (Shepheard, 2020). More recently, the UK government has set the world's most ambitious climate change target for 2035 (Gov. uk, 2021). The new legal obligation which will become law before July 2021, will set the aim to reduce emissions by 78% by 2035 compared to 1990 levels. Harrabin (2021) states that this will be a world-leading position, however plenty of work will need to be done in order to achieve the target.

Current context around the target

The UK Carbon Budgets briefing paper 2019, states that the first two carbon budgets were met, and the third budget (2018~2022) is set to be beaten. However, in the long run the UK is projected to underperform in budgets four and five by 5.6% and 9.6% respectively (Priestly, 2019). A report by Evans (2021) suggests that the UK is halfway to meeting the legally binding net zero target. This was a result of total greenhouse gas emissions falling by 8.9% in 2020 (Department for Business, Energy & Industrial Strategy, 2021). However, recent progress has been due to lower activity as a result of the pandemic and therefore a rebound is expected once all restrictions are lifted. The report also highlights a cause of concern which is that one of the largest annual increases in UK emissions came in 2010, as the economy bounced back after the financial crisis. The UK have to avoid following this trend as we recover from the pandemic as there is now a net zero emissions target to achieve. Additionally, new technology and more knowledge around carbon reduction measures should stop a similar rise. An article by Forster (2021) concluded that although the pandemic has paused and reduced emissions globally, the effect it will have in tackling rising temperatures in the short and long term is a lot less than is anticipated. In fact, several models suggest that the impact that the pandemic will have on global temperatures is a fall of 0.01°C, deeming its impact negligible (Forster, 2021). Therefore, key stakeholders in the UK cannot be lured into a false sense of security by short-term positive results and instead need to continue finding solutions and finding new ways to continue moving the UK to a net zero future.

What role does the commercial construction sector play in achieving the target?

The UK construction industry plays a pivotal role in achieving the net zero target set by the government for 2050 as the built environment contributes significantly to emissions in the UK. A report by LETI (2020) states that within the UK, buildings contribute 49% to annual carbon emissions. The UKGBC (2020) agree with the idea that the UK building sector contributes a large amount to the UK's total carbon emissions, however they believe the contribution is slightly lower at around 40%. Despite varying figures from leading climate change councils and networks, is it clear that the industry has a pivotal role to play if the UK are to achieve the 2050 target?

Within the built environment, the commercial construction sector is the sector which requires the most improvement in order to push the industry to a net zero future and to ensure emissions from the built environment reach net zero. This is highlighted in WSP's "Delivering Net Zero" report, which states that more action needs to be taken to improve energy efficiency in commercial buildings. It suggests that the lack of incentives is discouraging businesses to invest in energy efficiency (WSP, 2020). This is supported by the CCC "Reducing UK Emissions" progress report 2020 which recommends that net-zero policy packages must be implemented before the end of the year to tackle commercial buildings in particular (CCC, 2020). However, although both reports briefly mention commercial buildings in the UK, the reports are general, and the sectors contribution is not reviewed in much depth.

According to the Building Energy Efficiency Survey (BEES) between 2014~2015 offices and retail were the two largest sectors in terms of energy consumption at 17% each (Department for Business, Energy & Industrial Strategy, 2016). Although the survey is over half a decade old, the contribution by the sectors remain similar as the landscape within the sectors, and design specifications on projects have not changed significantly since the survey was conducted. The results of the survey highlight the importance of improving the efficiency and ultimately reducing carbon emissions within the sector in order to contribute positively to the UK government's 2050 goal. Brooks (2019) supports the idea that achieving net zero emissions in commercial properties is key for the industries transition to a more sustainable future. Although Brooks recognises the requirement to reduce emissions in all buildings, both new and old, he states that the value of early carbon savings in commercial properties is much more than in other construction sector. Therefore, with the sector contributing a large amount towards total emissions within the UK built environment, and with the value of achieving net zero within the sector being the greatest, it could be argued that the commercial construction sector has the biggest influence on the UK 2050 target within the construction industry.

What has been achieved to date within the construction sector?

The CCC submitted their "Reducing UK Emissions" progress report 2020 to parliament and suggested that despite building emissions falling 14% between 2008 to 2018, the sector will struggle to transition to a net-zero future (CCC, 2020). The report also suggests that the buildings sector has not decarbonised at the pace of the rest of the economy and has remained relatively flat over the latter half of the 2010s. This is a direct result of the lack of legislation and policy which can help incentivise achieving net zero emissions within the industry. This is evident when comparing the UK buildings sector to the power sector. Policy appears to have been the key driver for decarbonisation since with strong policies in place the power sector saw emissions fall by 67% from 2008 to 2019 (CCC, 2020). Despite worries about the progress being made within the construction sector, the industry has been praised by the business secretary Kwasi Kwarteng who said: "The UK's construction and engineering industries are leading the world in the drive to cut emissions, and I am thrilled to see so many businesses from all over the globe share this ambition." (Kwarteng, 2021)

Race to Zero, a global campaign which commits members to achieving net zero emissions by 2050, has seen 58 UK construction companies sign up, which is a positive sign as it shows that the industry has recognised it has a key role to play and large businesses such as Lendlease and Mace want to drive change.

Another significant achievement within the industry is the development of the UK's first net zero carbon concrete solution, Virtua. This product developed by CEMEX, is a game changer as concrete contributes a worryingly large amount towards embodied carbon emissions within the UK built environment. According to Lehne and Preston (2018) cement makes up around 8% of global carbon emissions. Therefore, a net zero alternative to traditional concrete which cuts the carbon footprint of the material by up to 70% and offsets the remainder is a positive step forward. The low carbon concrete solution has been used on the high speed 2 project, which showcases its potential within the UK construction industry. The challenge is now rolling out the use of the material across all projects in the UK where possible as there is no use for an innovative low carbon material if the industry is unwilling to adapt and use it to its full potential.

What obstacles does the industry face in achieving the target?

The main obstacle that the industry faces in helping achieve the government's target is the lack of policy and legislation to guide the industry towards a net zero future. A report by LETI (2020) states that the legislation within the construction and wider built environment industry is lagging behind in terms of the decarbonisation trend and therefore jeopardising the future of our planet. The key progress report from the CCC (2020) goes further by saying heating is an area in particular which needs major policy reforms and significant policy gaps still remain within the commercial construction sector. This is significant as this report will influence the next steps the government will take in order to achieve long term emissions targets. The fact that commercial buildings have been singled out highlights the importance of ensuring this sector is at the forefront of the net zero drive within the construction sector.

Another key obstacle is that most existing buildings will require zero-carbon retrofitting in order to decarbonise the sector. Alarmingly, 80% of the buildings that will exist in 2050 have already been built and so decarbonisation of existing buildings is key (UKGBC, 2020). This view is highlighted by the UKGBC's 2020 report "Energy performance targets for net zero carbon offices". The findings showed that the majority of key stakeholders within the industry believed that the emissions target is achievable for new buildings, however-

er the challenge lies in improving the performance of existing buildings, due to factors such as the complexity of retrofitting (UKGBC, 2020). A recent survey carried out by WSP (2020) further emphasises why retrofitting is a major obstacle for the industry as 85% of respondents believe there is not enough policy pushing owners to invest in retrofitting existing building stock. With a large scale nationwide retrofit campaign required but no policy to push it, the UK will struggle to reach net zero emissions within the built environment.

The report by Watkins and Hochlaf (2021) outlines that another major obstacle the industry is facing is the aging workforce. The proportion of workers aged 50 and above are 35% and only 20% of the total construction workforce are aged below 30. Another damning statistic highlighted in the report is that around 750000 UK construction workers will retire or be on the verge of retirement before 2035. The skills gap in the industry and the ageing workforce means that the industry might not have the workforce required to achieve the ambitious net zero targets. A report by UK FIRES (2019) states that education can play a major role in reconfiguring society to adopt methods to move towards a net zero carbon future, as was the case in the 1970's when education was used to influence the new generation of engineers to use more steel frames instead of concrete frames. This further highlights the issue of an aging workforce as if there is not enough young, highly educated talent with knowledge of net zero solutions and alternatives joining the industry, how will the industry shift towards a net zero future? The circa 20% of workers below 30, probably employed in lower positions within companies will struggle to influence key decisions which could help decarbonise the sector.

Chapter Summary

The literature review carried out by the researcher has informed the researcher of the various studies carried out to date around the chosen topic area and has highlighted a major gap in the research. The research shows that despite progress being made since the legislation was implemented, the UK risks not meeting the 2050 net zero target if improvements are not made.

Firstly, Priestly (2019) expects the UK to underperform in carbon budgets four and five by 5.6% and 9.6% respectively. This highlights the worrying fact that if the current landscape around net zero within the UK continues at the same trajectory, the UK will fall short of meeting the 2050 target. Further more, the research shows the built environment has a large role to play, as two reports, one by LETI (2020) and one by the UKGBC (2020), both highlight the fact that the built environment contributes to over 40% of annual UK carbon emissions. Within the built environment, the BEES 2014~2015 showed that offices and retail were the two largest sectors in terms of energy consumption (Department for Business, Energy & Industrial Strategy, 2016). The value and importance of reducing carbon emissions within the commercial construction sector is further emphasised by Brooks (2019) as he suggests that the value of

early carbon savings in commercial properties is greater than in other construction sector. Therefore, with there being a large contribution by the sector towards carbon emissions within the UK, and the value of reducing it being the most compared to other sectors within construction, the need and importance of this study is highlighted.

To date, there has not been any entry level research carried out around net zero in the UK and the achievability of the 2050 target. There is also a gap in research in regards to the UK commercial construction sector which is an important gap which this study aims to fill as the commercial sector is a large contributor to carbon emissions and is a sector lagging behind in terms of progress as shown in the literature review.

3 Objectives

To critically investigate the rationale behind the net zero emission targets being set by the UK government, and the current context around the targets, including what has been achieved to date within the construction sector.

To identify the obstacles that the UK construction industry needs to overcome, in order to achieve the net-zero emission targets by 2050.

To analyse the ideas of stakeholders within the UK commercial construction sector on how achievable the 2050 targets are.

To critically analyse a case study to determine the feasibility of moving the UK commercial construction sector towards net zero emissions and to identify any obstacles in order to inform recommendations.

To recommend what further measures need to be implemented by UK construction firms and the government, in order to achieve the targets by 2050.

4　Methods

A mixed method approach combining both quantitative and qualitative was be executed by the researcher in order to"triangulate" the data to improve its credibility and validity (Noble and Heale, 2019). "Triangulation of Data" refers to the use of several methods of research or data sources to form a comprehensive understanding of the research area or phenomenon (Patton, 1999). There are four types of triangulation as highlighted by Carter et al (2014); Method triangulation, investigator triangulation, theory triangulation and data source triangulation. The researcher elected to use method triangulation, by using multiple methods of data collection in order to improve confidence in the research findings (Noble and Heale, 2019).

Semi-structured, one-to-one interviews were used to allow the researcher to probe specific areas and raise particular queries as highlighted by Naoum (2013). Three participants were carefully selected by the researcher from different professions within the industry to bring varying perspectives on the net zero target and its achievability, as shown in Tab. 1.

Interviewees Tab. 1

Interviewee Reference	Job Role	Type of Company	See Appendix
A	Senior Cost Consultant (MRICS)	Consultancy	A
B	Head of Sustainability/CIBSE Vice President	Engineering Consultancy	B
C	Commercial Director	Main Contractor	C

The participants are all London based and were chosen by the researcher as they are all very experienced industry professionals who have a leading role to play within their organisations in contributing to the net zero drive within the industry. In addition, despite the limited sample size, all three interviewees together provide a comprehensive view of key stakeholders within the construction industry to help achieve the objectives mentioned.

The researcher chose to adopt a single case study analysis approach where design changes and subsequent cost implications on an anonymised UK commercial project were critically analysed. This highlighted where potential obstacles could lie in achieving net zero on commercial projects. Due to strict client confidentiality, detailed information for an entire net zero project was unattainable, however an anonymised commercial office scheme identified in the UKGBC "Building the Case for Net Zero" study (UKGBC, 2020) was used for analysis. The in-depth data in the UKGBC study consists of two designs of the same commercial office building. One achieving net zero and one at the baseline, and the cost implications of aiming for net zero.

4.1 Results—Interviews

In terms of the progress made within the construction sector, the results from interview C supports comments made by Kwarteng (2021), which is that the construction sector is leading the global drive for net zero. With large global contractors setting their own, inhouse net zero targets and achieving them, it shows that good progress is being made within the construction sector. Both interviewees A and C are in agreement on the fact that the UK commercial construction sector is seeing increased demand for net zero buildings from clients, particularly for office space. As mentioned by interviewee A, this in turn is driving developers to create and offer net zero offices for potential tenants in order to "Be more competitive and offer something that differentiates them from other developers". This is encouraging as it shows a positive trend towards a net zero future within the sector, which is seen as key by Brooks (2019).

The results from interview B highlight two encouraging aspects of progress within the industry. The lack of certification of net zero buildings is something that is mentioned as a major barrier by all participants. Therefore, if as mentioned by interviewee B, a certification process is ready before the end of 2021, this will be a huge step in the right direction for the UK construction industry. Finally, the mention of a new concept being used within the industry of carbon being used a currency in a new variation of value engineering is also indicative of good progress being made towards a net zero future. Cost needs to be considered alongside carbon savings and this is a process labelled as "True Value Engineering" by interviewee B. It should be common practice on all projects as soon as possible if the industry to achieve net zero emissions by 2050.

With regards to achievability of net zero emissions within the commercial construction sector, Interviewees B and C are in agreement that the 2050 net zero target is achievable if progress is made fast. Interviewee B goes on to mention legislation and regulation being key drivers and an important area which needs swift action to see the target achieved which supports the views of LETI (2020). Interviewee believes the 2050 target is very demanding, especially if carbon emissions from the manufacturing of materials is included in any industry targets or potential certification. The interviewee believes that the decarbonisation of the "Processes of getting the raw materials to finished products to get into the construction phase of a project" is where the difficulties lie.

A 2030 target for all new developments to be net zero is mentioned by both interviewees B and C. This target is not legally binding for the industry, it is simply a proposal being put forward by the UKGBC in their "Building the Case for Net Zero" report (UKGBC, 2020). Results from both interviews B and C suggest that this potential target is overly ambitious and may be difficult to achieve unless progress speeds up within the industry. Interviewee B does believe the target is achievable within the commercial sector due to aspirations of project funders, however the interviewee claims that the achievability of the goal within the public and residential sectors "depends on the willingness and aspirations of the project funders".

All three participants have identified a variety of obstacles that the UK construction industry needs to overcome in order to achieve the 2050 net zero target. Firstly, interviewees A and C agree that there is not a clear definition of net zero. Interviewee C stated that "There is not enough clarity" and that "It is all very slow and we risk being left behind as an industry". This is a major issue because it causes difficulties for clients and contractors when designing a building to be net zero. This also links in with the view from all three interviewees which is that a lack of certification is a major barrier for the industry. There is no clarity regarding what will allow clients to call their buildings net zero buildings and the only way this can be clarified is through a certification programme.

Another key obstacle which is highlighted by all three interviewees is the lack of legislation in place. This is a dominant theme across the results from all three interviews which highlights its importance. Interviewee B even goes on to state that it is the main obstacle facing the industry in achieving net zero by 2050. Interviewees A and C highlight the issue of liability. There is no legislation or industry standard as highlighted by interviewee C and so it causes confusion around the issue of which party is responsible for offsetting emissions. This needs to be overcome with legislation.

The results from the three interviews also indicate that the increased cost associated with achieving net zero emissions on a construction project is a major barrier. Interviewee A believes the additional cost is the main reason "That is stopping clients from choosing a net zero strategy and design to deliver projects". They then go on to explain that this is causing clients to choose between achieving either net zero in construction or net zero in operations and not both. Interviewee B states that one way around this could be to blend low-cost and high-cost carbon reduction methods instead of purely aiming for high-cost solutions which is unrealistic and not feasible for clients. Interviewee C believes that clients need to understand that a "slightly higher initial cost will be made up in the long run as the buildings will be so much more efficient".

Retrofitting is also seen as a key obstacle by the participants. All three interviewees share the view that retrofitting the existing building stock will be a huge obstacle for the industry. Interviewee A believes that this will take longer than the 2050 target and this could be explained by a worrying statement made by interviewee C which is that large projects are still being built currently, which will require retrofitting to meet net zero standards in the future. Data from all three interviews supports the view from the UKGBC's 2020 report "Energy performance targets for net zero carbon offices" in which it was found that the majority of key stakeholders within the industry believed that the emissions target is achievable for new buildings, however the challenge lies in improving the performance of existing buildings, due to factors such as the complexity of retrofitting (UKGBC, 2020).

Finally, the skills gap within the industry is also highlighted as a major obstacle by interviewees B and C. A lack of industry skills on net zero and what it means to achieve it and how it can be done is an overarching theme. Both interviewees share the view that it is not just a problem for people already working in construction but also the new wave of talent joining the industry. This is a major worry as UK FIRES (2019) states that education can play a major role in reconfiguring society to adopt methods to move towards a net zero carbon future. If the education is non-existent around net zero, the industry faces a huge task in achieving the 2050 target.

Regarding further action that is needed, all three interviewees are in agreement that multiple further measures need to be targeted and implemented in order for the UK construction industry to reach net zero emissions by 2050. A common theme across the results from all three participants is that every contractor should commit to their own internal net zero targets. Interviewee B suggests that this could be done by signing up to the World Green Building Councils net zero carbon declaration which: Challenges business, organisations, cities, states and regions to reach net zero carbon in operation for all assets under their direct control by 2030, and to advocate

for all buildings to be net zero carbon in operation by 2050 (WGBC, 2021).

Interviewee C supports this idea by stating that "Every contractor in the UK needs to set their own net zero target for around 2025 to 2030".

Whilst agreeing that further legislation needs to be implemented by the UK government, interviewees A and B have differing views on what areas should be target in any potential new legislation. Interviewee A believes that both operational and embodied carbon needs to be targeted for legislation to be effective. In contrast, interviewee B believes that the focus should be on more demanding building regulations, around areas such as air tightness. Both interviewees support the view of the CCC (2020) progress report, which is that significant policy gaps need to be filled.

Interviewee C further highlights the need for certification for net zero buildings in the UK, which is encouraging because it is a key measure that needs to be implemented by interviewee B, this will be introduced in 2021. This shows that progress is already being made to implement further measures to help achieve net zero emissions within the UK construction industry.

4.2 Results-Case Study Analysis

A study carried out by the UKGBC in the "Building the Case for Net Zero" report (UKGBC, 2020) uses a commercial project which is representative of new developments in the UK to examine what it will take to meet net zero targets. The study identified the key changes required in terms of design, delivery and operations for buildings to achieve net zero emissions. The cost implications were then analysed by the researcher to assess the feasibility of achieving net zero and any potential obstacles which could arise.

Project overview

16 storey office building.

Situated on an urban infill site.

Typical new city office building with ambitious environmental targets.

Initial design includes retail space, but this has been excluded from the study.

The three design scenarios and the net zero targets being compared are as shown in Tab. 2.

Design scenarios and net zero performance targets(UKGBC, 2020)　　Tab. 2

		Baseline Scenario	Intermediate Scenario	Stretch Scenario
Office	Operational Energy (kWh/m² (GIA)/year)	225 (RIBA-business as usual)	90 (UKGBC-2025 target)	70 (UKGBC-2030 target)
	Upfront Embodied Carbon (LCA module A kgCO$_2$e/m²)	1000 (LETI-business as usual)	600 (LETI-2020 target*)	350 (LETI-2030 target)

Cost results analysis

Design

For the purpose of the study, the design economics for the three scenarios are highlighted in Tab. 3. The stretch design is the scenario which has been designed to achieve the UKGBC's proposed 2030 net zero operational energy target for all new developments and so this is the scenario which will be analysed for the 2050 net zero target as the 2030 target was set with the aim to aid the achievement of the UK's 2050 net zero goal.

Design economics for three design scenarios (UKGBC, 2020)　　Tab. 3

	Baseline	Intermediate	Stretch
Gross Internal Area (m²)	28516	26975	24650
Net Internal Area (m²)	19391	17997	16035
NIA:GIA Overall Efficiency	68%	67%	65%
Total Floors (excl. roof)	17	16	15
Above Ground Floors (excl. roof)	16	15	15
Below Ground Floors	1	1	0
Slab to Slab Height (m)	3.60	3.87	3.85
Structural Frame	Steel Frame and Composite Floor	Steel Frame and CLT Floor	Glulam Frame and CLT Floor

Cost implications

The cost implications of design changes towards net zero are shown in Tab. 4.

Cost change by building element (£/m² GIA) for office design scenarios (UKGBC, 2020)　　Tab. 4

	Baseline	Intermediate		Stretch	
	(£/m²)	(£/m²)	Change from Baseline	(£/m²)	Change from Baseline
1. Substructure	£325	£365	12%	£160~185	-50%~-44%
2. Frame, Upper Floors & Stairs	£450	£625	39%	£730~820	63%~82%

continued

	Baseline	Intermediate		Stretch	
	(£/m²)	(£/m²)	Change from Baseline	(£/m²)	Change from Baseline
3. Roof	£75	£75	—	£75	—
4. External Walls, Windows & Doors	£495	£445	−10%	£495~550	0~11%
5. Internal Walls & Doors	£95	£95	—	£95	2%
6. Finishes & Fittings	£235	£235	—	£235	1%
7. Mechanical, Electrical & Plumbing(MEP)	£730	£745	2%	£745~820	2%~12%
8. Lifts	£110	£120	9%	£130~140	17%~27%
9. Preliminaries; Overheads & Profit; Design & Build Risk	£610	£610	—	£700~775	15%~27%
Total Shell & Core	£3125	£3320	6.2%	£3370~3660	8%~17%

Higher cost is a major obstacle.

As seen in Tab. 4, the overall cost increase for the shell and core of the office design used for the study is between 6.2% to 17% depending on which design scenario is aimed for. This is a considerable increase, especially when considering there is currently no incentive for clients to aim for net zero. However, the increased capital cost should not act as a deterrent for clients as it was found in the data from the interviews conducted, that tenants are currently seeking net zero offices. This increases the feasibility for clients as increased demand for net zero offices will ultimately lead to an increase in achievable rent income. Although a higher initial investment will be required the trends in the market mean it could easily be recovered in the long run.

The stretch scenario is not feasible in today's market due to the lack of incentives and regulation. A cost increase between 8% ~ 17% is simply too high as there is no certainty over what impact a net zero building can have on potential revenue. Interview data collected highlights the fact that it is unrealistic for clients to go for a high-cost net zero design in the current context as there is no reward or certification which guarantees the building will be sought after within the market. However, it should be considered once a certification process and standard is implemented, as potential tenants will want to use operating out of net zero spaces to improve their sustainability image.

Whole life costing makes the numbers look more attractive.

Although higher capital costs initially decrease the feasibility of achieving net zero on commercial projects, the data shown in Tab. 5 shows that a higher initial investment can bring long term financial gains. The data highlights the importance of assessing whole life costing when assessing the feasibility of net zero projects as mentioned by interviewee B who highlighted that there is not enough whole life costing taking place and because the sector evaluates capital costs only, it is having a negative impact on the net zero agenda. Interviewee C also stated that clients need to understand a slightly higher initial cost will be recovered in the long run due to the increased efficiency of the buildings.

A limited whole life costing analysis undertaken for three building elements **Tab. 5**

Baseline Scenario	Stretch Scenario	Cost Change (Over 30 Years)
Gas Boiler	Air Source Heat Pump	Saving of 30%~40%
Suspeneded Ceiling	Exposed Soffit	Saving of 50%~60%
Raised Access Flooring	Solid Timber Flooring	Increase of 3.5%

Unrealistic stretch design

The researcher has the view that the stretch design is unrealistic due to the exclusion of a basement. In the commercial office sector in busy cities such as London, space is at a premium and strict planning regulations mean total floors can sometimes be capped. Therefore, developers simply have no choice but to have multiple basement levels for plant and sometimes even amenities such as changing rooms and showers. The stretch scenario is therefore deemed as highly unachievable. This is a barrier because net zero solutions needs to be realistic and incorporate methods and designs which can be achieved in the UK construction industry.

Conclusion

The data analysis has highlighted the obstacles that the UK construction industry

needs to overcome in order to achieve the net zero targets by 2050. Evidence gathered in chapter 2 suggests that one of the main obstacles facing the industry in terms of reaching the 2050 net zero target is the lack of legislation and policy. This is supported by the data gathered from all three interview participants and therefore is deemed the most prominent obstacle which needs to be overcome.

Another key obstacle which was identified by the interviewees is the lack of net zero certification within the UK. This results in a lack of direction and clarity for developers when considering the development of a net zero built asset as there is no set criteria to aim for.

The data gathered from the interviews also suggests that the 2050 target is achievable within the UK commercial construction sector if progress gathers pace. However, problems could arise if carbon emissions from the manufacturing of materials is included in any industry targets or potential certification.

The data analysis from the case study has highlighted the issue of higher costs being an obstacle facing developers when considering net zero. Although the cost increase may not seem large, as there is currently no incentive to achieve net zero, the increased cost is a deterrent for some clients. However, the data also suggests that when considering the whole life cost of a project, clients can in fact recover higher initial costs in the long run as the increased efficiency of the buildings will bring monetary savings.

5 Conclusions

In conclusion, whilst there has been significant improvement within the commercial construction sector in particular towards achieving net zero emissions on projects, a lot more needs to be done within the industry in order for the 2050 target to be achievable. A reoccurring area of improvement is to further education around net zero and how it can be achieved not only for current industry professionals but the new crop of talent that will enter the industry over the next 10 years. This education will then help clients understand why net zero buildings need to be built as soon as possible and therefore drive the industry to a net zero future. Partnerships between the UKGBC and universities could be formed, and modules could be introduced to push the net zero agenda. Most construction discipline courses have some learning centred around sustainability, and so these could be updated with net zero being the focus of the material being taught.

Another key area is legislation. The UK government need to put legislation in place not only for operational net zero but also for embodied carbon to achieve net zero. There needs to be industry wide collaboration, where the government work with bodies such as the UKGBC and LETI to plug the legislation gaps. The professional bodies and councils should then work with clients, contractors and the supply chain to give the industry a nudge in the right direction with the backing of legislation. The main reason why progress is slow and the target is becoming less achievable by the day is due to the lack of any real incentive. Whether that is certification, legislation or higher rental income or grants from local authorities, there is simply not enough pressure for clients to build net zero buildings. The National Australian Built Environment Rating System UK (NABERS UK) should be utilised as the Australian system has a carbon neutral certification process. If the Building Research Establishment (BRE) and NABERS UK can work together to develop similar certification for UK buildings and in particular commercial spaces, it will go a long way in helping the UK as a whole achieve the 2050 net zero target.

The researcher would recommend further research to be carried out on the same topic with similar objectives, but with a larger sample size in order to improve validity. Similar research could be carried out within alternative sectors within construction such as the residential sector to gauge which sector needs the most focus and pressure to achieve the net zero 2050 target.

Acknowledgements

The author would like to thank all interview participants and Hua Zhong of Nottingham Trent University for her time and support.

References

[1] Annan K. Kofi Annan: We must challenge climate-change sceptics who deny the facts. [Online] the Guardian. Available at: < https://www.theguardian.com/environment/2015/may/03/kofi-

annan-interview-climate-change-paris-summit-sceptics>[Accessed 7 November 2020]. 2015.

[2] Bale T. The Conservative Party : From Thatcher to Cameron. [Place of publication not identified]: Polity Press, 2016.

[3] Ca. Why are carbon savings from commercial properties so important?. [Online] Better Buildings Partnership. Available at: <https://www.managingagentspartnership. co. uk/why-are-carbon-savings-commercial-properties-so-important>[Accessed 3 May 2021]. 2019.

[4] Carter and Childs. Friends of the earth as a policy entrepreneur: the big ask campaign for a UK Climate Change Act. [Online] Taylor & Francis. Available at: <https://www. tandfonline. com/doi/full/10. 1080/09644016. 2017. 1368151? scroll = top&needAccess = true> [Accessed 3 May 2021]. 2017.

[5] Carter and Clements. From "greenest government ever" to "get rid of all the green crap": David Cameron, the conservatives and the environment. [Ebook] Available at: < https://link. springer. com/content/pdf/10. 1057/bp. 2015. 16. pdf> [Accessed 13 August 2021]. 2015.

[6] Carter et al. [ebook] Available at: <https://onf. ons. org/onf/41/5/use-triangulation-qualitative-research>[Accessed 20 August 2021]. 2014.

[7] CCC. UK climate action following the Paris Agreement —Climate Change Committee. [Online] Climate Change Committee. Available at: < https://www. theccc. org. uk/publication/uk-action-following-paris/> Accessed 17 May 2021]. 2016.

[8] CCC. Net Zero The UK's contribution to stopping global warming. [Ebook] CCC. Available at: <https://www. theccc. org. uk/wp-content/uploads/2019/05/Net-Zero-The-UKs-contribution-to-stopping-global-warming. pdf> [Accessed 6 February 2021].

[9] CCC. Reducing UK emissions: 2020 progress report to parliament. [Ebook] CCC. Available at: <https://www. theccc. org. uk/publication/reducing-uk-emissions-2020-progress-report-to-parliament/> [Accessed 8 November 2020]. 2020.

[10] CCC. About the Climate Change Committee [Online] Climate Change Committee. Available at: < https://www. theccc. org. uk/about/> [Accessed 9 May 2021]. 2021.

[11] Davies. The big ask: the story behind the Climate Change Act. [Online] Linkedin. com. Available at: <https://www. linkedin. com/pulse/big-ask-story-behind-climate-change-act-nicholas-davies/> [Accessed 4 May 2021]. 2018.

[12] Department for Business, Energy & Industrial Strategy, 2016. BEES.

[13] Department for Business, Energy & Industrial Strategy. [Online] Assets. publishing. service. gov. uk. Available at: <https://assets. publishing. service. gov. uk/government/uploads/system/uploads/attachment_data/file/972583/2020_Provisional_emissions_statistics_report. pdf> [Accessed 9 May 2021]. 2021.

[14] Downs. p and Down with Ecology-the "Issue Attention Cycle". [Online] Fbaum. unc. edu. Available at: <https://fbaum. unc. edu/teaching/articles/Downs_Public_Interest_1972. pdf> [Accessed 3 May 2021]. 1972.

[15] Evans. Analysis: UK is now halfway to meeting its "net-zero emissions" target | Carbon Brief. [Online] Carbon Brief. Available at: <https://www. carbonbrief. org/analysis-uk-is-now-halfway-to-meeting-its-net-zero-emissions-target > [Accessed 8 May 2021]. 2021.

[16] Forster. Coronavirus lockdown will have "negligible" impact on the climate - new study. [Online] The Conversation. Available at: <https://theconversation. com/coronavirus-lock-down-will-have-negligible-impact-on-the-climate-new-study-143503> [Accessed 7 May 2021]. 2021.

[17] GOV. UK. UK enshrines new target in law to slash emissions by 78% by 2035. [Online] GOV. UK. Available at: <https://www. gov. uk/government/news/uk-enshrines-new-target-in-law-to-slash-emissions-by-78-by-2035> [Accessed 3 May 2021]. 2021.

[18] Guterres A. Guterres: The IPCC Report is a code red for humanity. [Online] United Nations Western Europe. Available at: <https://unric. org/en/guterres-the-ipcc-report-is-a-code-red-for-humanity/> [Accessed 15 August 2021]. 2021.

[19] Harrabin. Climate change: UK to speed up target to cut carbon emissions. [Online] BBC News. Available at: <https://www. bbc. co. uk/news/uk-politics-56807520> [Accessed 5 May 2021]. 2021.

[20] Hawking S. 30 of the most impactful Climate Change Quotes—Curious Earth | Environment & Climate Change. [Online] Curious Earth | Environment & Climate Change. Available at: < https://curious. earth/blog/climate-change-quotes/> [Accessed 11 November 2020]. 2007.

[21] Hill. Tighter policies necessary if the UK is to achieve net-zero by 2050. [Online] Property-week. Available at: <https://www. propertyweek. com/analysis-resi-and-data/tighter-policies-necessary-if-the-uk-is-to-achieve-net-zero-by-2050/5111018. article> [Accessed 13 November 2020]. 2020.

[22] House of Commons Library. UK carbon budgets. 2019.

[23] House of Commons. Meeting UK energy and climate needs. [Online] Google Books. Available at: < https://books. google. co. uk/books? id = t1tmCsydB _ 0C&dq =%22International＋Symposium＋on＋the＋Stabilisation＋of＋greenhouse＋gas＋concentrations＋-＋Report＋of＋the＋International＋Scientific＋Steering＋Committee%22&source = gbs _ navlinks _ s> [Accessed 3 May 2021]. 2006.

[24] Institute for government. The Climate Change Act (2008). Instituteforgovernment. 2020.

[25] IPCC. About—IPCC. [Online] Ipcc. ch. Available at: < https://www. ipcc. ch/about/> [Accessed 1 May 2021]. 2021.

[26] Jacobs. [Online] High pressure for low emissions: how civil society created the Paris climate agreement. Available at: <https://onlinelibrary. wiley. com/doi/abs/10. 1111/j. 2050-5876. 2016. 00881. x> [Accessed 3 May 2021]. 2016.

[27] Kwarteng. [Online] Available at: <https://www. pbctoday. co. uk/news/energy-news/construction-net-zero/93226/> [Accessed 8 May 2021]. 2021.

[28] Lehne and Preston. Making concrete change: innovation in low-carbon cement and concrete. [Online] Chatham house-international affairs think tank. Available at: <https://www. chathamhouse. org/2018/06/making-concrete-change-innovation-low-carbon-cement-and-concrete > [Accessed 4 May 2021]. 2018.

[29] LETI. LETI-climate emergency design guide. LETI. 2020.

[30] LSE. What is the 2008 Climate Change Act? — Grantham Research Institute on climate change and the environment. [Online] Grantham Research Institute on climate change and the environment. Available at: <https://www. lse. ac. uk/granthaminstitute/explainers/what-is-the-2008-climate-change-act/> [Accessed 6 November 2020]. 2020.

[31] Naoum S. Dissertation research & writing for construction students. London: Routledge. 2013.

[32] NASA. Carbon Dioxide Concentration | NASA Global Climate Change. [Online] Climate Change: Vital Signs of the Planet. Available at:

< https://climate. nasa. gov/vital-signs/carbon-dioxide/> [Accessed 8 August 2021]. 2021.

[33] Noble and Heale. Triangulation in research, with examples. [Ebook] Available at: <https://ebn. bmj. com/content/22/3/67. long> [Accessed 4 August 2021]. 2019.

[34] Patton. Enhancing the quality and credibility of qualitative analysis. [Ebook] Available at: <http:// europepmc. org/backend/ptpmcrender. fcgi? accid=PMC1089059&blobtype=pdf> [Accessed 4 August 2021]. 01999.

[35] Priestley S. A house of commons library briefing paper on the UK carbon budgets. [Ebook] Available at: <https://commonslibrary. parliament. uk/research-briefings/cbp-7555/> [Accessed 6 November 2020]. 2019.

[36] Shepheard. UK net zero target. [Online] Instituteforgovernment. org. uk. Available at: < https://www. instituteforgovernment. org. uk/explainers/net-zero-target> [Accessed 1 May 2021]. 2020.

[37] UK FIRES. Absolute zero. [Ebook] Available at: < https://ukfires. org/absolute-zero/> [Accessed 1 August 2021]. 2019.

[38] UKGBC. Climate change -UKGBC -UK Green Building Council. [Online] UKGBC -UK Green Building Council. Available at: <https://www. ukgbc. org/climate-change/> [Accessed 10 November 2020]. 2020.

[39] UKGBC. Energy performance targets for net zero carbon offices. [Ebook] Available at: <https:// www. ukgbc. org/ukgbc-work/net-zero-carbon-energy-performance-targets-for-offices/> [Accessed 5 November 2020]. 2020.

[40] UN. Sustainable development goals | UNDP. [Online] Available at: <https://www. undp. org/content/undp/en/home/sustainable-development-goals. html> [Accessed 8 November 2020]. 2020.

[41] UNFCCC. [Online] UNFCCC. Available at: <https:// unfccc. int/process-and-meetings/the-paris-agreement/the-paris-agreement> [Accessed 7 November 2020]. 2020.

[42] Watkins and Hochlaf. Skills for a green recovery: A call to action for the UK construction industry. [Ebook] Available at: < https:// www. ippr. org/research/publications/skills-for-a-green-recovery > [Accessed 9 August 2021]. 2021.

[43] WGBC. The Net Zero Carbon Buildings Commitment | World Green Building Council. [Online] World Green Building Council. Available at: < https://www. worldgbc. org/thecommitment > [Accessed 14 August 2021]. 2021.

[44] WSP. Delivering Net Zero. [Online] WSP. Available at: < https://www. wsp. com/en-GB/news/2020/bright-blue-and-wsp-net-zero> [Accessed 5 November 2020]. 2020.

Planning for the Future: the Rise of Robotics and Artificial Intelligence in UK Towns and Cities

April Baker[1]   Hua Zhong[2]

(1. Graduate Project Manager, Arcadis UK, London EC3M 4BY, United Kingdom;

2. School of Architecture, Design, and the Built Environment, Nottingham

Trent University, Nottingham, UK)

【Abstract】 Technology is changing the fabric of UK towns and cities. This study centres on how robotics and artificial intelligence are currently being used in an urban context and how these will impact our "utopian" settlements of the future. Robotics and artificially intelligent technologies can revolutionise UK towns and cities with their utilisation within urban infrastructure systems and smaller localised construction projects. The government have recognised the importance of these technologies within the construction industry and have produced numerous industry related publications to boost utilisation. Recent government publications have been scrutinised identify if these government publications are aligning with current academic literature and current industry practice. RAI technologies are likely to become engrained within our urban spaces and transform these evolving ecosystems of human, metal and machine. If the government embraces the current digital revolution and paves the way for the construction industry through funding, legislation, and best practice on public sector projects real transformative change will be achievable in our dynamic urban spaces.

【Keywords】 Robotics and Artificial Intelligence; Digital Construction; Urban Infrastructure; Industrial Strategy

1 Introduction

Industry 4.0 is creating a world in which virtual and physical systems are intwined through a complex network of neural and electronic systems. This has started to reshape the towns and cities in which we live. To fully enable and integrate the technological transition we are undergoing it is important that the government lays the necessary foundations to enable implementation. Academics have deemed robotics and artificial intelligence (RAI) to be key technologies in the development of smart cities[33]. This research aims to address the lack of academic works focusing on how industrial practice can be shaped by UK government policy.

1.1　Background to the Research

Technology is rapidly transforming today's society and already plays an integral part of our everyday lives. Despite the dawn of the "fourth industrial revolution", the UK construction industry has seen a systemic absence of innovation[24]. There has been a slight change in recent years as a small number of RAI technologies have started to be successfully implemented[12]. Academia has revealed that the construction industry has begun to benefit from technological developments; examples from the field of robotics and engineering include the emergence of advanced robotics in off-site manufacturing[19] the development of additive manufacturing and 3D printing and the use of unmanned aerial vehicles (UAVs) for the purposes of progress monitoring, site surveillance and surveys[36]. In addition to this connected and autonomous vehicle (CAV) technology has been used in the construction of plant systems[5]. Furthermore, advances in the field of AI such as the Internet of Things (IoT) has been used by the construction industry to create the Industrial Internet of Things (IIoT). AI has been utilised effectively in numerous towns and cities across the world to improve efficacy and alleviate temporal restraints on construction[23, 34]. Despite the UK being a recognised leader in technological innovation its cities are far

behind global competitors. Academics have stated that this pitfall is predominantly due to a lack of physical infrastructure[9, 31].

1.2　Research Rationale

For the UK to revitalise and reinvent its towns and cities through the implementation of RAI technologies it is crucial that industrial professionals and the government realise and work towards shared goals. The aim of this study is to uncover whether there is currently an alignment in thinking by scrutinising the prevailing trends from the aforementioned sectors. This research will uncover if government policy is rooted in physical reality and has the possibility to insight real change on a grand scale in UK towns and cities.

1.3　Justification for the Research

Research has shown that the implementation of digital technologies in the UK could lead to an additional 80000 jobs and an increase of £232 billion to our GDP by the year 2030[3]. Despite RAI technologies having the potential to revolutionise the construction industry and with it the fabric of our towns and cities, the farmer review[14] revealed a worrying prognosis on the current state of affairs. The independent review urged the construction profession to address its numerous pitfalls; including poor productivity levels and a negative industry reputation.

The UK government introduced an industrial strategy in 2017 which detailed their plans to increase investment into the R&D of RAI technologies. Despite the government committing £100 billion from 2017~2021 into this study area, there was a lack of clarity regarding how this funding was used in reality on the ground to benefit both industry and society[1, 2]. This research comes at a pivotal moment in 2021 when the government published several topical policy papers. An analysis of key UK policy documents will follow to determine if these plans give specific details on the implementation of RAI technologies.

1.4 Research Aims and Objectives

The following statement provides the aim of this research project:

"To assess key UK government policies regarding the construction sector and future UK urban infrastructure relating to RAI. The result of this will be to identify gaps in the existing literature between current knowledge and levels of implementation and stipulate how this will affect the 'utopian' settlements of the future."

To achieve the aim of this research, the following objectives have been set:

(1) Undertake a review of UK government policy and highlight their long-term strategy for the implementation of RAI.

(2) Identify areas where UK government policy is misaligned with current academia through a systematic quantitative analysis of the data.

(3) Uncover the enablers and barriers of RAI technologies through interviews with construction industry professionals.

(4) Make recommendations to the UK government and industry professionals and propose areas for additional research to account for any identified shortcomings.

2 RAI Implementation in UK Towns and Cities

There is an array of RAI technologies that are currently being used and developed to aid urban infrastructure and construction projects in the UK[21]. Academics have revealed the key technologies coming to the fore which include construction robotics, AI, additive manufacturing, BIM, CAVs, virtual reality, UAVs and IIoT[12, 17]. Recent studies have shown that the utilisation and adoption of BIM and autonomous vehicles in the UK seems to be progressing at a faster rate when compared to the other RAI technologies[30, 35]. This is largely due to the government backed funding these sectors have recently received. Government funding and policies appear to drive best practice within the construction sector as their initiatives inspire other landowners and developers to invest and adopt these technologies. There also appears to be a geographic disparity between the north and south of the UK along with one between that of the town and the city regarding RAI adoption levels.

London is undoubtedly the most prominent site for investment and the utilisation of RAI technology and is well on its way to becoming a smart city[10, 26, 28]. Edinburgh, Reading, and Manchester are also other key cities of note for their RAI advancements[7, 15]. Urban spaces in the UK are not advancing at the same pace as large cities attract greater financial investment and also have the advantage of

"Brain Gain" as skilled workers are drawn to these sites for employment opportunities. This development process seems to be stuck in a negative feedback loop as the cities with the most advanced existing infrastructure are evolving at an even quicker rate than their counterparts. For UK towns and cities to equally reap the benefits of RAI the government needs to be focusing on improving the existing services within these localities.

3 UK Government Policy

The Construction 2025 strategy was published by the Department for Business, Innovation and Skills[13], and marks one of the government's initial commitments to revolutionise the UK construction industry through innovation, research, digital design, new technologies, advanced materials, and smart construction. Following on from this, the National Infrastructure Delivery Plan (2016) (NIDP) was announced by the Infrastructure and Projects Authority[2] and specified their long-term plans until 2021 this year. The policy detailed the UK wide strategy for infrastructure and government investment over this timeframe. Independent analysis of the NIDP reveals that RAS technology is cited only once within the document and RAI related investments, or strategies are not mentioned at all within this body of work. The UK Industrial Strategy[3] followed these previous publications and aimed to boost employment and earning power through investing in industries, skills, and infrastructure[16]. Amongst the policies "Four Grand Challenges"—artificial

intelligence and data economy were highlighted as key areas of focus[4]. This policy specifically refers to AI a total of 65 times and robotics five times. This inquiry highlights how RAI technologies gained increasing attention within UK policy documents. This publication saw the government specifically relaying its strategies for the use of RAI technologies in the future.

However, during the Covid-19 pandemic the Government replaced the 2017 strategy with the National Infrastructure Strategy (NIS) (2020)[16] and Build Back Better: Our Plan for Growth (2021)[32]. These address the new socio-economic and political landscape of the UK in the aftermath of Brexit and bolstered national climate change goals. Whilst both strategies highlight the importance of technological innovation on our road to recovery RAI technologies are mentioned only in passing once in the NIS and not at all in the other policy document. Build Back Better (2021) does note that the government will be supporting "Digital Infrastructure" to improve towns and cities across the UK. Fig. 1 depicts how investment by the government in several local infrastructure project will benefit different regions; the paper does not however mention if or how RAI technologies will be used on these projects. These government policies although focussed on digital infrastructure clearly lack specific direction regarding the key technologies that will bring about urban development and growth. There seems to be a vast disparity between the promises made in earlier policy documents and what is currently being implemented and proposed in the future. Given the current circumstances, it is unsurprising that the government's post-pandemic policies lack clarity however elaboration on how they will achieve their RAI technological ambitions are much needed so that a clear plan for growth along with sufficient funding can be established.

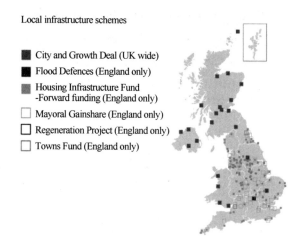

Local infrastructure schemes

■ City and Growth Deal (UK wide)
■ Flood Defences (England only)
■ Housing Infrastructure Fund
  -Forward funding (England only)
□ Mayoral Gainshare (England only)
□ Regeneration Project (England only)
□ Towns Fund (England only)

Fig. 1　Highlights local infrastructure schemes
the government intend to support
（HM Government 2021）

4　Long-term Government Strategy

The government have produced numerous policy documents over the last 10 years relating to RAI implementation. Whilst these clearly reveal a long-term commitment to RAI implementation in the construction sector in the years to come the details of how these technologies will be rolled out in our towns and cities are not specified. National RAI implementation is an undertaking that will require years of foreplaning and coordination between numerous stakeholders. We are on the precipice of an RAI revolution, and it is critical that the government leads the way by pro-

viding clear strategic direction during this technological transition.

5　Research Methodology

The methodology detailed below has been designed in a way that allows for a sequential flow from one research objective to another. This research was developed using the widely established and acclaimed research onion model to produce a coherent and robust methodology[29]. This research philosophy was based on pragmatism. This stance accepts that there are many ways to understand and interpreting the world in which we live and hence undertaking research. Right from the outset this allows for research to be grounded as one amongst many true realities[25]. Pragmatism enables value-driven research and means that problems, practice and relevance are central to the research design[18].

5.1　Covid-19 Response

The research strategy was significantly modified as a result of the working conditions due to the covid-19 pandemic. Finding new industry contacts was made extremely challenging and hence the research participants consisted of existing professional colleagues and contacts. This still allowed for a valid range of results to be collated from experienced industry professionals, but the possible sample size was reduced as social-distancing rules meant that all interviews were restricted to virtual online communication.

5.2　Research Strategy

This mixed methods research follows a

sequential explanatory research design. This was deployed as a tool to examine change, by allowing initial findings to influence the questions and answers that followed[11]. This study commenced with a content analysis of key UK government policy documents. This led to insights which were then questioned through an initial quantitative survey, which in turn lead to key themes being identified which were further unpicked through qualitative semi-structured interviews.

5.3　Government Policy vs Academic Literature

Analysis of key UK government policy documents was carried with the following key terms being searched for: construction robotics, AI, additive manufacturing, BIM, UAVs, VR, CAVs and IIoT. Following this, a database search was carried out using the RAI-specific technologies noted above to locate academic papers that had "construction" in the title and keywords. This was to reveal the volume of academic literature published since 2015 on the related technologies in the construction industry. These two quantitative stages were designed to be easily and accurately compared against one another. Academics speak highly of the use of such techniques for it allows real data-driven conclusions to be drawn from live information[8, 20, 22]. It also bridges the gap between traditional narrative methods and meta-analysis[27].

5.4　Online Survey

To contextualise the initial findings from the government policy and academic literature analysis, an online survey was produced with a blend of open and closed questions. This afforded respondents the opportunity to elaborate on some of their answers which allowed for greater insight into the research topic.

The survey questions were centered around uncovering the main drivers for the implementation of these technologies and aimed to highlight the factors that facilitate RAI adoption in the construction of UK towns and cities.

5.5　Semi Structured Interviews

Upon identifying key themes from the online survey, a set of semi-structured interviews were used as a framework for further development. The interviews were based on a set of predetermined questions, but the conversations were kept flexible to allow for points of interest to be discussed in greater detail. This self-expression is said to lead to more nuanced understandings being wrought[6]. By allowing the interviewee to dictate the flow of conversation within the remit of the research topic key themes can be brought to light to enrich research findings.

Three semi-structured interviews were held with industry professionals who were based in the city of London. To achieve a range of perspectives from the limited sample group the individuals where all from different compagnies with diverse professional backgrounds. Interviewee one was a director at an international design and consultancy firm who has 20 years' experience as a project manager in high-end and built-to-rent commercial development. Interviewee two held the position of Digital Transformation Lead and had been at their multi-national firm for nine years. This individual worked in the commercial cost management team but also aided the wider company in developing the use of cloud-based digital tools. The final interviewee was a Business Development Partner at a property development and investment company based in the UK. This individual had worked for their firm for six years and provided niche insights from a developer's perspective.

6 Research Findings

6.1 RAI Research and Government Policy

Initial content analysis of Construction 2025 (2013), National Infrastructure Strategy (2020) and Build Back Better (2021) high-lighted how key RAI technologies are referenced only a handful of times. The terms "Robotics" and "Artificial Intelligence" are mentioned only once respectively in the NIS and not at all in the other two papers. Tab. 1 shows a quantitative breakdown for RAI terms used per paper. To contextualise these findings against current academic literature on the subject an article search engine was used to identify papers published internationally since 2015 for relevance. This was done to identify trends between current academia and UK government policy documents. Note that papers were not filtered to only UK-published papers as it was assumed that all recent literature on RAI could have a beneficial influence on UK government policies.

Summary of content analysis of three UK government policy documents. Word count has been taken for each keyword and made a percentage of the total word count (author's own) Tab. 1

	Construction 2025 (2013)	National Infrastructure Strategy (2020)	Build Back Better (2021)
Total Word Count	17056	34050	26071
Technology	23 (0.0013%)	23 (0.0007%)	27 (0.0010%)
Robotics	0 (0)	1 (0.0001%)	0 (0)
Artificial Intelligence (AI)	0 (0)	1 (0.0001%)	0 (0)
Additive Manufacturing	12 (0.0007%)	11 (0.0003%)	3 (0.0001%)
BIM	25 (0.0015%)	0 (0)	0 (0)
Virtual Reality	1 (0.0001%)	0 (0)	0 (0)
UAVs	0(0)	0(0)	1(0.0001%)
CAVs	0 (0)	1 (0.0001%)	1 (0.0001%)
IIoT	1 (0.0001%)	1 (0.0001%)	0 (0)
Total	62 (0.0036%)	38 (0.0011%)	32 (0.0013%)

It became clear that robotics and AI are areas of significant academic focus. Fig. 2 highlights how researchers believe these to be important technologies for the construction industry now and in the future. The other RAI technologies identified also received academic attention—around $1\% \sim 2\%$ each of the total number of papers published during the period of 2015 to current; robotics and AI take up 10% and 5% respectively.

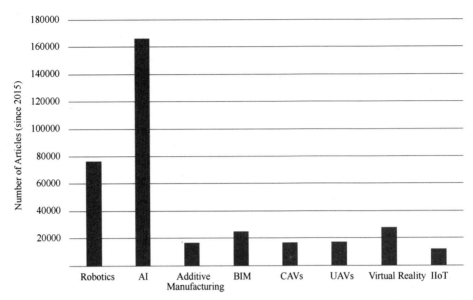

Fig. 2　Number of construction—related articles published since 2015 on RAI technologies（author's own）

When plotted against one another there was shown to be no correlation between the academic interest given to certain RAI technologies and the validation given in UK government policy papers（Fig. 3）. This means that the government haven't successfully aligned themselves with recent academic research which will ultimately impede RAIs potential for adoption. Despite the government recognising the importance of RAI and pledging investments into this emerging sector there are no strong policies in place to ground their aspirations. If the UK government truly want to support the digitisation of the construction industry and with it our towns and cities, they need to act produce more robust, coherent, and specific policies.

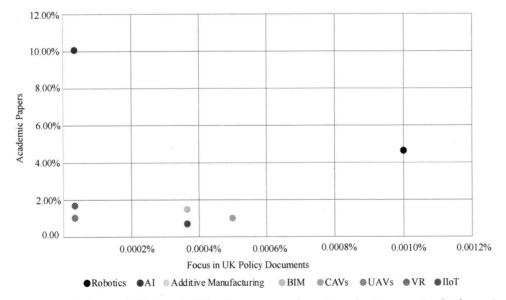

Fig. 3　Graph showing RAI focus in UK policy papers vs focus in academic papers（author's own）

6. 2　Investigative Survey

The profile of respondents for the online survey shows that an array of results was collected from individuals with differing industry experience. Tab. 2 highlights that 39% of respondents had between one—and five—years' experience in the industry. Almost 20% of the sample group had over 20 years' experience which meant that a large wealth of information could be drawn upon and translated into insightful survey responses.

Breakdown of respondents by the years of experience in the construction industry (author's own)　Tab. 2

Time in the Industry	Respondents	Percentage (%)
1~5 Years	11	39
6~10 Years	8	29
11~15 Years	2	7
16~20 Years	2	7
Over 20 Years	5	18
Total	28	100

Respondents also came from a range of professional backgrounds. Project managers formed the largest part of the sample pool, followed by quantity surveyors and subsequently directors. By having an array of individuals from distinct disciplines within the construction industry niche perspectives and understandings could be explored.

Upon analysing the data from the online survey responses clear quantitative patterns emerge. Fig. 4 portrays a breakdown of the technologies that have been used by respondents on the construction projects they have worked on. Additive manufacturing has been the technology most widely used by the sample pool. Interestingly additive manufacturing is the most frequently mentioned term out of all the three policy documents analysed. This could imply that increasing government attention and initiatives lead to higher levels of RAI implementation in the construction industry.

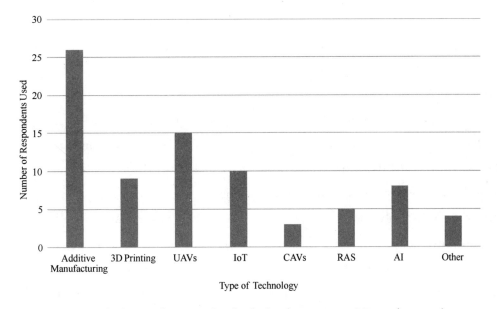

Fig. 4　Graph plotting the types of technologies the survey participants have used on
UK construction projects (author's own)

Fig. 5 displays what the participants believe to be the most important factor regarding the implementation of RAI technologies. It is clear that "Client Objectives" are seen to be the main driver behind RAI implementation. This indicates that the client ultimately dictates the level of RAI used on individual construction projects. "Capacity within the Project Team" and "Knowledge" are seen as having the least impact on technological implementation. This suggests that the construction industry in the UK already has the capabilities to utilise these technologies.

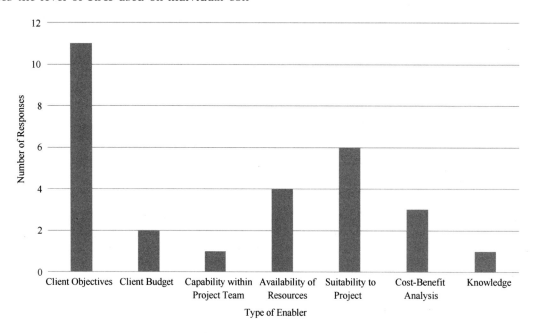

Fig. 5   Graph displaying the factor that survey respondents believe affect RAI implementation (author's own)

6.3 Qualitative Data

Following on from the quantitative analysis, thematic analysis, was deployed to allow for meaningful qualitative results to be produced. Questions relating to key UK government legislation revelled that not a single participant was aware of any of the main publications regarding RAI. Only one participant was able to name any legislation, and this related to the adoption of BIM on government funded projects. The survey revealed that every participant believed that the government was not doing enough to support the adoption of these new technologies in the construction industry. Exerts from the survey below highlight this: "I think the government could do more. The construction industry does seem to be waking up to the benefits of RAI but more public projects often lead the way in demonstrating new technologies. This isn't currently happening."

"The government as the largest client of the construction industry should ask the constructor to implement these technologies."

"There could be more grants to encourage innovative working."

The survey also exposed how many construction professionals believe that RAI implementation is occurring at a higher rate in UK towns and opposed to cities. Almost all the research participants work for construction firms based in London. This suggests that these head offices are mostly dealing with projects in large cities as opposed to towns. This could be due to a host of reasons but is likely to be linked to the fact that large cities generally attract more investment due to receiving more public and private sector funding.

An analysis of the online survey revealed several key themes: the client is key, insufficient government legislation, uneven RAI adoption and the city trumps the town.

6.4 Semi Structured Interviews

Uneven RAI Adoption

The online survey portrayed that RAI technologies are currently being adopted at different rates across the UK construction industry. Interviewees were asked to elaborate on what the most widely adopted RAI technology currently used in the construction industry was. Interviewee one answered:

"It's got to be BIM; ever since the government made a minimum of level 2 BIM mandatory on all their public construction projects its use has sky-rocketed. This legislation definitely inspired other developers to adopt this best practice. All the private sector projects I'm currently working on make use of this technology."

This finding was supported by interviewee two and three who also cited BIM and additive manufacturing as being the most widely adopted technologies currently used in the UK.

Interviewee one noted that their firm had recently produced a new digital strategy detailing the company's commitment to researching and adopting key RAI solutions. Interviewee one commented:

"I think the industry is at a crunch point now where companies that start to adopt these technologies will gain a critical edge over competitors."

This reinforces the modernise or die mentality. If large corporations in the UK commit to adopting these digital solutions this could spark other industry players to adopt new business strategies that incorporate the much—needed use of RAI.

Client is Crucial

All participants corroborated that client objectives were a central factor in driving RAI implementation. Interviewee three who is client side commented:

"All clients ultimately have their own objectives; (x) strive to deliver innovative and impactful projects which aim to increase the user's wellbeing and contribute to the social values of the communities we operate in." (x - blanked out corporation's name).

It was also highlighted that:

"We are constantly looking for process improvement and as we produce more complex projects more complex digital tools are needed."

Interviewee three also noted that their company has a rigorous internal system, with each potential project being submitted to their main board for review before full financial backing is given to a project. Each project had to demonstrate that certain criteria has been met such as sustainability commitments, quality standards and financial parameters. It seems that clients have started to recognise the intrinsic and extrinsic benefits digital tools bring both as stand-alone tools for innovation and as technologies that aid the achievement of other business objectives. Clients should continue to endorse the use of RAI technologies as this is likely to start a snowball effect within the industry.

Insufficient Government Legislation

The analysis of government policy documents along with the online survey clearly revealed that the current legislation regarding RAI adoption in the UK construction industry is lacking. When elaborating on the main enablers and barriers regarding RAI adoption interviewee two stated:

"Government legislation is critical in getting the construction industry to adopt their digital agenda. However, with increasing emphasis on sustainability and reducing our carbon footprint at some point companies need to take it on themselves to drive these innovations."

It seems that technological innovation requires multiple stakeholder input for full transformation to be achieved. Interviewee one and three also noted that whilst clearer government legislation is needed, government funding is also essential.

Interviewee two noted:

"Investment into robotics and AI is yet to be seen on the macro level which makes the industry speculative about using them in a day-to-day situation."

Participants believed that the most critical RAI enabler was client objectives. It was noted that cost—benefit analysis should also be seen as key for ultimately the technology needs to be financially viable before it is considered for use on projects. Technological barriers appear to stem from inadequate government legislation, a lack of cohesion between urban construction stakeholders, and an unwillingness from UK construction companies to drive RAI implementation.

If the government wish for RAI to be effectively rolled out within the UK construction industry updated legislation is needed. This needs to define what funding opportunities are available and denote what exact RAI adoption standards are to be achieved by the industry.

The City Trumps the Town

The interviewees all noted that RAI adoption was not happening at even rate across towns and cities in the UK. Interviewee two stated:

"The disparity in profit margins between large infrastructure companies and SMEs is creating a divide between city infrastructure development and general construction works across the country."

This participant highlighted that large construction corporations often win leading projects where the client is more likely to entertain technological innovations due to the sheer size and complexity of the project. This trend is being seen in cities over towns as this is where more investment is being placed in the UK. Interestingly all the correspondents believed that the covid-19 pandemic had highlighted the importance of these emerging technologies. It was suggested that SMEs have been forced to adopt new technologies to survive economic uncertainty.

Clearly if the government wish to develop the town in tandem with the city more needs to be done to support both investment and development into these urban settlements.

The Future Urban Landscape of the UK

All participants agreed that RAI technologies will undoubtably change the UK towns and cities of the future. Many noted the numerous benefits RAI would bring to urban infrastructure including heightened health and safety, increased clash detection and higher productivity. It was discussed that future"Utopian" settlements are a long-term goal and must be planned and prepared for. The very fabric of the town and city is likely to be transformed as RAI is ingrained within the living, functioning city and not just in urban construction methods. How our cityscapes will physically and virtually work is still to be determined, however if true positive implementation is to occur the government, construction professionals, and clients need to align their visions and buy-in to this new shared reality.

7 Recommendations

In light of the research findings displayed above the following recommendations are made:

UK Government

Produce an updated, clear, and detailed policy document for the construction industry that gives critical attention to each emerging RAI technology. This document should implement specific legislation relating to RAI adoption targets based on the size of construction corporation.

Adopt a long-term perspective for the urban development of towns and cities and produce a UK roadmap that pinpoints how differing geographical locals will receive RAI investment and development.

Collaborate with academics and the construction industry so that future RAI policies can be based on robust and practicable knowledge and research.

Construction professionals

Individuals in the field should work to upskill themselves and gain knowledge regarding the use of RAI so they can understand, foster, and boost the use of these technologies.

Professionals where possible should educate uninformed clients on the benefits of RAI and support and recommend their use.

The construction industry should champion this technological shift toward innovation and recognise the intrinsic and extrinsic values of RAI.

Additional Research

Undertake surveys and interviews with the public to gauge widespread perceptions relating to RAI technologies and see how adoption rates relate to customer demand for these services.

Measure the uptake of RAI technologies amongst UK SMEs against large corporations to uncover if the covid-19 pandemic has impacted adoption rates.

Follow up this research by looking at how the UK compares to international competitors working to advance RAI capabilities. This will highlight if the UK are maintaining a pole position in the race for digitalisation.

8 Conclusions

It was highlighted how there appears to be no correlation between the amount of relevant academic literature produced on key RAI technologies and current government policies. This knowledge gap if left unchecked is likely to affect the development of our "Utopian" settlements of the future. These settlements are likely to incorporate unprecedented levels of technological innovation. This urban transition must however be seen a long-term goal that requires careful planning and coordination from stakeholders. These technologies have the potential to revolutionise UK towns and cities as RAI start to be deployed within urban infrastructure systems and not just isolated construction projects. Devices such as CAVs are likely to become an engrained part of our urban spaces and form a new component to these transformative ecosystems of metal, human, and machine. The future is not certain and how our cityscapes and town plans will look, and function in reality is still to be determined. What is clear is that if the government lay down the necessary foundations for this digital revolution and pave the way through funding, legislation, and best practice on public sector projects real transformative change will be possible.

8. 1 Key Findings

Client is crucial

The qualitative research conducted showed that many industry professionals believed that the clients of construction projects were the biggest factor dictating the use of RAI in the industry. Findings revealed that clients had begun to acknowledge the intrinsic and extrinsic benefits of these technologies. It was suggested that if clients continue to endorse the implementation of RAI technologies on construction projects a wider seismic shift could occur within the industry. This best—practice through a top-down approach could lead to these transformative technologies becoming more engrained in mainstream construction processes.

Insufficient government legislation

One of the key findings from this research has been that the government to date have not produced adequate or clear legislation to fully support the role out of RAI technologies in UK urban infrastructure in towns and cities. Government publications only stated the "what" and failed to detail the "how" and "where" these technologies will be implemented. The government cannot be held fully accountable for these issues, the construction industry must also start to push for these changes from within. However, the government must be seen as the main external factor dictating chance and hence need to address these shortcomings.

Uneven RAI Adoption

The mixed methods deployed during the data analysis portrayed how RAI technologies are not currently being rolled out at an even pace across the UK construction industry. One size fits all approach will not be sufficient in supporting the role of RAI in towns and cities across the UK. Construction firms were advised to take matters into their own hands and start driving technological innovation from within the industry. Bespoke management plans from key industry players should be produced and used to drive RAI implementation in areas where their businesses are lacking. This is likely to have a knock-on effect within the industry as other firms would be inspired to follow suit and smaller firms would be encouraged to upskill in order to stay competitive within the market.

City Trumps the Town

It was highlighted that UK cities are developing and making use of RAI technologies at a faster rate than towns. It was seen that large construction corporations tend to win complex and costly projects based in cities. The nature of these projects meant that not only was the client more willing to adopt RAI technologies but increased budgets and capabilities within the project team allowed for greater innovation. It was suggested that the covid-19 pandemic had highlighted the importance of RAI and that small and medium sized construction companies had been forced to adopt new technologies in order to survive economic uncertainty.

8.2 Research Contribution

This work has highlighted how government policy is a vital tool that can be used for real chance. The government not only act as a public sector role model but have the power to also drive innovation in the industry through legislation, R&D and funding opportunities. This work has revealed that the government need to work to align themselves with the construction industry and academics so that a unified plan for chance can be established. Buy-in from all stakeholders is needed for RAI implementation to be rolled out in the construction of UK towns and cities.

8.3 Limitations

It must be noted that data collection methods were largely dictated and at times hindered by the covid-19 pandemic. The literature that has been appraised came predominantly from online. Efforts were made to collect original primary data at a time when social interaction was stunted. As a result, the survey and semi-structured interviews consisted of a smaller sample size than planned. None the less existing contacts with a variety of industry professionals helped gain a variety of results which culminated in new and insightful conclusions.

References

[1] Armitt J. Letter to Rt Hon Philip Hammond. 2019.

[2] Authority I and P. National infrastructure delivery plan 2016-2021. 2016.

[3] BEIS. Industrial strategy: building a Britain fit for the future. 2017.

[4] BEIS. Post-pandemic economic growth: industrial policy in the UK. House of Commons, Lon-

don，UK，2021.

[5] Bo Yu，Fan Bai，Dressler F. Connected and Autonomous Vehicles. 2018,22:4-5. doi：10. 1109/MIC. 2018. 032501510.

[6] Brinkmann S，Kvale S. InterViews：Learning the Craft of Qualitative Research Interviewing，3rd ed. SAGE Publication Ltd. ，London，UK,2014.

[7] Chui M，Manyika J，Miremadi M. What AI Can and Can't Do (Yet) For Your Business. 2018，34:1-4.

[8] Collins JA，Fauser BCJ. Balancing the strengths of systematic and narrative reviews. 2005，11：103-104. doi：10. 1093/humupd/dmh058.

[9] Cotton B. Why are UK cities finding it so hard to become smart? | Business Leader News. https://www. businessleader. co. uk/why-are-uk-cities-finding-it-so-hard-to-become-smart/69011/. Accessed 10 Jul 2021.

[10] Cousins S. NHS nightingale hospital，East London——how BIM played its part. http://www. bimplus. co. uk/projects/nhs-nightingale-hospital-east-london-how-bim-playe/. Accessed 10 Jul 2021,2020.

[11] Creswell J. Research design：qualitative，quantitative，and mixed-method approaches，3rd ed. SAGE Publications Ltd. ，London，UK,2009.

[12] De Schutter G，Lesage K，Mechtcherine V，et al. Vision of 3D printing with concrete—technical，economic and environmental potentials. 2018(112):25-36.

[13] Department for Business I&S. Construction 2025. HM Government，London，UK,2013.

[14] Farmer M. Modernise or die：the farmer review of the UK construction labour model. Construction Leadership Council，London，UK,2016.

[15] Garcia F. UK artificial intelligence analysis 2020. https://thedatacity. com/insight/uk-artificial-intelligence-analysis-2020/. Accessed 10 Jul 2021.

[16] Government H. The UK's industrial strategy. https://www. gov. uk/government/topical-events/the-uks-industrial-strategy. Accessed 21 May 2021.

[17] Guerra BC，Leite F，Faust KM. 4D-BIM to enhance construction waste reuse and recycle planning：case studies on concrete and drywall waste streams. 2020 (116)：79-90. doi：10. 1016/j. wasman. 2020. 07. 035.

[18] Kelemen M，Rumens N. Anoe introduction to critical management research，1st ed. SAGE Publications Ltd. ，London，UK,2008.

[19] Küpper D. Advanced robotics in the factory of the future，1st ed. Boston Consulting Group，Boston：MA,2019.

[20] Lee A，Kamler B. Bringing pedagogy to doctoral publishing. 2008(13):511-523. doi：10. 1080/13562510802334723.

[21] Macrorie R，Marvin S，While A. Robotics and automation in the city：a research agenda. 2021(42)：197-217. doi：10. 1080/02723638. 2019. 1698868.

[22] McGrail MR，Rickard CM，Jones R. Publish or perish：a systematic review of interventions to increase academic publication rates. 2006(25)：19-35. doi：10. 1080/07294360500453053.

[23] Media O. O'reilly artificial intelligence conference 2019-London，United Kingdom. O'Reilly Media，Inc，London,2019.

[24] Miettinen R，Paavola S. Reconceptualizing object construction：the dynamics of building information modelling in construction design. 2018(28):516-531. doi：10. 1111/isj. 12125.

[25] Molina-Azorin JF. Mixed methods research：an opportunity to improve our studies and our research skills. 2016(25):37-38. doi：10. 1016/j. redeen. 2016. 05. 001.

[26] Nesta. Flying high：the future of drone technology in UK cities. Nesta，London，UK,2018.

[27] Pickering C. Systematic quantitative literature review,2020.

[28] Place Tech. VU. City's virtual reality model of square mile launched. https://placetech. net/news/vu-citys-virtual-reality-model-of-square-mile-launched/. Accessed 10 Jul 2021,2020.

[29] Saunders M, Lewis P, Thornhill A. Research methods for business students, 6th ed. Pearson, London, UK,2012.

[30] Srivastava A. The UK self-driving vehicles startups on cusp of transport revolution. https://www. uktech. news/news/the-uk-self-driving-vehicles-startups-on-cusp-of-transport-revolution-20210408. Accessed 10 Jul 2021.

[31] Stone S. Key challenges of smart cities & how to overcome them. https://ubidots. com/blog/the-key-challenges-for-smart-cities/. Accessed 10 Jul 2021.

[32] Treasury H. Build back better: our plan for growth. 2021.

[33] UK-RAS. UK-RAS white paper: urban robotics and automation-critical challenges, international experiments and transferable lessons for the UK, 1st ed. EPSRC UK-RAS Network, London,2017.

[34] Vodă AI, Radu L-D. Artificial intelligence and the future of smart cities. 2018(9): 110-126. doi: 10. 5281/zenodo. 1249819.

[35] Williamson J. What are the UK's top 20 internet of things cities? https://www. themanufacturer. com/articles/what-are-the-uks-top-20-internet-of-things-cities/. Accessed 10 Jun 2021.

[36] Zaychenko I, Smirnova A, Borremans A. Digital transformation: the case of the application of drones in construction. 2018(193): 5066. doi: 10. 1051/matecconf/201819305066.

An Investigation into the Appropriateness of the UK Government's Response to "The Future Homes Standard 2019 Consultation" (FHS), with Consideration Given to the 2050 Net-Zero Emissions Target

Matthew Hourigan[1]   Hua Zhong[2]

(1. Beadmans LLP, London, United Kingdom;

2. School of Architecture, Design and the Built Environment, Nottingham Trent University,

Nottingham NG1 4FQ, United Kingdom)

【Abstract】 This paper aimed to evaluate the trajectory of the government's response to the future homes standard consultation, which will put the United Kingdom housing stock on to meet the "net-zero" emissions target for 2050. A systematic review was conducted of 12 responses of governmental bodies, Professional bodies and environmental organisations. The paper will show that responses were unsatisfied with the government's approach in relation to its targets and it also identifies areas where the government failed to adequately respond to these issues. The data collected from interviews gave this research a unique contribution. The data gathered was primarily from architects, engineers and surveyors who were grouped together in the government's summary as the largest group to respond to the consultation. The data collected from this primary and secondary research, cumulates to suggest that the United Kingdom government is not on track to meet a net-zero target. The paper identifies that this is due to key factors being ignored which would help achieve the target. For example, embodied and whole-life carbon, actual energy consumption including unregulated energy, incentives for developers, modular construction and the use of technology to reduce demand. It is also suggested that in future consultations summaries, the government should give the percentage of people that gave each reason for the answers as this would allow a more quantifiable analysis to be undertaken. More research is recommended with larger sample sizes to get a boarder view. An analysis was also made by comparison of published responses to the Future Homes Standards, the government's response and interviews with industry professionals. Then to evaluate the trajectory this will put the United Kingdom housing stock on towards meeting the net-zero target for 2050.

【Keywords】 Future Homes Standards; 2050 Net-Zero Emissions; Ventilation Airtightness; Zero Carbon Housing; Embodied Carbon

1 Introduction

In 2019, the UK agreed to a net-zero emissions target for 2050. In 2018 ~ 2019, 19% of the UK's CO_2 emissions came from the residential sector. This shows the impact low carbon homes can have on the net-zero target.

The FHS is a consultation on parts L and F of the building regulations was published by the government on October 1st, 2019. 69 questions were asked and in total 3310 responses were received. On January 1st 2021, the government summarised the received responses and replied with plans for the next 5 years. This research project compared and evaluated the government's response with published consultation responses and interviews with professionals.

This research benefits the net-zero goal by holding the government accountable. In the past 10 years, low carbon housing legislation has been subject to policy dismantling which has caused the industry to lose faith in the government's ability to achieve carbon targets. This study has critiqued the government's response to the consultation and evaluated its ambition and achievability. The results of this study can be used to assess whether the industry responses were sufficiently considered in the making of the new legislation. Moreover, it has obtained the views of a small sample of industry professionals on the government's response to allow for a more valid conclusion to be drawn.

The aim of the research was to analyse published responses to the FHS; and to compare these with the government's response and interviews with industry professionals. Then to evaluate the trajectory this will put the UK housing stock on towards meeting the net-zero target for 2050.

Why is there a net-zero emissions target for 2050?

Carbon dioxide emissions make up 76% of global man-made greenhouse gas emissions (EPA, 2017) and 81% of UK emissions. Greenhouse gases (GHG's) cause global warming by trapping heat in our atmosphere. Global warming is having negative effects across our planet, such as rising sea levels causing flooding, heat waves, and loss of habitats and carbon sinks such as forests.

In 2016, the paris agreement has been signed by 189 counties and is a legally binding document aiming to keep global warming "well under" 2℃ more than it was prior to the industrial revolution, although the target is to keep it below a 1.5℃ increase (UNFCC, 2018).

In June 2019, the UK agreed to a net-zero emissions target for 2050 as recommended by the Climate Change Committee (CCC), improving upon the previous target set in the Climate Change Act 2008 to reduce GHG emissions by 80% of what they were in 1990 by 2050 (institute for government, 2020).

Net-zero emissions means balancing the amount of GHG's emitted into the atmosphere and the amount removed from the atmosphere. GHG's are removed from the atmosphere though carbon sinks such as forests and oceans.

A net-zero carbon building is defined by the World Green Building Council as a "Building that is highly energy efficient and fully powered from on-site and/or off-site renewable energy sources" (WGBC, 2019).

In 2018~2019, 19% of the UK's CO_2 emissions came from the residential sector (Beis, 2020). This shows the impact low carbon homes can have on the net-zero target. Energy use in homes accounts for 14% of the UK's total GHG emissions (CCC, 2019). This shows that increasing how energy efficient a home is in the construction stage will have a large effect on reaching the net-zero target.

The need for this research is emphasised by the government's record of repeatedly neglecting their responsibilities to meet greenhouse gas reductions. In 2007 the government published "Building a Greener Future" (DCLG, 2007), this claimed that by 2016 all new homes would be net zero carbon. The Code for Sustainable Homes (the code) was also published in 2007, it aimed to encourage housebuilders to voluntarily improve the energy efficiency of new-build houses. However, the higher levels of the code were rarely achieved, especially by private housebuilders, Heffernan (2015) argued that this shows a resistance from the private housebuilders to act voluntarily.

This research benefits the net-zero carbon goal by holding the government accountable for the regulations they published in the past 10 years, low carbon housing legislation has been subject to policy dismantling which has potentially caused the industry to lose faith in the government's ability to achieve carbon targets. This study will critique the government's response to the consultation and evaluate its ambition and achievability. The results of this study can be used to assess whether the industry responses were sufficiently considered in the making of the new legislation; moreover, it will obtain the individual views of a small sample of industry professionals on the government's response to allow for a more valid conclusion to be drawn.

The research paper will aim to:

• Identify consumer's opinions on energy efficient / low carbon housing.

• Analyse and compare the responses of professional bodies, government bodies and energy efficiency organisations; to the government's own response to the future homes standard 2019 consultation.

• Investigate a sample of construction industry professionals' views on the government's response to the consultation.

● Evaluate the trajectory that the government's response to the future homes standard consultation will put us on towards meeting the net-zero carbon target by 2050.

● Evaluate the key factors that need consideration to help achieve zero carbon homes in the UK.

2　Literature Review

The Future Homes Standard consultation (MHCLG, 2019) gave two options within it, these were:

● Option 1 (Future Homes Fabric): 20% improvement on the 2013 part L standard, delivered through very high fabric standards.

● Option 2 (Fabric Plus Technology): 31% improvement on the 2013 part L standard, delivered through "minor" increase to fabric standards and low-carbon heating and/or renewables e. g. PV (Photovoltaic) panels.

Option 2 was stated as the government's preferred choice. This suggests that the fabric of homes was not their priority; however, it delivers more carbon savings annually and costs less in bills for homeowners/occupiers, although it does have higher build costs. The consultation also states that "It would help to prepare supply chains for heat pumps and in-crease the number of trained installers" (MHCLG 2019, p. 7).

However, the government's response to the consultation insists that the package being delivered "will ensure a fabric first approach is at the heart of all new homes alongside a low carbon heating system" (MHCLG 2021, p. 5), this was a response to the concerns raised in responses to the consultation that underperforming fabric requirements would result in homes that add to the retrofit burden. The Currie & Brown and Aecom report for the CCC (2019) shows that the cost to improve the fabric now is marginal compared to the cost of retrofitting later(See Tab. 1).

Tab. 1 shows that it costs £124 to save a tonne of CO_2 via fabric improvements compared to £525 per tonne of CO_2 saved via renewable technology. This proves that fabric improvements are the most cost effective method of improving CO_2 emissions in homes.

The importance of the fabric first approach is emphasised by the Carbon Trust[2] who state, "Taking a 'fabric first' approach is fundamental to the energy performance of a building". They indicate the advantages of such an approach as being lower capital costs; increasing the buildings value; better thermal control and comfort for occupiers; and compliance with regulations meaning that costly retrofitting later is less likely.

NHBC foundation review-main findings-costs and CO_2 savings　　　Tab. 1

Scenario	Initial 20-year CO_2 Saving (Tonnes)	60-year CO_2 Saving (Tonnes)	Immediate Cost (£)	60-year Cost (£)	Cost per Tonne of CO_2 Saved(£)
Built-fabric Improvements Only	163(10.9%)	488(10.9%)	47787	60546	124

continued

Scenario	Initial 20-year CO_2 Saving (Tonnes)	60-year CO_2 Saving (Tonnes)	Immediate Cost (£)	60-year Cost (£)	Cost per Tonne of CO_2 Saved(£)
Renewables Only	180(12%)	540(12%)	101198	283593	525
Built-fabric Improvements and Renewables Combined	343(22.9%)	1028(22.9%)	148984	344138	335

This is supported by the NHBC (2012) whose findings show that the cost savings that can be made from improving the fabric efficiency cost less per tonne of CO_2 than using renewable energy does (see fig. 1). However, the figures used in this 2012 research may now be outdated. The table shows that the cost per tonne of CO_2 saved is less when fabric improvements are made than using renewable energy technologies such as PV panels. This makes it more cost effective to improve a home's efficiency with a fabric approach and then to improve upon this with the more costly renewable technologies and therefore this supports the "Fabric First Approach". Minimising the energy demand of dwellings through high fabric standards ensures LZC sources are used in the most efficient way (Zero carbon hub, 2009).

This fabric first approach is supported by Lowe and Oreszczyn (2008) who argue that too much investment is put into low and zero carbon technology (LZC) without full consideration of the importance of improving the buildings fabric to maximise its performance.

The fabric first approach which the government have adopted can save maintenance costs in comparison to LZC's and remove the consumer variables (Carbon futures, 2021), as it does not rely on the occupier to use it in an effective way for it to be efficient. Smith (2021) stated "In contrast to renewable technologies, a building's fabric can't be meddled with by tenants, in any significant way. Built correctly it will continue to perform as intended for decades and easily have other technologies added in future if required." This also reduces the performance gap which is commonly found in homes between the designed energy efficiency performance and the achieved performance is potentially much lower; however, this is dependent on the proper quality control and installation of the necessary building fabrics which is a consistent cause of the performance gap (Taylor, Counsell and Gill, 2013).

A case study (buildenergy, 2013) to demonstrate this is H + H's Ashford council project which focused on achieving above the building regulations standards with "Excellent U-Values and Airtightness" but primarily through the efficiency of the fabric and not through renewable energy.

There is no mention of a performance gap in this case study. However, because the y-value calculations were based on the accredited construction details it does not consider the possibility of poor installation quality. Moreover, the accredited construction details have now been archived which leaves their reliability unclear. there is no mention of further testing of the thermal bridging elements which means a performance gap is possible in this case study.

3 Research Strategy

This research project makes use of primary and secondary research which resulted in both qualitative and quantitative data. The primary research methods include a survey, a systematic review and standardised open ded interviews. The secondary methods include reports, articles and case studies(See Fig. 1).

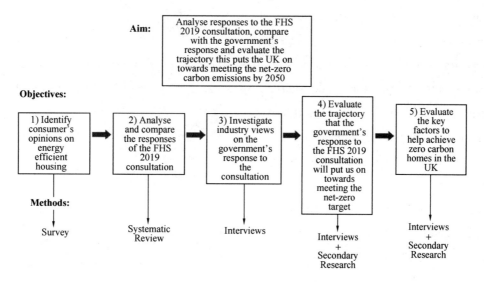

Fig. 1 Research methodology flowchart

3.1 Survey

15 questions were asked; 11 targeted at answering the objective, and 4 questions to identify the population of the respondents (Fig. 2). The questions written for this research had not been used before. When wording the questions, care was taken to ensure that the questions were written with "Common Sense" in mind (Peterson, 2000), this allowed issues such as EPC ratings, PV panels, the UK's net-zero target and the planning and energy Act 2008; to be investigated despite their specialist nature. Sinclair (1975) emphasises the importance of short clear questions using language understood by all the whole population, to achieve this he suggests a pilot study.

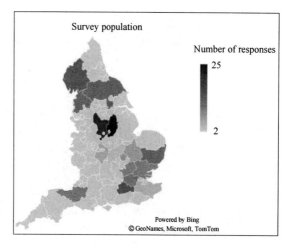

Fig. 2 Survey population

Two pilot studies took place with 10 respondents each, the first found that the wording of two questions was overly ambiguous and not adequately defined for those without specialist knowledge of the subject to understand. The questions were changed and a second pilot study with a new batch of respondents confirmed that the questions were now suitable for the sample. The questions asked the respondents about their views, this increased the reliability of the questionnaire by keeping the respondents engaged (Sinclair 1975).

Chain referral sampling through social media was used. The link to the questionnaire was shared on several social media accounts. People were asked to fill it out and share it with their own social media audiences this resulted in snowball sampling (Johnson, 2014). This resulted in a sample population of 200 across England.

Fig. 3 and Fig. 4 suggest that energy efficiency is important to home buyers. However, Fig. 5 shows that when they have to prioritise the factors they would consider when

buying a house, it performs poorly compared to other factors.

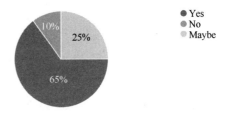

Fig. 3   Would you consider how energy efficient a home is whe renting/purhasing it?

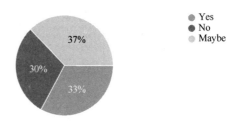

Fig. 4   Would you consider paying more for a home that produces zero-carbon emissions?

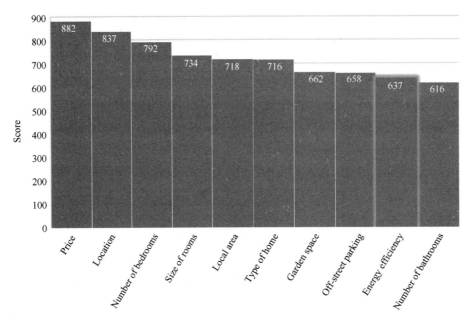

Fig. 5   Pleas rank these factors in the order from 1 to 10, that you would take them into consideration when buying a home

3.2 Systematic Review

A Systematic review was used to collate all the responses from the sample and compare them. Of the 3310 total responses to the consultation this research found 12 that met the eligibility criteria(Fig. 6).

The systematic review is a repeatable method，by using the same eligibility criteria and protocol another research team should be able to replicate the results.

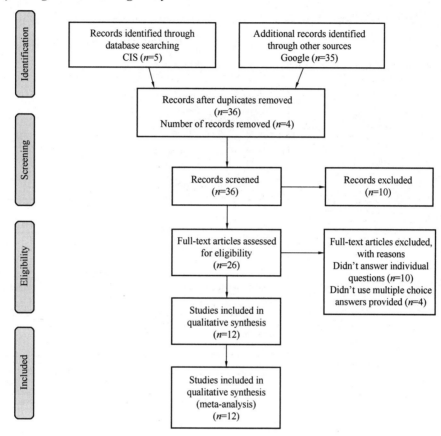

Fig. 6　Systematic review process

The eligibility criteria：

（1）A published response to the consultation.

（2）Published between the 1st October 2019 and 1st January 2021.

（3）By a professional body / a government body / energy efficiency organisation.

（4）Respond to at least 3 questions.

（5）Pick multiple choice answers.

The final sample was 12 external responses in addition to the government's own response to the consultation.

（1）4 professional body responses：

● RIBA，Royal Institute of British Architects.

● CIBSE，Chartered Institute of Building Services Engineers.

● REA，The Association for Renewable Energy and Clean Technology.

● ALEO，Association of Local Energy Officers.

（2）2 government body responses：

● MOL，Mayor of London.

● RTPI，Royal Town Planning Institute.

（3）6 energy efficiency organisation responses：

● REGEN.

● ATS，Independent Airtightness Testing Scheme.

- CREDS.
- UKGBC，UK Green Building Council.
- TEIGN.
- NEA，National Energy Action.

Protocol：

Following the ranking criteria(Fig. 7)：

- 19 questions were "Highly contested".
- 11 questions were "Mildly contested".
- 35 questions were " not sufficiently contested" to require further inquiry.
- 4 questions were omitted due to a lack of data.

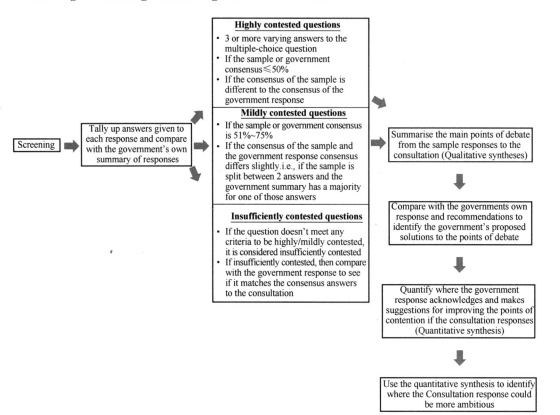

Fig. 7　Protocol

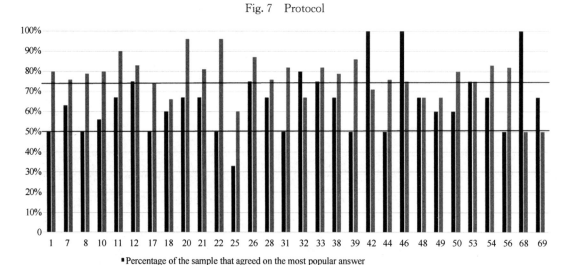

■Percentage of the sample that agreed on the most popular answer
■Percentage of the total consultation responses that agreed on the most popular answer

Fig. 8　Contested questions percentage of agreement on most common answer：sample/full consultation

Fig. 9 shows the proportion of questions where the sample had a different most common answer to the total consultation response. These 17 questions make up more than half of the contested questions, showing that the sample responses often had different opinion to the rest of the responses.

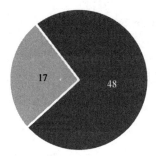

■ Different most common answer ■ Same most common answer

Fig. 9 Result

3.3 Interviews

The systematic review was used to find the areas of greatest contention within the FHS. The questions in the consultation that were significantly "contested" were then formed into questions which were asked to a sample of construction professionals.

The sample was:

(1) 2x Architects (both a part of the Architects Climate Action Network-ACAN).

(2) A project planning committee member.

(3) 2x senior RICS chartered quantity surveyors.

(4) An engineering services cost and project management consultant.

(5) 2x building services engineers (both chairs of respective CIBSE groups).

(6) A head of sustainability at a top 5 UK housebuilding contractor.

(7) The director of a smart building systems company.

The data gathered via these interviews has allowed this research to analyse the largest group of respondents to the consultation "Designer, Engineer, Surveyor" who made up 44% of the total responses. These interviews allow for this groups opinion's to be separated and compared with each other. They also provide a mixture of perspectives from senior quantity surveyors and contractors who focused on the time and cost implications of this standard, to engineers and architects who prioritised the reduction of emissions and the importance of regulating whole life carbon, embodied carbon and in-use energy.

Every interview participant acknowledged that the standard was going in the right direction but had suggestions as to where it falls short in one form or another. The sole point that every interviewee returned to was that there has been no plan put in place for retrofitting the existing housing stock.

4 Use of the Data Gathered from the Systematic Review, Interviews and Secondary Research to Analyse the Appropriateness of the Government's Response

4.1 Embodied Carbon/Whole Life Carbon Impacts

For example, embodied carbon and whole life carbon assessments were absent from the FHS and the government's subsequent response, despite being a recurring theme in the samples responses to the consultation and interview responses. Particularly the UKGBC and the architects interviewed were passionate advocates for projects being required to disclose whole life carbon impacts and embodied carbon being accounted for. The case was put forward that in other countries such as Denmark national databases were being introduced to calculate embodied carbon and that there is already a British standard that is accepted by the EU and has an RICS guide on how it works. The quantity surveyors on the other hand were sceptical of how this would be practically introduced, especially with smaller contractors.

The initial consultation proposed the removal of the Fabric Energy Efficiency Standard (FEES), this was widely rejected by the industry due to risk of the fabric standards of new homes following this being potentially worse than under the 2013 part L standard which led to FEES being retained.

4 metrics were proposed in the government's response to the consultation, the principal metric being primary energy followed by CO_2 targets, fabric energy efficiency targets and minimum standards for fabric and fixed building services. CIBSE held the position that primary energy should not be the principal metric because, like the CO_2 target it is a system level metric that does not represent the actual performance of the building. They instead called for an in-use metric that relates to consumer bills so as to be familiar to them, this view was supported by both building services engineers. An in-use energy metric was also suggested by the Mayor of London as

well as ALEO and CREDS. The RIBA suggested that the principal metric be actual energy consumption as measured by Energy Use Intensity in $kWh/m^2/yr$, this would be calculated at the design stage and then reviewed via Post Occupancy Evaluation (POE) to ensure the home is working as intended.

4.2 Heating Systems

A large part of the consultations impact is to do with how homes are heated. The movement towards air source heat pumps was largely well received in consultation responses and in interviews. However, concerns were expressed regarding the change in consumer behaviour it would require, it was also repeated in interviews that proper assessments should be done to ensure heat pumps are the most efficient choice for individual homes. The RIBA, CIBSE and Mayor of London all mentioned that if installed or operated incorrectly it could result in high bills for the end user. This skill shortage was demonstrated further in the interviews with the housebuilder and a quantity surveyor who both expressed apprehensions regarding the dramatic supply chain increase required for these new low-carbon heating technologies as currently there are only tens of thousands of heat pumps and the government wants to scale this up to 600000 per year by 2028.

4.3 Performance Gap

The performance gap between design and actual homes is a major obstacle to achieving a truly net-zero housing stock. The interviewees were asked: how closely should, predicted building energy use and monitored building energy use map. There was a consensus that some allowance is necessary to account for unpredictable occupant behaviour and the issue was raised that, if there is an allowable percentage, what happens if a dwelling has more than the allowable performance gap? Would the contractors have to return at the cost of the client?

The quantity surveyors suggested financial incentives would be best to reduce performance gaps and the planning committee member suggested increased off-site construction would be the solution.

The MoL suggested that operational data could be used to bridge the performance gap and suggest mandatory operational ratings like that which are proposed for commercial buildings in the future building standard.

4.4 Minimum Airtightness Levels

CIBSE state that $5m^3/m^2/h$ air permeability "is far from world class" and agree with the UKGBC who suggested that past $3m^3/m^2/h$ at 50Pa homes should have MVHR to ensure air quality. Both building services engineers and the cost consultant / project manager agreed with this and the reason was given that inhabitants need ventilation year round to ensure good air quality and in the winter windows cannot be opened without increasing the heating use so MVHR should be implemented at high levels of airtightness. Although a quantity surveyor argued that this is very general and it should perhaps be decided on a case-by-

case basis because there are other factors such as population density, they also proposed that in this scenario developers would make their homes less airtight than the requirement for MVHR because mechanical ventilation costs more.

5 Conclusion

This research found several areas where the responses were unsatisfied with the government's approach to the standard and where the government failed to adequately respond to these issues. These areas were found to be: regulating actual in-use energy consumption, no plan for retrofitting the existing housing stock, measuring embodied carbon and whole life carbon impacts, inadequate changes to the systematic problems with SAP and finally the performance gap was addressed but not enough was done to tackle it.

The results of the interviews showed that, there was generally a lot of uncertainty about how the supply chains would be built up by 2025, there was also disappointment among some groups regarding the exclusion of embodied and whole life carbon impacts, most commonly by architects and engineers. The surveyors and developers argued that contractors and developers need financial incentives to make the suggested improvements, otherwise it will cut their profit margins and the consumers will suffer the financial consequences. These interviews were valuable; however, they are the views of a few and so cannot be generalised as widely as the government's summary of total responses or as much as the systematic review samples views.

These interviews revealed that 100% of the sample believed that 2050 was an unrealistic target for the housing stock to reach net-zero. The most common perception found in this research is that new homes may reach net-zero carbon emissions, but that this should account for unregulated energy, embodied carbon and whole life carbon which the 2025 standard will not regulate. There was agreement that the existing housing stock is too large to be tackled by 2050.

This research also found that professionals think there should be more support for; modular / off-site construction for homes, innovation into technology which can reduce energy demand and incentives for developers and homeowners to encourage and support retrofitting the existing housing stock.

References

[1] BEIS. 2019 UK greenhouse gas emissions, provisional figures. ONS. 2020.

[2] Carbon trust. Building fabric. Energy saving techniques to improve the energy performance of buildings. Carbon trust, 2018:4.

[3] CCC. UK homes unfit for the challenges of climate change, CCC says-Climate Change Committee. [Online] Climate Change Committee. Available at: <https://www.theccc.org.uk/2019/02/21/uk-homes-unfit-for-the-challenges-of-climate-change-ccc-says/> [Accessed 10 May 2021].

[4] CIBSE. MHCLG Consultation on part L and part F 2020 and the Future Homes Standard CIBSE Response. CIBSE,2020.

[5] CREDS. Future homes standard: changes to part L and part F of the building regulations for new dwellings-comments from CREDS. [Online] Available at: <https://www.creds.ac.uk/wp-content/uploads/PartL-consultation-CREDS-response-final.pdf> [Accessed 10 May 2021].

[6] DCLG. Building a greener future: policy statement. [Online] Available at: <https://www.rbkc.gov.uk/PDF/80% 20Building% 20a% 20Greener% 20Future% 20Policy% 20Statement% 20July%202007.pdf> [Accessed 10 May 2021].

[7] EPA. Inventory of U.S. Greenhouse Gas Emissions and Sinks | US EPA. [Online] Available at: <https://www.epa.gov/ghgemissions/inventory-us-greenhouse-gas-emissions-and-sinks> [Accessed 10 May 2021]. 2017.

[8] Gov. The ten point plan for a green industrial revolution. London: HM Government, 2020:20-22.

[9] Heffernan E, et al. Zero carbon homes: perceptions from the UK construction industry. Energy Policy, 2015(79):23-36.

[10] iATS. iATS draft response to: the future homes standard: Chapter 5 Airtightness. [Online] iATS. Available at: <https://iats-uk.org/wp-content/uploads/2019/12/Consultation-Response-iATS.pdf> [Accessed 10 May 2021], 2019.

[11] Instituteforgovernment. org. uk. UK net zero target. [Online]Available at: <https://www.instituteforgovernment.org.uk/explainers/net-zero-target>[Accessed 10 May 2021].

[12] Kvale S. Inter view: an introduction to qualitative research interviewing. 1996:45

[13] NHBC. Foundation facts. [Online] NHBC. Available at: <https://www.nhbcfoundation.org/wpcontent/uploads/2016/06/Foundation-Facts-2010.pdf> [Accessed 10 May 2021]. 2012.

[14] Mayor of London. Future homes standard: changes to part L and part F of the building reg-

ulations for new dwellings response from the Mayor of London. [Online] Available at: <https://www. london. gov. uk/sites/default/files/fhs_consultation_response. pdf> [Accessed 10 May 2021]. 2020.

[15] MHCLG. The future homes standard 2019 consultation on changes to part L (conservation of fuel and power) and part F (ventilation) of the building regulations for new dwellings. 2019.

[16] MHCLG. The future homes standard: 2019 consultation on changes to part L (conservation of fuel and power) and part F (ventilation) of the building regulations for new dwellings sum-

mary of responses received and government response. 2021.

[17] Peterson R. Constructing effective questionnaires. Thousand Oaks, Calif. Sage Publications,2000.

[18] Sinclair M A. Questionnaire design. Applied ergonomics, 1975,6(2):73-80.

[19] Unfccc. int. What is the Paris Agreement? [Online] Available at: < https://unfccc. int/process and meetings/the-paris-agreement/the-paris-agreement> [Accessed 10 May 2021]. 2018.

Evaluation of Domestic Thermoelectric Cogeneration in Regions with Temperate Climate

Xiaofeng Zheng[1,2]　Yuying Yan[1]　Hoy-yen Chan[3]

(1. Department of Architecture and Built Environment, Faculty of Engineering,

University of Nottingham, University Park, Nottingham NG7 2RD, UK;

2. School of Built Environment, Engineering and Computing, Leeds Beckett University,

Leeds LS1 3HE, UK;

3. Invite Green Consultancy, Lintang Pantai Jerjak 3, 11700, Gelugor, Penang, Malaysia)

【Abstract】 Research in thermoelectric material and application has been extensively performed due to its various advantages. However, its poor economic viability limits the applications to niche industrial areas due to the low conversion efficiency given by current commercially available thermoelectric modules. In the absence of fundamental breakthrough in materials performance, its wide and feasible application remains challenging. A thermoelectric cogeneration of electrical power and hot water designed for residential houses is proposed and tests have been carried out on a bench scale prototype to investigate its performance. Its viability has been discussed based on the availability of domestic boiler waste heat and solar radiation. The economic and environmental benefit of employing thermoelectric generators in domestic environment has been evaluated on the basis of experimental studies to the usage in the regions with temperate climate. The analysis shows that the proposed system concept potentially offers a reasonable cost recovery period (less than 5 years).

1 Introduction

In the most of developed countries with temperate climate, space heating and domestic hot water is often provided by burning gas or oil in a domestic appliance such as boiler or furnace. Current condensing boilers are highly efficient achieving thermal efficiencies in excess of 90% with combustion gases rejected at dew point temperature. Despite a growing market for condensing boilers, a large number of old boilers are still being used with the seasonal efficiency as low as 55%, resulting in large energy loss [1].

Thermoelectric modules, consisting of pairs of p-type and n-type semiconductor materials forming a pair of thermocouple, generate electricity when a temperature difference is established across the module. Thermoelectric devices have found a wide range of applications including power generation, heat recovery and thermal sensing[2]. These applications span a wide range of industries such as transport [3,4], process industries[5~7], medical[8,9] and space[10]. Many efforts have been made in the development of advanced thermoelectric materials with high conversion efficiency to enhance the potential of a wider range of applications[11~14].

The main advantages that boast thermoelectric devices are being static, compact, and low maintenance cost. However, low energy conversion efficiency has confined their application to specialised niche markets. One of these markets is generating electrical power in buildings located in inaccessible remote areas. For instance, back in 1996, the Swedish Royal Institute of Technology[15] developed a thermoelectric stove to provide small amounts of power to residential houses in the remote northern areas of the country where grid connection is prohibitively expensive. Recent works on thermoelectric stoves include those of Champier[16], Nuwayhid [17] and Mastbergen[18], which recover waste heat from cooking stove to generate electricity to power fan or lamp.

Because domestic boilers rely on grid connection for operation, Daniel[19,20] attempted to develop a selfpowered domestic boiler using thermoelectric generators by integrating the thermoelectric modules between the combustion chamber and the water channel in the boiler enclosure. Thermoelectric generators were also regarded as direct contenders to replacing diesel gensets for power generation in off-grid buildings, eliminating noise, and high maintenance that characterize internal combustion engines. A simple design arrangement of such a thermoelectric generator could be presented by Fig. 1. It shows that the thermoelectric modules are placed between a heat source and a heat sink.

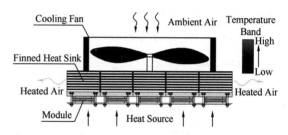

Fig. 1　Cooling fan at the heat sink

The thermal performance of the thermoelectric devices for power generation-only applications remains, however, poor with maximum efficiency below 5%[21]. This means 95% of the fuel energy content is rejected as low grade heat. Therefore, recovering rejected heat for useful utilisation for space heating and domestic hot water as part of a cogeneration thermoelectric system would make thermoelectric more attractive as overall thermal efficiency can be increased up to 80%, which was shown in another studies carried out by the author [21,22]. Gao[23] proposed a symbiotic application which uses the rejected thermal energy to improve the combustion process efficiency by pre-heating the air/fuel mixture to higher temperatures. Qiu[24] developed a thermoelectric power generation system which generates electricity and hot water by burning natural gas in a furnace. Relying on the supply of natural gas, its operation is suitable for the applications which are purposefully designed for using the natural gas as the primary fuel. A cogeneration system which uses the heat from boiler exhaust and solar power in the UK residential houses has been proposed and experimentally studied in[21]. A potential benefit has been prospected for supplementing the domestic energy need and improving the energy efficiency in the UK domestic sector on the basis of the available heat from boiler waste and solar energy.

This paper introduces a domestic thermoelectric cogeneration application to improve the energy utilisation efficiency by recovering the waste heat and using the solar energy and evaluates the economical and environmental benefit that can be achieved by adopting this system in a residential house in regions with temperate climate.

2　Fundamentals and Applications

The story of discovering the thermoelectric effect originates from a deflected compass needle which was placed near a closed loop formed by two dissimilar conductors, one of which was heated. Seebeck, the person who found this phenomenon, attributed it to an interaction of the earth's magnetism with the temperature difference between the equator and the poles. Based on the current knowledge, the magnetic phenomenon discovered by Seebeck was because of the electricity genera-

ted in the presence of temperature difference across a thermocouple. Following this discovery, after 12 years a complementary effect was discovered by Peltier, who observed temperature changes in the vicinity of the junction between dissimilar conductors when a current passed. This is how the well known Peltier effect came to our awareness. Thomson (Lord Kelvin) predicted the existence of a third thermoelectric effect (known as Thomson effect), which he subsequently observed experimentally on the basis of establishing a relationship between Seebeck and Peltier effect. However, since the discovery of these thermoelectric effects, the application of them didn't attract much interest due to many more exciting discoveries were made during that time. Until 1850 when the interest was focused on all forms of energy conversion, attentions were tilted to thermoelectricity, which was considered in 1885 by Rayleigh who tried to calculate the efficiency of a thermoelectric generator. Following rayleigh, Altenkirch gave a satisfactory theory of thermoelectric generation and refrigeration with a conclusion that a large seebeck coefficient should be possessed by good thermoelectric materials which also need to show low thermal conductivity[25].

The theory of thermoelectric transport is essentially the domain of transport in semiconductors[26]. All the essential features and complications of transport phenomena in solids are shown in semiconductors. It involves a flow of electric charge or energy or the combination of both. Due to the "forces" caused by external causes such as an electric field or temperature gradient, these "flows" or "transportation" occurs. The various transport coefficients that characterizes the electrons and phonons in the material are defined by the relationships between various "forces" and "flows"[27~29]. Basically, the flow can be driven by any force. A linear relationship can be obtained between "forces" and "flows" on the assumption that the electron and phonon systems depart only slightly from their equilibrium distributions. Assuming a solid semiconductor contacting with two reservoirs, one is energy and the other is electrons. In the steady state, a steady flow is maintained through the solid as are the differences in electrochemical potential (p) and temperature (T) between the two ends.

A good choice of the forces are grad (μ/T) and grad ($1/T$), the components of the flows of electrons ($\vec{j}$) and of energy ($\vec{w}$) are given by Eq. (1) and Eq. (2)

$$-j_i = \sum_{k=1}^{3} L_{ik}^{(1)} \frac{\delta}{\delta x_k}\left(\frac{\mu}{T}\right) + \sum_{k=1}^{3} L_{ik}^{(2)} \frac{\delta}{\delta x_k}\left(\frac{1}{T}\right)$$

(1)

$$w_i = \sum_{k=1}^{3} L_{ik}^{(1)} \frac{\delta}{\delta x_k}\left(\frac{\mu}{T}\right) + \sum_{k=1}^{3} L_{ik}^{(4)} \frac{\delta}{\delta x_k}\left(\frac{1}{T}\right)$$

(2)

The total energy flow can be written as a sum of W_e and W_p where e and p refer to the electron and phonon systems, respectively. For example, $L^{(3)} = L_e^{(3)} + L_p^{(3)}$. The set of coefficients L_{ik} provide a complete description of the transport properties of the solid.

The electric current density ($\vec{i} = -e\vec{j}$) and thermal current density ($\vec{w}$) can be expressed by Eq. (3) and Eq. (4) in terms of grad (μ) and grad(T).

$$i_i = \frac{1}{e}a \cdot \mathrm{grad}\mu - a \cdot a \cdot \mathrm{grad}T \quad (3)$$

$$\vec{w} = \left(\pi - \frac{\mu}{e}\right)\vec{i} - \lambda \cdot \mathrm{grad}T \quad (4)$$

Where, a, π and λ are second order tensors which are related to the coefficient, $L^{(n)}$.

The theory of electronic transport has been described by Single Spherical Band Model, Two-Band Conduction, Multi-valley Effects and Intervalley Scattering. More details are given in the CRC handbook of thermoelectrics [25].

Thermoelectric materials can be used for either heat pump or power generation, shown in Fig. 2. Its construction consists of arrays of N & P type semiconductors in which, by applying a heat source on one side and a cooler heat sink to the other side, electric power is produced and vice versa. Namely, electric power can be converted to cooling or heating by reversing the current direction.

Despite the low conversion efficiency of a-

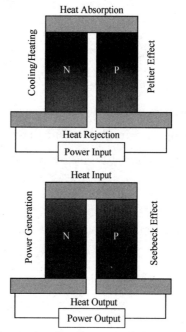

Fig. 2   Cooling/Heating and power generation

thermoelectric heat engines

round 10% when used as power generators, they are strongly advantageous as they have no moving parts and are therefore both more reliable and durable compared to conventional energy technologies. Apart from that, they are scalable without releasing any pollutant to the environment during the operations. Hence they would be ideal for applications in many areas at different scales replacing the traditional cooling and power generation methods.

In typical TE (thermoelectric) devices, the N & P materials are electrically connected in series and thermally connected in parallel in the form of flat arrays called modules, as shown in Fig. 3.

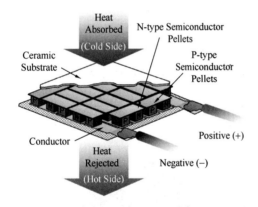

Fig. 3 Typical thermoelectric module construction

Thermoelectric materials are evaluated by the figure-of-merit ZT, it is defined in terms of intrinsic material properties of both the N and P type materials and determined by three physical properties—seebeck coefficient (S), electrical conductivity (σ), and thermal conductivity (λ). It can be related to the physical properties by Eq. (5):

$$ZT = \frac{\sigma S^2}{\lambda} T \qquad (5)$$

Where, T is the absolute temperature, the figure-of-merit ZT serves as a dimension-

less parameter to evaluate the performance of a thermoelectric material. The larger the value of ZT, the better is the thermoelectric material. Obviously, the materials with higher electrical conductivity and lower thermal conductivity have larger value of Z which contributes more to the enhancement of conversion efficiency η. It is given by Eq. (6):

$$\eta = \frac{\Delta T}{T_1}\left(\frac{\sqrt{ZT+1}-1}{\sqrt{ZT+1}+1-\frac{\Delta T}{T_h}}\right) \qquad (6)$$

The conversion efficiency of electrical power to cooling is given in terms of COP, defined as:

$$COP = Q_c / P_{input} \qquad (7)$$

Where, P_{input} is the electric power input and Q_c is the cooling thermal power produced by TE module. The COP of TE module measures the cooling effectiveness of thermoelectric cooler. In an ideal assembly, the optimum cooling effectiveness can be expressed by Eq. (8):

$$\beta_c = \frac{T_c}{\Delta T}\left(\frac{\sqrt{Z\overline{T}+1}-1-\frac{\Delta T}{T_c}}{\sqrt{Z\overline{T}+1}+1}\right) \qquad (8)$$

Where, T_c, $\overline{T}$ and ΔT are the cold side temperature, average temperature of hot and cold sides, and temperature differential.

Supposing a load of resistance R_{ex} is connected across the thermocouple at A and C shown in Fig. 4, a heat source is supplied at the rate Q so as to maintain a temperature difference ($T_1 - T_2$) between the junctions. The produced voltage by the generator is ($\alpha_p - \alpha_n$)($T_1 - T_2$) and this yields useful power across the load given by:

$$W = \left[\frac{(\alpha_p - \alpha_n)(T_1 - T_2)}{R_{ex} + R_m}\right]R_{ex} \qquad (9)$$

Among the supplied heat by the heat source, most of the heat is conducted to the sink through the thermocouple branches, some is used to balance the Peltier effect which is associated with the current flow just as for the case of thermoelectric refrigeration, half of the Joule heating in the arms finds its way to the source. It can be described by Eq. (10):

$$Q = K(T_1 - T_2) + (\alpha_p - \alpha_n)IT_1 - I^2R/2 \tag{10}$$

Where the current I equals to $(\alpha_p - \alpha_n)(T_1 - T_2)/(R_{ex} + R_m)$.

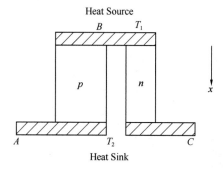

Fig. 4 Thermocouple for power generation

The efficiency q is equal to W/Q and its value depends to some extent on the way that the load is matched to the resistance of the module. The condition for maximum power transfer is obtained if R_{ex} and R_m is made equal to one another. However, if this condition is satisfied, the efficiency can never exceed 50% of the ideal thermodynamic value $(T_1 - T_2)/T_1$. Therefore, it is assumed that the load resistance is chosen so as to yield maximum efficiency. If the ratio R_{ex}/R_m is denoted by m, it is required that $\mathrm{d}q/\mathrm{d}m = 0$.

3 System Description

In this concept shown in Fig. 3, one primary heat source is waste heat from the boiler exhaust. The waste heat is recovered by the hot side heat exchanger which is attached to the hot side surface of TE module. A cooling fluid is then circulated through a compact heat exchanger on the cold side of the TE module to establish temperature difference to generate electricity and produce hot water or heating for the building[30]. The system is also capable of using solar energy collected by a solar collector mounted on the roof of a building whereby solar energy is absorbed and supplied to the thermoelectric module hot side. This can be achieved by direct utilisation or indirect utilisation of solar energy, depending on the heat exchanger design. The heat absorbed from the two aforementioned heat sources can be used to generate "free" electricity and attendant heat is used for heating purposes, increasing the overall energy utilisation efficiency of the boiler. However one of the challenging designs is to overcome limitations of heat transfer to and from the thermoelectric module.

Fig. 5 shows the concept of domestic thermoelectric cogeneration system which absorbs heat from boiler waste and solar energy using a hot side heat exchanger. The heat rejected from the cold side of thermoelectric modules is taken away by a cooling plate which is connected to the boiler to provide preheated water. The hot side heat exchanger is specifically designed for absorbing the heat from boiler waste and solar energy. Two concepts are shown in Fig. 6. Fig. 6(a) shows the indirect use in which heat transfer oil absorbs the heat from boiler waste and solar energy.

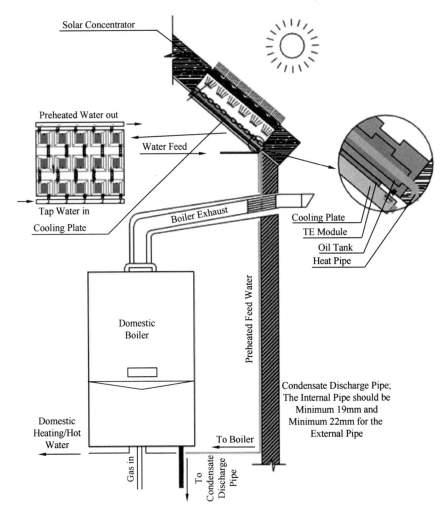

Fig. 5　Concept diagram of domestic thermoelectric cogeneration system

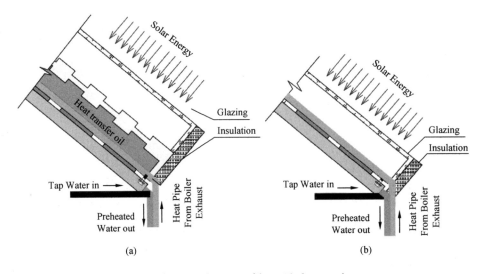

Fig. 6　Schematic diagram of hot side heat exchanger

The oil is heated up by the absorbed energy and flows through the TE module to proceed with the energy conversion and water preheating. Fig. 6(b) shows the direct use where the solar energy and boiler waste heat is used directly to heat the TE hot side. The selection can be made according to the availability of heat sources and requirements on the system response time. Solar concentration measures can also be taken to promote the temperature level at the heat source side.

4 Discussion

This system basically has two products, electrical power and preheated water. The amount of power generation is determined by the temperature difference across the thermoelectric generators. The temperature level of the preheated water is determined by the heating period in the cooling plate. It can be adjusted according to the need by varying the inlet velocity in the economic range. When the temperature level of the preheated water needs to be higher, the feed water into the cooling system can be supplied at a smaller flow rate, vice versa. As shown in the results of the previous experimental study, the power output hardly had any change when the flow rate of feed water reduces. This is because the temperature difference across the module does not have obvious fluctuation when the cold side temperature changes. However, the heat output is increased when the water feed velocity is decreased.

Due to the characteristic of this domestic TCS, it largely replies on the heat. Its opera-tion can be fulfilled by using the available heat in the domestic environment. The available heat includes the waste heat from domestic facilities, solar power or purposely designed heat source. It can either be used as a parasitic application which recovers the waste heat given out by other facilities, or as a main power generator by using the purposely designed heat source. The former concept is suitable for the areas or regions where the electricity supply is not an issue. It can be used as a supplemental power generation method to partially supply the electrical power. The parasitic application concept helps improve the energy efficiency of the host facilities. The latter concept is suitable for the regions where lacks of electricity or no electricity supply. A facility, relying on the local sufficient sources, can be built to provide heat for power generation, e. g. remote mountain areas where the wood is of great abundance.

The purpose of this system is to recover the waste heat existing in domestic environment and use solar energy at the same time. More boiler waste or solar energy means more electrical power and thermal energy can be produced. This does not necessarily mean a more inefficient boiler or domestic heating device is rewarding. The goal is to improve the energy utilisation efficiency by recovering the waste heat in residential house and meanwhile utilise the solar energy to reinforce the system viability in domestic environment.

4.1 Economic Analysis

There are a large amount of existing boilers are operating highly inefficiently in the UK, leading to massive energy waste. When we discuss about the potential methods of solving this issue, the boiler retrofit always seems to be on top of the list. However, it is very expensive and large, requiring a huge amount of upfront capital. It does not meet the ROI (return of investment) thresholds for energy efficiency retrofit, although it saves a considerable amount of money over the expected lifetime of the boiler. The economic viability of domestic TCS is discussed in this section to demonstrate its economic viability.

When considering the energy saving benefit of using thermoelectric generators, it is important that the amount of energy produced by the thermoelectric generators during their life time should be larger than the amount of energy for fabricating them. Fig. 7 shows energy recovery period for various methods of electrical power generation [31]. In the case of thermoelectric power generation by a Bi-Te-based module of 200℃ type, the energy recovery period is 0.85 year which shows sufficient competitiveness among other methods.

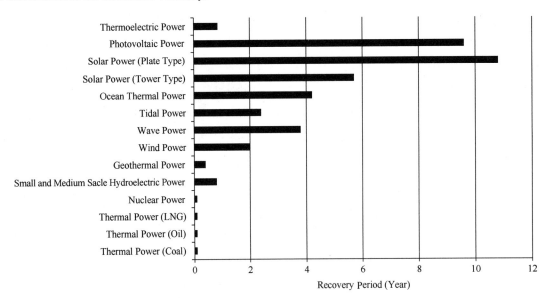

Fig. 7   Comparison of energy recovery Period

The energy recovery period is calculated against the application which generates electricity only. The aforementioned 0.85 year energy recovery period for the 200°C module type can be reduced further when the thermoelectric generator is used in the cogeneration concept introduced in this research.

To accurately evaluate the economic benefit that the use of this system concept can bring

to the residential house which uses conventional domestic or old inefficient boiler, the amount and availability profile of the heat sources along the timeline of whole year and the temperature of boiler flue should be understood.

For the UK, a comprehensive study conducted on the solar energy arriving at the surface of the Earth has been carried out by the Institute for Environment and Sustainability of European Commission[32]. It is cited here to show the solar availability in different regions and countries. The amount of available solar energy depends on the geographical variability, weather conditions and time dynamics. The analysis to the availability helps us under- stand the contribution that TCS could make to the improvement of domestic power condi- tions.

Taking a two-bed bungalow in Northampton as an example, the average daily solar irradiance of each month is shown in Fig. 8.

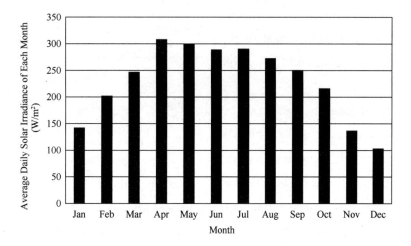

Fig. 8　Average daily solar irradiance of each month in Northampton

The profile of boiler waste heat is deter- mined by the outdoor temperature and the us- er demand. The heating demand of the two- bed bungalow in the whole year is shown in Fig. 9 based on the operating time in each hour of the day in the whole year. The data is based on the air source heat pump, whose op- erating pattern should be able to represent that of a domestic boiler if it was used in this bungalow. Assuming the used domestic boiler is a new conventional boiler with 78% effi- ciency (24kW capacity) and the flue gas is ex- hausted at 180°C, the operating temperature at the hot side of the system can be estimated in each hour of the day. Assuming 5% of the energy is lost from radiation and conduction,

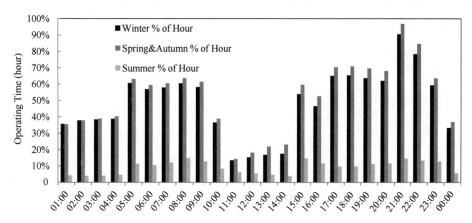

Fig. 9　Hourly heating demand of a UK two-bed bungalow in a year

then 17% of the energy escapes to the environment via boiler flue gas, which is about 4.08kW. Assuming the effectiveness of the flue heat exchanger is 60%, then approximately 2.45kW thermal energy can be absorbed by the flue heat exchanger. A thermoelectric cogeneration system with 16 modules is used to convert the absorbed thermal energy to electrical power and domestically useable heat.

Taking one set of experiment as an example, the heat input is given at the rate of 93W by cartridge heater which simulates the heat source. The power output and open voltage of a single cell thermoelectric cogeneration system under different operating temperature difference is shown in Fig. 10. When the temperature difference across the thermoelectric generator is 130℃, the power output and heat output of the system is 2.5W and 59.4W. The conversion efficiency and thermal efficiency is 4% and 67%, respectively. The conversion efficiency η and thermal efficiency η_t can be calculated by Eq. (11).

$$\eta = \frac{P}{Q_{input}}, \quad \eta_t = \frac{P + Q_{output}}{Q_{input}} \qquad (11)$$

Where, P, Q_{output} and Q_{input} are power output, heat output and heat input, respectively. P and Q_{output} can be calculated by Eq. (12) and Eq. (13):

$$P = S^2 \frac{(T_1 - T_2)^2}{R_{ex}} \qquad (12)$$

$$Q_{output} = c\rho g (T_{outlet} - T_{inlet}) \qquad (13)$$

Where, S, T_1, T_2, R_{ex}, T_{inlet} and T_{outlet} are the Seebeck coefficient of used TE generator, temperature at the hot side surface of gen-erator, temperature of cold side surface, external load resistance, water temperature at the inlet and outlet of cold side heat exchanger.

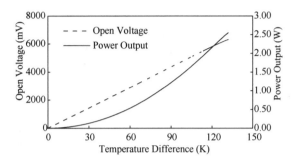

Fig. 10　Open voltage and power output of a single cell thermoelectric cogeneration system

The heat output and open voltage of a single cell thermoelectric cogeneration system under different operating temperature difference is shown in Fig. 11. The abrupt fluctuation in the curve of heat output could be caused by equipment error and can be omitted.

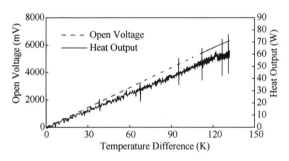

Fig. 11　Open voltage and heat output of a single cell thermoelectric cogeneration system

For a domestic TCS using 16 thermoelectric modules, the cost recovery and energy recovery period is estimated on the assumption that the same conversion efficiency and thermal efficiency is achieved by adopting the reproducible assembly technique and configuration used in the single thermoelectric cogeneration system.

The economic impact is analysed as follow: The users are normally concerned about the time of recovering the system cost when they justify the economic benefit of turning to utilising this system. Here the cost recovery period will be evaluated to provide the information for assessments. The cost for establishing this system is estimated and shown in Tab. 1 as below.

Cost estimation of a TCS with 16 thermoelectric modules　　　　Tab. 1

Component	Modules	Htf Oil	Hot Side x	Cold Side x	TIM	Installation and Operation	Total Cost
Quantity	16	0.4litre	1	1	1	1	£467
Cost	£200	£1.20	£2	£22	£12	£150	
Type	First (Electricity 668kWh/ Gas 148kWh)						Rest
Electricity	22.765p						9.88p
Gas	6.621p						3.05p

For electricity and gas, the rates are 22.765p and 6.621p per unit for the first 668 kWh and 148 kWh, the rest is 9.879p and 3.05p per unit, respectively

Let's assume there is a 20℃ temperature drop from the flue gas to the module hot side surface and the average temperature of cold side is 30℃, then the operating temperature difference is 130℃. With the conversion efficiency and thermal efficiency is 4% and 67% at 130℃ temperature difference, a running 24kW domestic boiler can produce 98W electricity and 1640W useable heat. To include the contribution from the solar energy, the availability of solar energy based on a 1m² solar collector is overlapped with that of boiler waste heat, shown in Tab. 2.

One of the tariffs of the electricity and gas from British Gas is shown in Tab. 1. Then the cost recovery period of this system can be calculated according to this tariff scheme. The annual saving by the electrical power and gas is £80.8 and £186.5, respectively. The labour hours are comprised of the time spent on fabricating the hot side and cold side heat ex-

changer, which are 10 hours and 16 hours, respectively. The labour cost is calculated as £650 in total using the pay rate of technicians in the Faculty of Engineering of the University of Nottingham at £25/hour. The total cost of building the system can be calculated as £1117. Therefore, the cost recovery period can be calculated as £1117/(£267/year) = 4.18 years. Here a clarification needs to be made that the costs are estimated in research laboratory environment, where the material cost and labour cost is higher than that could be realized in industrial environment. The reasons mainly lie in the higher material cost due to small purchase and likely ineffective material sourcing as well as lower efficiency in material use due to weak integration in laboratory environment. Namely, in the commercial case, the total cost can be reduced further which also means the cost recovery period could be less than 4.18 years.

Hourly availability of solar energy and boiler waste heat in the two-bedroom bungalow of each month in a whole year in Northampton (kW)   Tab. 2

Time	01:00	02:00	03:00	04:00	05:00	06:00	07:00	08:00	09:00	10:00	11:00	12:00	13:00	14:00	15:00	16:00	17:00	18:00	19:00	20:00	21:00	22:00	23:00	00:00
W	0.36	0.38	0.38	0.39	0.61	0.57	0.58	0.60	0.58	0.37	0.13	0.15	0.17	0.17	0.54	0.47	0.65	0.65	0.64	0.62	0.90	0.78	0.59	0.33
Jan	2.45	2.45	2.45	2.45	2.45	2.45	2.45	2.56	2.56	2.56	2.56	2.56	2.56	2.56	2.56	2.56	2.45	2.45	2.45	2.45	2.45	2.45	2.45	2.45
Feb	2.45	2.45	2.45	2.45	2.45	2.45	2.61	2.61	2.61	2.61	2.61	2.61	2.61	2.61	2.61	2.61	2.61	2.45	2.45	2.45	2.45	2.45	2.45	2.45
Mar	2.45	2.45	2.45	2.45	2.45	2.65	2.65	2.65	2.65	2.65	2.65	2.65	2.65	2.65	2.65	2.65	2.65	2.65	2.45	2.45	2.45	2.45	2.45	2.45
S/A	0.36	0.38	0.39	0.40	0.63	0.59	0.61	0.64	0.61	0.39	0.14	0.18	0.22	0.23	0.60	0.53	0.70	0.71	0.70	0.68	0.97	0.85	0.64	0.37
Apr	2.45	2.45	2.45	2.45	2.45	2.70	2.70	2.70	2.70	2.70	2.70	2.70	2.70	2.70	2.70	2.70	2.70	2.70	2.70	2.70	2.45	2.45	2.45	2.45
May	2.45	2.45	2.45	2.45	2.69	2.69	2.69	2.69	2.69	2.69	2.69	2.69	2.69	2.69	2.69	2.69	2.69	2.69	2.69	2.69	2.69	2.45	2.45	2.45
Jun	2.45	2.45	2.45	2.45	2.68	2.68	2.68	2.68	2.68	2.68	2.68	2.68	2.68	2.68	2.68	2.68	2.68	2.68	2.68	2.68	2.68	2.45	2.45	2.45
Oct	2.45	2.45	2.45	2.45	2.45	2.45	2.62	2.62	2.62	2.62	2.62	2.62	2.62	2.62	2.62	2.62	2.62	2.62	2.45	2.45	2.45	2.45	2.45	2.45
Nov	2.45	2.45	2.45	2.45	2.45	2.45	2.56	2.56	2.56	2.56	2.56	2.56	2.56	2.56	2.56	2.56	2.45	2.45	2.45	2.45	2.45	2.45	2.45	2.45
Dec	2.45	2.45	2.45	2.45	2.45	2.45	2.45	2.53	2.53	2.53	2.53	2.53	2.53	2.53	2.53	2.53	2.45	2.45	2.45	2.45	2.45	2.45	2.45	2.45
S	0.04	0.04	0.04	0.05	0.11	0.11	0.12	0.15	0.13	0.08	0.06	0.05	0.05	0.04	0.14	0.12	0.10	0.10	0.11	0.12	0.15	0.13	0.13	0.06
Jul	2.45	2.45	2.45	2.45	2.68	2.68	2.68	2.68	2.68	2.68	2.68	2.68	2.68	2.68	2.68	2.68	2.68	2.68	2.68	2.68	2.68	2.45	2.45	2.45
Aug	2.45	2.45	2.45	2.45	2.45	2.67	2.67	2.67	2.67	2.67	2.67	2.67	2.67	2.67	2.67	2.67	2.67	2.67	2.67	2.67	2.45	2.45	2.45	2.45
Sep	2.45	2.45	2.45	2.45	2.45	2.65	2.65	2.65	2.65	2.65	2.65	2.65	2.65	2.65	2.65	2.65	2.65	2.65	2.65	2.45	2.45	2.45	2.45	2.45

* W, S/A and S represent winter, spring/autumn and summer, respectively. The number in the same row represents the period of operation in each hour. The gold part represents the period when both the solar energy and boiler waste heat are available.

The hourly availability of solar energy and boiler waste heat is shown in Tab. 2.

With the aforementioned operational condition, the power output of this system can be predicted according to the availability of heat source in Tab. 2.

Hourly output of electrical power of each month in a whole year in Northampton (kWh)　　Tab. 3

W	01:00	02:00	03:00	04:00	05:00	06:00	07:00	08:00	09:00	10:00	11:00	12:00	13:00	14:00	15:00	16:00	17:00	18:00	19:00	20:00	21:00	22:00	23:00	00:00	Subtotal
Jan	34.9	37.0	37.7	38.0	59.4	55.7	56.6	62.0	59.6	37.5	13.8	15.7	17.2	17.9	55.2	47.7	63.8	64.0	62.5	60.8	88.7	76.8	58.1	32.6	35.8
Feb	34.9	37.0	37.7	38.0	59.4	55.7	60.4	63.2	60.7	38.2	14.0	16.0	17.5	18.2	56.2	48.6	68.0	64.0	62.5	60.8	88.7	76.8	58.1	32.6	36.2
March	34.9	37.0	37.7	38.0	59.4	60.2	61.2	64.0	61.5	38.7	14.2	16.2	17.7	18.5	57.0	49.3	68.9	69.2	62.5	60.8	88.7	76.8	58.1	32.6	36.7
S/A																									
Apr	34.8	37.1	38.2	39.6	61.9	64.2	65.4	68.7	66.3	41.9	15.3	19.6	23.7	24.9	64.3	56.7	75.9	76.6	75.1	73.5	94.9	83.03	62.41	36.22	40.3
May	34.8	34.8	34.8	34.8	38.2	38.2	38.2	38.2	38.2	38.2	38.2	38.2	38.2	38.2	38.2	38.2	38.2	38.2	38.2	38.2	38.2	34.8	34.8	34.8	27.7
Jun	34.8	37.1	38.2	39.6	67.8	63.8	65.0	68.3	65.9	41.7	15.3	19.5	23.6	24.8	63.9	56.4	75.4	76.1	74.7	73.1	104	83.0	62.4	36.2	40.6
Oct	34.8	37.1	38.2	39.6	61.9	58.3	63.6	66.8	64.5	40.8	14.9	19.1	23.1	24.2	62.5	55.2	73.8	74.5	68.3	66.8	94.9	83.0	62.4	36.2	39.2
Nov	34.8	37.1	38.16	39.59	61.93	58.29	62.06	65.22	62.93	39.81	14.56	18.59	22.5	23.7	61.0	53.8	68.9	69.6	68.3	66.7	94.9	83.0	62.4	36.2	38.6
Dec	34.8	37.1	38.2	39.6	61.9	58.3	59.4	64.6	62.3	39.4	14.4	18.4	22.3	23.4	60.4	53.3	68.9	69.6	68.3	66.8	94.9	83.0	62.4	36.2	38.4
S																									
July	4.15	3.80	4.06	4.45	12.0	11.3	12.9	15.9	13.7	9.03	6.57	5.72	4.95	4.10	15.5	12.4	10.3	10.3	12.0	12.5	15.7	13.0	12.5	5.53	7.2
Aug	4.15	3.80	4.06	4.45	11.0	11.3	12.8	15.8	13.7	8.98	6.54	5.69	4.93	4.07	15.4	12.3	10.2	10.2	11.9	12.4	14.3	13.0	12.5	5.53	7.1
Sep	4.15	3.80	4.06	4.45	11.0	11.2	12.7	15.7	13.6	8.93	6.50	5.65	4.90	4.05	15.3	12.2	10.1	10.2	11.8	11.4	14.3	12.5	12.5	5.53	7.0
																							Total		355

* W, S/A and S represent winter, spring/autumn and summer, respectively.

Hence, the annual output of electrical power of this system is 355kWh. The annual heat output can be calculated as 5941 kWh.

The carbon emission from natural gas and grid electricity is 0.18523kg per kWh and 0.53909kg per kWh. The saving of carbon emission in the use of electricity and gas can be calculated as 65.8kg and 3202.7kg, respectively. Hence, the total saving of carbon emission is 3268.5kg.

The real factor that will determine market success is the minimal acceptable cost. It was concluded that boilers need 700 watts of power to achieve the goal of self-powering, which came down to 100 watts reported in a later report done Gasunie research[33]. It also concluded that thermoelectric costs would have to come down to below £6.3/watt to achieve the self-powering goal.

4.2 Environmental Impact

Bismuth telluride is a compound of bismuth and tellurium, a gray powder. It is a semiconductor which is an efficient thermoelectric material for refrigeration or power generator. Bismuth telluride comprises some of the best performing room temperature thermoelectrics. The single crystalline bismuth telluride compounds can be grown by using the Czochralski method. They can be obtained with directional solidification from melt or powder metallurgy processes.

It is prepared by sealing a sample of bismuth and telluride metal in a quartz tube under vacuum and heating it to 800℃ in a muffle furnace. Exposure (at 15mg/m³ for 6 hr/day, 5 days/week) to the dust of undoped bismuth telluride (0.4micron) does not impose adverse influence on animals like dogs, rabbits and rats[34]. For human being, exposure to undoped bismuth telluride can occur through inhalation, eye/skin contact and ingestion. Acute exposure to undoped bismuth telluride can cause local irritation of eyes and skin[35]. No signs or symptoms of chronic exposure to the undoped bismuth telluride have been reported. However, suggestions have been given on avoidance and rescue in case of exposure to it[36].

When it is used in residential house, the long lifespan of continuous operation of thermoelectric generator (up to 11 years) makes sure the system operate for a considerably long time. The ability of recovering the waste heat from the boiler flue gas improves the fuel utilisation efficiency of domestic boiler and at the same time utilise the free solar energy. The thermoelectric cogeneration system partially offsets the need of electrical power and thermal energy.

5 Conclusion

Thermoelectric generators have found wide applications in different areas due to reliable operation, nomoving parts and compact structure. The experimental results prove that the thermoelectric cogeneration concept is able to deliver much higher energy utilisation efficiency by generating electrical power and producing thermal energy. This research has been looking into the issues associated with adopting the thermoelectric technology in an economically viable way by studying the single cell thermoelectric cogeneration system.

The cost recovery period of deploying this

system in a residential house installed with a 24kW boiler and 1m² solar collector has been evaluated. Based on the conversion efficiency and thermal efficiency is 4% and 67% at 130℃ temperature difference, the house can produce 98W electricity and 1640W useable heat when the boiler is running and the contribution from the solar energy is included. It takes less than 4. 2 years to recover the system cost.

References

[1] Boiler efficiency database. http://www. sedbuk. com/.

[2] X F Zheng, C X Liu, Y Y Yan et al. A review of thermoelectrics research-recent developments and potentials for sustainable and renewable energy applications. Renewable and Sustainable Energy Reviews, 2014(32):486-503.

[3] Hsu C T, Huang G Y, Chu H S, et al. Experiments and simulations on low-temperature waste heat harvesting system by thermoelectric power generators. Applied Energy, 2011 (88): 1291-1297.

[4] Love N D, Szybist J P, Sluder C S. Effect of heat exchanger material and fouling on thermoelectric exhaust heat recovery. Applied Energy, 2012(90):322-328.

[5] Kajikawa T, Ito I M, Katsube I,et al. Development of Thermoelectric power generation Utilising Heat of Combustible Solid Waste. In Proc. 13th int. conf. on thermoelectrics, Kansas City, USA, 1994:314-318.

[6] Kajikawa T. Thermoelectric power generation systems recovering heat from combustible solid waste in Japan. In Proc. 15th int. conf. on thermoelectrics, Pasadena, USA, 1996:343-351.

[7] Gou X L, Xiao H, Yang S W. Modelling, experimental study and optimization on low-temperature waste heat thermoelectric generator system. Applied Energy,2010(87):3131-3136.

[8] Gueler N F, Ahiska R. Design and testing of a microprocessor-controlled portable thermoelectric medical cooling kit. Applied Thermal Engineering,2002(22):1271-1276.

[9] Chatterjee S, Pandey K G. Thermoelectric cold-chain chests for storing/transporting vaccines in remote regions. Applied Energy, 2003 (76): 415-433.

[10] Rowe D M. Applications of nuclear-powered thermoelectric generators in space. Applied Energy,1991(40):241-271.

[11] Boukai A I, Bunimovich Y, Tahir-Kheli J, et al. Silicon nanowires as efficiency thermoelectric materials. Nature, 2008(451):168-171.

[12] Hochbaum A I, Chen R K, Delgado R D, et al. Enhanced thermoelectric performance of rough silicon nanowires. Nature, 2008(451):163-167.

[13] Heremans J P, Jovovic V, Toberer E S, et al. Enhancement of theremoelectric efficiency in Pbte by distortion of the electronic density of states. Science,2008(321): 554-557.

[14] Hsu K F, Loo S, Guo F,et al. Cubic AgPbmSbTe2+m: Bulk thermoelectric materials with high figure of merit. Science, 2004 (303): 818-821.

[15] Killander A, Bass J. A stove-top generator for cold areas. 15th Int. Conf. on Thermoelectrics, Pasadena, USA,1996:390-393.

[16] Champier D, Bedecarrats J P, Rivaletto M, et al. Thermoelectric power generation from biomass cook stoves. Energy 0, 2010 (35): 935-942.

[17] Nuwayhid R Y, Rowe D M, Min G. Low cost stove-top thermoelectric generator for regions with unreliable electricity supply. Renewable Energy,2003(28):205-222.

[18] Mastbergen D, Willson B, Joshi S. Producing light from stoves using a thermoelectric generator. Report of the Energy Conversion Lab Colorado State University (2005). Available at: www. vrac. iastate. edu/ethos/files/ethos2005/pdf/mastbergen. pdf.

[19] Allen D, Wonsowski J. Thermoelectric self-powered hudronic heating demonstration. In Proc 16th int conf on thermoelectric, Dresden, Germany, 1997:571-574.

[20] Allen D, Mallon W. Further development of "self-powered boilers". In Proc 18th int conf on thermoelectrics, Baltimore, USA, 1999:80-83.

[21] X F Zheng, C X Liu, R Boukhanouf. Experimental study of a domestic thermoelectric cogeneration system. Applied Thermal Engineering, 2014. https://doi. org/10. 1016/j. applthermaleng. 2013. 09. 008.

[22] Zheng X F, Yan Y Y, Simpson K. A potential candidate for the sustainable and reliable domestic energy generation-thermoelectric cogeneration system. Applied Thermal Engineering, 2012. doi: 10. 1016/j. applthermaleng. 2012. 03. 020.

[23] Min G, Rowe D M. Symbiotic application of thermoelectric conversion for fluid preheating/power generation. Energy Conversion and Management, 2002(43):221-228.

[24] Qiu K, Hayden A C S. A natural-gas-fired thermoelectric power generation system. Journal of Electronic materials, 2012 (38): 1315-1319.

[25] Rowe D M. CRC handbook of thermoelectrics. CRC Press, 1995.

[26] G Bennett. Space applications Chapter 41 CRC Handbook of Thermoelectrics. Ed. D. M. Rowe, CRC Press.

[27] Ziman JM. Electrons and phonons. London: Cambridge University Press.

[28] Callen, HB, Phys. Rev. , 73, 1349, 1948.

[29] Drabble J R, Goldsmid H J. Thermal conduction in semiconductors. Oxford:Pergamon Press,1961.

[30] X F Zheng, C X Liu, Y Y Yan. Investigation on an oriented cooling design for thermoelectric cogenerations. Journal of Physics: Conference Series,2012,395(1):012062.

[31] Sano S, Mizukami H, Kaibe H. Development of high-efficiency thermoelectric power generation system. Komai'su Technical Report, 2003,49(152).

[32] M Suri, T A Huld, E Dunlop. Potential of solar electricity generation in the European Union member states and candidate countries. Solar Energy,2007(81):1295-1305.

[33] Auxiliary power generation for gas heating systems: evaluation of the business potential. Gas Research Institute,1995, GRI-95/0343.

[34] Cralley L J, Cralley L V. Patty's industrial hygiene and toxicology. 2nd ed. New York, NY: John Wiley and Sons,1985(3).

[35] NJDH. Hazardous substance fact sheet: bismuth telluride. Trenton, NJ: New Jersey Department of Health,1985.

[36] U. S. Department of health and human services. Occupational safety and health guideline for bismuth telluride, undoped, available at: http://www. cdc. gov/niosh/docs/81-123/pdfs/0056. pdf.

Simulation-based Method for Determing Parameters of Material Price Adjustment Model in Large-scale Hydropower Development in China

Lijing Yang [1]   Zhenyuan Liu[1,2]   Xi Chen[3]

Yaodi Fan[3]   Hualin Chen[1]

(1. School of Artificial Intelligence and Automation, Huazhong University of Science and Technology, Wuhan 430074, China;

2. Key Laboratory of Education Ministry for Image Processing and Intelligent Control, Wuhan 430074, China;

3. Yalong River Hydropower Development Company LTD, Chengdu 610051, China)

【Abstract】 In the construction of large-scale hydropower projects in China, enormous amounts of materials are consumed. Influenced by various factors, fluctuation of materials price is very common during the construction. To protect the benefits of the demander and the supplier, reasonable material price adjustment methods need to be set in the procurement bidding documents. A price adjustment model is established considering price fluctuation stratifying and risk sharing. A simulation-based method is proposed for determing parameters in the model, and the simulation process is optimized with the optimal computing budget allocation algorithm. Based on the judgment of engineers and real data of a hydropower project in Sichuan, China, and simulation cases are generated and the experimental results are analyzed with the TOPSIS. The results show that the simulation-based method provides scientific and reasonable support for determing parameters of material price adjustment model, and the optimal computing budget allocation algorithm plays an optimizing role.

【Keywords】 Large-scale Hydropower Development; Material Price Adjustment Parameters Determing; Simulation; Optimal Computing Budget Allocation; TOPSIS

1 Introduction

The construction cycle of large-scale hydropower projects is long and the investment is huge in China. The cost of bulk materials often accounts for $50\% \sim 70\%$ of the total project investment (Yu et al., 2013). Influenced by various factors, materials price often fluctuates (Yang et al., 2010). Due to the huge consumption, slight fluctuation in price during the contract period could lead to a settlement amount change of up to millions of yuan. To protect the interests of both the supplier and the demander, corresponding material price adjustment method is usually set in the procurement bidding.

Many researchers have studied how to formulate material price adjustment method. Zhang (2012) pointed out that comprehensive adjustment method is used widely, the core of which is to measure the actual market price fluctuation of materials by the change of information price index published by authoritative organizations, and there are many related applications and researches. Based on the comprehensive adjustment method, Zhang (2005) found that there was a certain linear relationship between the price fluctuation ratio published by authoritative institutions and the price fluctuation ratio in the actual market, and proposed the corresponding solution. Based on the Delphi method, Choi et al. (2006) concluded that the deduction rate should be introduced into the price adjustment. Lam et al. (2007) found that the two parties to the material procurement contract need to share the risk of the price adjustment difference caused by price volatility. Nasirzadeh et al. (2014) proposed a comprehensive fuzzy system dynamics method for quantitative determination of the sharing proportion. Jin (2011) used simulation theory and neuro-fuzzy techniques. Embrechts et al. (2017) addressed the problem of risk sharing among agents using a two-parameter class of quantile-based risk measures, the so-called Range-Value-at-Risk (RVaR), as their preferences.

In general, in the field of material price adjustment, the theoretical framework of material price adjustment method and risk sharing of the price adjustment difference is studied to some degree, but there are no in-depth studies on the stratification of price fluctuation and the quantitative determination of specific parameters in the price adjustment method. Based on a large-scale hydropower construction project in China, methods of system simulation and statistics are applied to provide decision support for determing the price adjustment method in material procurement bidding from the perspective of the tenderee in this paper.

2 Selection of Material Price Adjustment Methods for Large Hydropower Projects

2.1 Problem Description and Model Establishment

In the management of large-scale project construction, the supplier of bulk raw materials needed for construction is usually selected by the owner through unified bidding. In the bidding contract, the comprehensive adjustment method is used as reference by the bid inviter to formulate corresponding price adjustment method, that is, to determine the price adjustment model and several parameters of it (Cao et al. ,2013).

In the procurement bidding of materials, the bidder is the material supplier, while the tenderer is the project owner. The purpose of the study is to provide decision support for the project owner to formulate a scientific and reasonable price adjustment method in the procurement bidding contract.

Based on the comprehensive adjustment method, fluctuation stratification is set in this paper, and so as the risk sharing parameters in corresponding fluctuation range.

The model is established as follows:

Information price fluctuation ratio in month j:

$$V_j = \frac{M_j - M_0}{M_0} \qquad (1)$$

Price fluctuation ratio in month j after risk sharing is considered:

$$U_j = \begin{cases} \beta_1 \cdot (V_j - \alpha_0), \alpha_0 \leqslant V_j < \alpha_1 \\ \beta_1 \cdot (\alpha_1 - \alpha_0) + \beta_2 \cdot \\ \quad (V_j - \alpha_1), \alpha_1 \leqslant V_j < \alpha_2 \\ \quad\quad \vdots \\ \sum_{k=1}^{n-1} \beta_k \cdot (\alpha_k - \alpha_{k-1}) + \beta_n \cdot \\ \quad (V_j - \alpha_{n-1}), \alpha_{n-1} \leqslant V_j \leqslant \alpha_n \end{cases} \qquad (2)$$

The monthly settlement price of materials in month j:

$$P_j = P_0 \cdot (1 + U_j) \qquad (3)$$

Total settlement amount for m months:

$$S = \sum_{j=1}^{m} Q_j \cdot P_j \qquad (4)$$

In fluctuation range k, the owner's risk-sharing proportion of price adjustment difference shall meet the constraint:

$$0 \leqslant \beta_k \leqslant 1 \qquad (5)$$

In the above model, the determination of authoritative price and base period parameters are relatively fixed, so they will not be discussed in this paper. The key is the selection and determination for parameters including the number n of fluctuation ranges, the upper bound α_k of each fluctuation range, and the risk-sharing proportion β_k ($k = 1, 2, \cdots, n$).

2. 2 Decision-making Methods for Determing Parameters of Material Price Adjustment Model

In the previous management, experience-based decision-making method and prediction-based decision-making method for determing parameters of engineering materials price adjustment model have been applied. However, the experience-based method relies heavily on the decision makers and lacks scientific rationality. While the prediction-based method cannot obtain credible results as the predicted situation is not general. Therefore, from the perspective of the bid inviter, a simulation-based decision-making method is proposed in this paper.

The process of simulation-based method is shown in Fig. 1. Firstly, the contract benchmark price, the authoritative price and several combinations of price adjustment parameters are selected. Then the decision makers provide a fluctuation range about future materials price based on analysis of historical data and their own experience. Then a large number of simulation cases are generated with the OCBA algorithm. Next, the sets of the monthly unit materials price under different combinations of price adjustment parameters are calculated respectively. Finally, suitable combinations of price adjustment parameters can be selected by the comprehensive evaluation analysis.

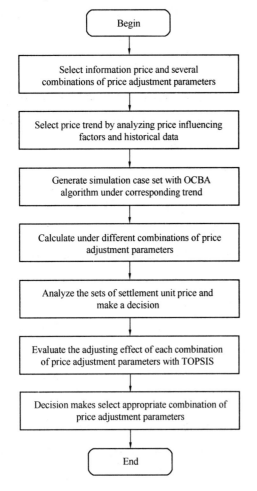

Fig. 1 Flowchart of simulation-based method for determining parameters of price adjustment model

Decision makers can estimate various kinds of materials price trends, and then choose the combinations of price adjustment

parameters appropriate for many trends, which is more reasonable and practical.

3 Key Components of Simulation-based Method for Determing Parameters of Material Price Adjustment Model

3.1 Selection of Price Trend

The problem of selecting price trend can be divided into two sub-problems: determing the number of segments for price trend and determing the fluctuation situation in each segment. After analyzing the relevant historical data of the procurement materials (cement) of a hydropower project in China, the following settings are made:

(1) According to the statistical analysis of historical data, distribution F_1 of price fluctuation ratio of material historical information price can be obtained after selecting the benchmark price, which is described by frequency distribution table.

(2) Based on historical data, the month-on-month fluctuation ratio distribution F_2 of materials price can be obtained.

3.2 Generation of Simulation Case Set

A large number of simulation cases need to be generated in the price trend estimated. Each case is a line graph composed of discrete points representing the monthly price fluctuation ratio V_j in the settlement month, and the points in the same segment share the same fluctuation range.

The generation of a simulation case is divided into two steps. First, generate the points in each segment and connect them according to the selected number of segments and the fluctuation range in each segment. Second, connect the curves generated in each segment to make up a complete simulation case.

3.3 Optimization for Generation of Simulation Case Set

In the case generation mentioned above, the overall distribution F'_1 of the final case set will be different from the historical price fluctuation ratio distribution F_1, because continuity of the case curve was preferentially considered. Therefore, the optimal computing budget allocation algorithm (hereinafter referred to as OCBA) (Chen et al., 2000) will be adopted to optimize the case generation. In each stage, the number of cases generated within each fluctuation range in the next stage is adjusted according to the distribution of the generated cases and the allocation rules of the computational load.

In this paper, the allocation rules of computational load are based on the following settings: a wide fluctuation range is subdivided into several subranges as different schemes, and the frequency of the subranges in the wide fluctuation range is regarded as the performance value of the scheme.

3.4 Evaluation Analysis of Simulation Results

Stability and marketability of each combination of price adjustment parameters are evaluated through the statistical analysis of the calculated results. The two qualitative indicators, stability and marketability, are characterized by the following four quantitative indicators. Quantitative data and intuitive graphics can provide decision support for decision makers to select the combination of price adjustment parameters.

(1) Deviation rate of settlement difference μ

$$\mu = \frac{\sum\limits_{j=1}^{m} Q_j \cdot P_j - \sum\limits_{j=1}^{m} Q_j \cdot P_0}{\sum\limits_{j=1}^{m} Q_j \cdot P_0} \tag{6}$$

μ denotes the difference between the total settlement amount after price adjusting and the total settlement amount without price adjustment. The smaller the value, the better the stability of the combination of price adjustment parameters.

(2) Goodness of fit R_1^2 between the adjusted price fluctuation ratio (U_j) and 0 (0 indicates that the price has not changed)

$$R_1^2 = \frac{\sum\limits_{j=1}^{m} (0-U)^2}{\sum\limits_{j=1}^{m} (U_j - U)^2} \tag{7}$$

U is the mean of U_j, the value of R_1^2 is between $[0,1]$, and the closer the value is to 1, the closer the settlement price is to the ex-factory price, which means better stability.

(3) Goodness of fit R_2^2 between the adjusted price fluctuation ratio U_j and the original fluctuation ratio V_j

$$R_2^2 = \frac{\sum\limits_{j=1}^{m} (V_j - U)^2}{\sum\limits_{j=1}^{m} (U_j - U)^2} \tag{8}$$

The value of R_2^2 is between $[0,1]$, and the

closer its value is to 1, the closer the settlement price generated is to the actual situation of the market, which means better marketability.

(4) Monthly settlement price distribution

In this paper, the box diagram shown in Fig. 2 is adopted to show the calculation results of a large number of cases under different combinations of price adjustment parameters. Each box in the horizontal direction represents the statistical distribution of the calculation results of a combination of price adjustment parameters.

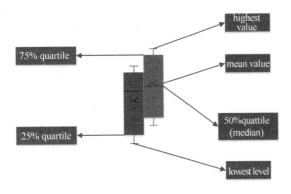

Fig. 2　Box diagram

The overall position of the box represents the monthly settlement unit price generated by corresponding combination of price adjustment parameters, and the length of the box represents the distribution range of the final monthly settlement unit price. The decision makers should choose the combinations of price adjustment parameters corresponding to appropriate box position and box length.

It is a multi-criterion decision problem to conduct the comprehensive selection of price adjustment parameter combination according to these four indicators. The TOPSIS method can be used to provide quantitative support for the selection of price adjustment parameter combination.

4　Project Material Price Adjustment Simulation Experiment

4.1　Design of Simulation Experiment

The experimental data comes from the relevant historical data of cement procurement of the hydropower project in Yalong river basin, Sichuan province. A simulation case is set to be composed of 36 discrete points representing monthly information price fluctuation ratio. According to the historical data, the fluctuation range of each situation is set by k-means as follows:

The situation of substantial increase: $[0.24, +\infty)$, that is, the material information price index M_j as increased by more than 24% relative to the benchmark price M_0;

The situation of moderate increase: $[0.06, 0.24)$;

The situation of slight increase: $[0, 0.06)$;

The situation of slight decline: $[-0.05, 0)$;

The situation of moderate decline: $[-0.16, -0.05)$;

The situation of substantial decline: $(-\infty, -0.16)$.

Decision makers make the following estimation: the trend is divided into three segments evenly. For the first segment (the first 12 points), the fluctuation situation includes situation of slight increase and the situation of slight decline, that is, the generated discrete points range in $[-0.05, 0.06]$. In the second and the third segment, the generated discrete points range in [−0.16, 0.24].

This calculation experiment generates two kinds of simulation case sets, one is set with 1000 cases generated directly; the other is set with 1000 cases generated with the OC-BA algorithm in five stages, which is used for comparative experiment.

In this experiment, five combinations of price adjustment parameters (hereinafter referred to as "CPAP") shown in Tab. 1 are selected for comparative calculation. The fluctuation ratio of the price set here will not exceed 1.

The weight vector can be obtained from the judgment matrix established by the decision makers. For convenience, goodness of fit R_1^2 and goodness of fit R_2^2 are replaced by GF1 and GF2 respectively, deviation rate of the settlement difference is replaced by DRSD in the following tables and diagrams.

4.2 Simulation Experiment and Decision Analysis

4.2.1 Effect Analysis of Generating Simulation Case Set

In the case generation stage, two case sets are generated, and the distribution of the generated two case sets and the historical data in the fluctuation range [−0.1, 0.1] is shown in Fig. 3. As can be seen that the overall distribution F_1' of the simulation case set generated by the OCBA algorithm is closer to the historical distribution F_1. More practical simulation cases can make the decision-making results are more reasonable and reliable.

Five CPAPs　　　　　　　　　**Tab. 1**

	$[\alpha_0 , \alpha_1)$	β_1	$[\alpha_1 , \alpha_2)$	β_2	$[\alpha_2 , \alpha_3)$	β_3
CPAP 1	[0,0.05)	1	[0.05,1]	1		
CPAP 2	[0,0.05)	0	[0.05,1]	1		
CPAP 3	[0,0.1)	0	[0.1,1]	1		
CPAP 4	[0,0.05)	0	[0.05,0.15)	1	[0.15,1]	0.5
CPAP 5	[0,0.1)	0	[0.1,0.2)	1	[0.2,1]	0.5

Note: CPAP = Combination of Price Adjustment Parameters.

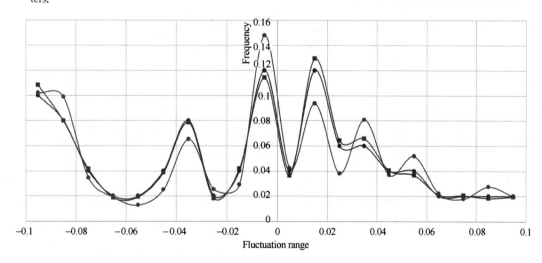

Fig. 3　Generated case distribution

4. 2. 2 Decision Analysis of Material Price Adjustment Parameters

1. Prediction-based Method

Referring to the actual management experience, a price fluctuation curve in the next three years is predicted by the simple moving average method in this experiment. The distribution of monthly settlement unit price under this price fluctuation curve of five CPAPs is shown in Fig. 4.

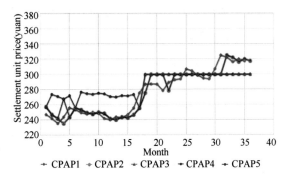

Fig. 4 Monthly settlement unit price display of
5 CPAPs by prediction-method

As can be seen that the distribution range of monthly settlement unit price of CPAP 1 is the largest, and the effect of price adjustment is unstable; the distribution ranges of CPAP 3, CPAP 4 and CPAP 5 are relatively small and the price adjustment effect is stable.

The predicted-based method can compare results under different CPAPs by data and graphs. However, there is no quantitative standard to measure the analysis results of the prediction-based method, and the overlapping of curves makes it impossible to compare and analyze the price adjustment effects of various CPAPs intuitively at the same time, so it cannot support decision well.

2. Simulation-based Method

Tab. 2 and Fig. 5 demonstrates the simulation results of the five CPAPs.

Evaluation results of 5 CPAPs   Tab. 2

Indicator	DRSD	R_1^2	R_2^2	Box Height	Box Length	D_i^+	D_i^-	C_i
Weight	0. 1589	0. 3350	0. 3089	0. 0846	0. 1117			
CPAP 1	0. 4746	0. 4468	0. 4513	1. 0000	0. 6026	0. 3605	0. 0563	0. 1350
CPAP 2	0. 5499	0. 4486	0. 4487	0. 0000	0. 5122	0. 1463	0. 3363	0. 6968
CPAP 3	0. 6145	0. 4520	0. 4407	0. 0000	0. 3954	0. 1611	0. 3396	0. 6783
CPAP 4	0. 2176	0. 4427	0. 4499	0. 0000	0. 3570	0. 0171	0. 3768	0. 9566
CPAP 5	0. 2176	0. 4459	0. 4454	0. 0000	0. 3013	0. 0048	0. 3802	0. 9875

Note: CPAP = Combination of Price Adjustment Parameters; DRSD = Deviation Rate of Settlement Difference.

As can be seen from Tab. 2 that CPAP 2 corresponds to the DRSD with relatively higher value, and GF2 with relatively higher value, which suggests that the effect of the CPAP 2 is not stable, and the monthly settlement price generated is closer to the actual market price. The data shown in Fig. 5 can verify this conclusion: the box diagram corresponding to the CPAP 2 is relatively longer, which indicates that the corresponding distribution range of monthly settlement unit price is wider and the effect is unstable.

For the comprehensive evaluation analysis, the TOPSIS method is adopted. The re-

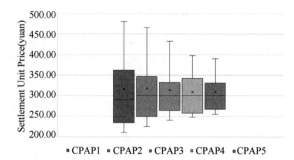

Fig. 5 Distribution of monthly settlement price after adjusted by 5 CPAPs

sults of evaluation and analysis of the combination of various adjustment parameters are shown in Tab. 2. As can be seen that CPAP 5 ranks first and CPAP 4 ranks second with close scores, which indicates that under the current price trend of the decision maker's choice, CAPA 4 and CPAP 5 consider both the marketability and stability of price adjustment well. The decision makers can choose the price trend again and the combination of price adjustment parameters with good effect under various price trends.

Through calculation and analysis of the example, we can get the following conclusion: compared with the prediction-based method, the simulation-based method for determing parameters of material price adjustment model can provide intuitive data comparison and show the price adjustment effect of various CPAPs with quantitative indicators. Therefore, it can provide better decision support for the decision makers to select the optimal CPAP.

5 Conclusion

In this paper, asimulation-based method for determing parameters of material price adjustment model is proposed. Through simulation and analysis, specific values are used to provide decision support for the determination of price adjustment parameters. The experimental results show that the proposed simulation-based method for determing parameters of material price adjustment model is more scientific and reasonable than other decision-making methods, not only of great practical significance to the material price adjustment of large-scale hydropower projects in China, but also of referable significance to the material price adjustment of other large-scale projects and even the material price adjustment in other fields.

However, there are still some problems to be studied. For example, only qualitative strategies are provided for parameters in the model, such as fluctuation stratification number and risk sharing proportion. Only some qualitative rules are given for price trend segmentation; the number of stages and the number of cases generated in each stage in simulation optimization have not been studied deeply. In the future, the theory and rules of quantitative determination of these problems can be studied to make the simulation-based method for determing parameters in material price adjustment model and its optimization method more scientific and effective.

Acknowledgements

This work is supported partially by the National Natural Science Foundation of China 71732001, the Science Foundation of Yalong

River Hydropower Development Company LTD, China, and the Fundamental Research Funds for the Central Universities, HUST: 2017KFYXJJ178.

Data Availability Statement

Some or all data, models, or code that support the findings of this study are available from the corresponding author upon reasonable request.

- Historical information price

Notation

The following symbols are used in this paper:

j = Month number in the contract period, $j = 1, 2, \cdots, m$;

k = Fluctuation stratification sequence number, $k = 1, 2, \cdots, n$ (here n denotes the stratification number of price fluctuation);

F_1 = Distribution of price fluctuation ratio of material historical information price;

F'_1 = Distribution of price fluctuation ratio of generated simulation case set;

F_2 = Distribution of month-on-month price fluctuation ratio;

M_0 = The benchmark price in the contract;

M_j = Information price index of materials published by authorities in month j (referred to as "authoritative price");

P_0 = Ex-factory price of materials in the contract;

P_j = The monthly settlement unit price in month j;

Q_j = Material consumption in month j;

S = The total settlement amount for m months;

U_j = Price fluctuation ratio in month j after adjustment;

V_j = Information price fluctuation ratio in month j;

α_k = The upper bound of the fluctuation range k;

β_k = The owner's risk sharing proportion of price adjustment difference in the fluctuation range k value in $[0, 1]$.

References

[1] Barbarosoǧlu G, Arda Y. A two-stage stochastic programming framework for transportation planning in disaster response. Journal of the Operational Research Society, 2004, 55(1): 43-53.

[2] Cao D Z, Sun Y, Guo H R. Optimizing maintenance policies based on discrete event simulation and the OCBA mechanism. In: 2013 Annual Reliability and Maintainability Symposium. Orlando: IEEE, 2013:1-6.

[3] Chen C H, Lin J W, Enver Y. Simulation budget allocation for further enhancing the efficiency of ordinal optimization. Discrete Event Dynamic Systems: Theory and Applications, 2000, 10(3): 251-270.

[4] Choi M, Kim J, Kim M. A study on the price escalation system in a construction contract. KSCE Journal of Civil Engineering, 2006, 10(4): 227-232.

[5] Dong J F, Hua K Q, Cai Y F. OCBA theory based on physical meaning and the perfection of OCBA theory of probability analysis. Automation and instrumentation (in Chinese), 2005, 21 (S1): 16-21.

[6] Embrechts P, Liu H Y, Wang R D. Quantile-Based Risk Sharing. Swiss Finance Institute Re-

search Paper Series, 2017.

[7] Horng S C, Yang F Y, Lin S S. Applying PSO and OCBA to minimize the overkills and re-probes in wafer probe testing. IEEE Transactions on Semiconductor Manufacturing, 2012,25 (3): 531-540.

[8] Ilbeigi M, Castro-Lacouture D. Effects of price-adjustment clauses on number of bidders and dispersion of bid prices in highway construction. 2017,33(4): 04017013.

[9] Jin X H. Model for efficient risk allocation in privately financed public infrastructure projects using neuro-fuzzy techniques. Journal of Construction Engineering and Management, 2011, 137 (11): 1003-1014.

[10] Lam K C, Wang D, Lee P T K. Modelling risk allocation decision in construction contracts. International Journal of Project Management, 2007,25(5): 485-493.

[11] Ma X. Research on resource allocation in simulation optimization. Dissertation for the Master Degree. Shanghai: Fudan University (in Chinese),2010.

[12] Nasirzadeh F, Khanzadi M, Rezaie M. Dynamic modelling of the quantitative risk allocation in construction projects. International Journal of Project Management, 2014,32(3): 442-451.

[13] Seneviratne P N. Costs and benefits of the price adjustment clause in FIDIC MDB. In: TRB 92nd Annual Meeting Compendium of Papers. Washington,2013:13-2312

[14] Yang G Sh, Huang G, Gao H. Research on the strategy of construction material procurement when the price fluctuation is considered. In: 2010 International Conference on Management and Service Science. Wuhan: IEEE, 2010:1-4.

[15] Yu X M, Geng D Y, Xie Z B. Research on Materials for Construction Projects Cost Control Taking a Project in Shijiazhuang as an Example. Applied Mechanics and Materials, 2013 (438-439): 1846-1850.

[16] Zhang M D. Application of price adjustment formula in international project contracting. China Science and Technology Information (in Chinese), 2012,24(12): 107-108.

[17] Zhang S. Study on the adjustment method of construction contract price when the price of building materials fluctuates (in Chinese). Dissertation for the Master Degree. Hangzhou, Zhejiang University (in Chinese),2005.

The Multi-period Immediate Delivery Problem with a Comprehensive Indicator for Order Selection

Xiaosong Ding[1,2]   Hu Liu[1]   Xi Chen

(1. International Business School, Beijing Foreign Studies University

100089 Beijing, China;

2. DECIDE Group, Department of Computer and Systems Sciences,

Stockholm University

Postbox 7003, 16407 Kista, Sweden)

【Abstract】 The immediate delivery problem is a variant of vehicle routing problem with stochastic demand, in which not all customers are known in advance, but are revealed while the system progresses. In this paper, we investigate a multi-period immediate delivery problem with a comprehensive indicator for order selection. In particular, we model the order selection as a sequential decision-making problem. Considering the immediacy of delivery tasks, the mismatch between demand and capacity, and the autonomy of the delivery staff, we propose herein a new comprehensive indicator consisting of distance, time pressure, and order revenue as the order selection criterion to help drivers choose the right order. Then the nearest neighbor algorithm is adopted to optimize routes dynamically. Finally, extensive simulation experiments show that the proposed indicator performs stable and better than the other single indicators.

【Keywords】 Vehicle Routing Problem; Dynamic Requests; Multi Period; Order Selection; Nearest Neighbor Algorithm

1 Introduction

Real-time logistics is a data-based delivery service that matches real-time demand and capacity through a global scheduling way(IResearch, 2020a). In recent years, with the rise of O2O (online to offline), online demand has achieved explosive growth, most notably in industries such as food delivery, fresh fruits and vegetables, retail convenience, and medical health (Meituan, CFLP, 2019). The current decline in the proportion of dine-in meals due to the COVID-19 has led to a rapid increase in demand for immediate consumption.

Meituan Dianping (Meituan, CFLP, 2019) pointed out that in 2020, Gross Transaction Volume (GTV) of its food delivery business increased by 24.5% to RMB 488.9 billion. The growth rate for the number of food delivery transactions continued to surge, with the daily average number of food delivery transactions increasing by 16.0% to 27.7 million. In terms of delivery and capacity, various emerging modes such as crowd-sourcing, full-time, and agent greatly ease capacity pressure, which provides diverse services for surging demand generated during the outbreak (IResearch, 2020). Therefore, in the face of large number of orders and multiple modes of capacity, how to design an effective just-in-time delivery system has become an important challenge for those instant logistics companies.

Two major decisions need to be considered in the instant logistics and delivery problem, namely order dispatch and routing optimization. The driver starts from the depot and delivers goods to the corresponding customer according to the planned route, and new potential orders appear after each delivery is completed. Considering his/her own constraints, a driver picks up the customers to be served from the potential orders and re-plans the service path dynamically. In order to improve order completion rates and overall customer satisfaction levels, traditional order dispatch is done through a centralized "platform-assign-order-to-driver" mode, i.e., using the "full-time" capacity model(Xu et al., 2018). However, order demand may be different at different times. When the rate of requests is too high, and the supply and demand are not in harmony, the centralized way may increase a driver's working hours and provide poor service quality. Facing the severe situation of the epidemic makes many riders choose to reduce their working times, coupled with the increase in demand during the Spring Festival, resulting in seriously inadequate capacity. In order to achieve a rapid improvement in delivery efficiency, crowd-sourcing will become the primary delivery mode in the post epidemic era (IResearch, 2020). Therefore, this paper will develop order selection algorithm from a driver's perspective.

Order dispatch, which aims to find the best matching between drivers and customers, directly affects the user experience and vehicle operational efficiency (Zhang et al., 2017). Glaschenko describes a multi-agent approach to real-time scheduling, especially applicable to taxi service (Glaschenko et al., 2009). The traditional order dispatch system employs the global positioning system (Liao, 2001) (GPS) or the latest positioning technology for matching passengers with the nearest available cabs (Liao, 2003), thus achieving greater productivity and customer satisfaction. Based on the nearest-coordinate method, Lee proposed an alternative dispatch system, i.e., the taxi assigned the booking job is the one with the shortest time path, reaching the customer in the shortest time (Lee et al., 2004). This dispatch ensures that customers are served within the shortest period of time and increases customer satisfaction. However, merely increasing individual customer satisfaction is a local endeavor, without considering the effects of the assignment on other awaiting customers in the request queue (Seow et al., 2009).

The crowd-sourcing mode emphasizes the driver's autonomy, i. e., drivers can select orders that are more in line with their conditions from a large number of potential orders at any time, thereby improving their satisfaction level. In addition to the traditional distance factor, time pressure and order revenue are also important factors that affect the driver satisfaction. Under the immediate delivery system, customers generally give drivers a relatively flexible delivery time, and order revenue can be estimated (Xu et al., 2018). Time pressure, also known as the time window, refers to the time slot that a customer needs to be served, reflecting the driver's delivery efficiency (Kallehauge et al., 2005). Order revenue refers to the amount of money the driver can get by serving the customer and directly reflects the driver's income and satisfaction level. As important factors in the logistics delivery system, time pressure and order revenue are commonly overlooked, so we propose an order selection indicator that comprehensively considers time pressure, delivery time, and revenues from the driver's perspective.

Beside order selection, routing optimization is also crucial. In order to improve the efficiency of delivery, it is necessary to solve the vehicle routing problem (Dantzig, Ramser, 1959; Korte, Vygen, 2009; Cordeau et al., 2002). The static vehicle routing problem generally assumes that customer information is fixed and known before routing (Zhang et al., 2010). In contrast, the dynamic vehicle routing problem emphasizes that information such as customer locations, demands, and driver's states is dynamically changing (Psaraftis, 1995; Bertsimas, 1992; Gendreau et al., 1996; Dror, Powell, 1993). Designing continuous real-time optimization strategy and corresponding algorithm are essential for solving the dynamic vehicle routing problem (Li et al., 2021). Greedy algorithm is a basic algorithm for solving combinatorial optimization problems. The nearest neighbor algorithm (Rosenkrantz et al., 1977) is a greedy algorithm widely used in traveling salesman problem (Flood, 1956) and vehicle routing problem (Bendall, Margot, 2006). As a constructive heuristic algorithm, the nearest neighbor algorithm performs well in solving the European traveling salesman problem (Gutin et al., 2002). Compared with other heuristic algorithms, it is easy to implement and efficient, so we will construct the driver's routes with nearest neighbor algorithm. In summary, the immediate delivery problem is a Vehicle Routing Problem with Time Window and Dynamic Requests (VRPTWDR). We first introduce a comprehensive indicator from the driver's perspective to solve the order selection problem and then use the nearest neighbor algorithm to optimize the routing problem. Finally, to verify the algorithm's effectiveness, we provide an overall comparison of the seven indicators. Experiments show that the indicator proposed in this work performs stable and better for solving such problems.

The paper is outlined as follows. We provide a brief overview of the background and related literature on order dispatch in section I. In section II, we define the problem and model it as a Markov decision process. We introduce the indicators that affect the order selection process in section III. The proposed algorithm is then presented in section IV. Examples and computational analysis are reported in section V and section VI. Section VII concludes the paper and discuss the future research.

2 Problem Description and Mathematical Formulation

We first illustrate the problem with a small example, and then model the multi-period immediate delivery problem as a Markov decision process.

2.1 Problem Description

In this paper, we study the multi-period immediate delivery problem, which is an variant of the Vehicle Routing Problem with Time Window and Dynamic Requests (VRPTW-DR), with the goal of maximizing total revenue. The setting of the problem can be described as follows. The driver starts from the depot and delivers the goods to the corresponding customers sequentially according to the planned route. Each time a customer is served, new potential orders generate, which indicates the beginning of a new decision epoch. In each period, two decisions need to be made: order selection and routing. The driver selects the orders following certain rules. The accepted orders must be served within its time window, and the rejected orders will be in the next round of redistribution. How to select orders, plan routes so as to maximize the total revenue of the driver is the focus of this work. The following example (Fig. 1) intuitively represent the problem studied in this paper.

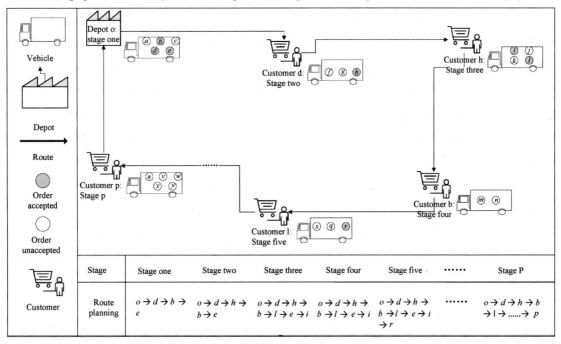

Stage	Stage one	Stage two	Stage three	Stage four	Stage five	......	Stage P
Route planning	$o \to d \to b \to e$	$o \to d \to h \to b \to e$	$o \to d \to h \to b \to l \to e \to i$	$o \to d \to h \to b \to l \to e \to i$	$o \to d \to h \to b \to l \to e \to i \to r$	......	$o \to d \to h \to b \to l \to ...... \to p$

Fig. 1 Illustration of order selection in immediate delivery service

During the first period, the driver encounters five orders a,b,c,d,e, selects three orders b,d,e, and plans the route for service. When the first customer d is served, the second period begins, and three new orders f,g,h appear. The driver selects customer h and adds it to the current path and re-plans the service path. The driver continues to serve the second customer h, after which the third period begins, adding new orders l,i to the current path. The driver takes no orders in the fourth period because the two new potential orders m,n violate availability constraints. In the fifth period, only order r satisfies the constraints. In all subsequent stages, the same patterns are followed, i. e., the driver continuously receives new orders while serving accepted orders and dynamically updates its service sequence until the total service time reaches the working time constraints of the day and all accepted orders are completed. The criteria to select orders results from a comprehensive consideration of driver's income, time pressure, and delivery time, which will be discussed in section Ⅲ. We use the nearest neighbor algorithm to optimize the service routes and our goal is to maximize the total revenue.

According to the problem description, we make the following assumptions：

（1）Rational person hypothesis. The driver is rational, i. e., every choice is based on the principle of maximizing self-interest；

（2）New requests can be observed at the beginning of a new decision-making cycle, and related information such as demand, time pressure, geographic location and vehicle speed remains unchanged；

（3）A driver's serviceability meets all needs of the orders received, and the demand of an order cannot be split.

2.2 Markov Decision Model

The problem described above is a stochastic and dynamic decision-making problem. It is stochastic because the set of customers to be served only becomes known at the beginning of each phase. It is dynamic because a sequence of decisions needs to be made at each stage. Furthermore, current decisions change the driver's location and therefore impact decisions in the future periods.

Markov Decision Process is an effective mathematical model for solving such sequential decision-making problems (Puterman, 2014). Each independent driver can be regarded as an agent (Agent) in the model to make a series of actions. In each period, a decision is made and the reward is obtained, followed by a stochastic transition that leads to the next decision state. The goal of the agent is to maximize the expected cumulative reward (Reward). In the following, We will discuss the details of states, action, reward, and transitions for our problem.

2.2.1 System State

The system state contains all necessary information to make a decision. We assume a driver serves customers over a sequence of periods $1,\cdots,P$. V_p represents the new potential customers encountered in period p. The set of orders that has been received but not yet serviced up to period p is denoted by N_p. In period p, the drivers can get the revenue r_{ip} by serving customer i, then the revenue vector is denoted by $R_p = (r_{ip})_{i=1,\cdots,v}$. Customer i must be serviced before the time t_{ip} in period p and the corresponding time window vector is $T_p = (t_{ip})_{i=1,\cdots,v}$. M_p represents the distance matrix of all customers that have appeared up to the current stage p. In general, we denote the system state in period p as $S_p = (V_p, N_p, M_p, T_p, R_p)$.

2.2.2 Action

In period p, a driver needs to perform two actions sequentially: the assignment of proper customers to the driver, which denoted by a_{p1} and the routing by a_{p2}. For simplicity, we assume that the driver's location at the beginning of period p is denoted by o_p. The order set including the depot is denoted $N_p^o = N_p \cup \{o_p\}$. Two actions a_{p1} and a_{p2} are feasible if the following constraints hold:

$$\sum_{i \in N_p^o} x_{ij} = 1, \forall j \in N_p \# \tag{1}$$

$$\sum_{j \in N_p^o} x_{o_p j} = 1 \# \tag{2}$$

$$\sum_{i \in N_p^o} x_{io_p} = 1 \# \tag{3}$$

$$\sum_{j \in N_p^o} x_{ji} - \sum_{j \in N_p^o} x_{ij} = 0, \forall i \in N_p \# \tag{4}$$

$$\sum_{i \in S} \sum_{j \in S} x_{ij} \leqslant |S| - 1, S \subseteq N_p \# \tag{5}$$

$$y_j \geqslant y_i + d_{ij} x_{ij} - M_{ij}(1 - x_{ij}), \forall i,j \in N_p^o \# \tag{6}$$

$$y_i \leqslant t_{ip}, \forall i = N_p^o \# \tag{7}$$

Where

$$x_{ij} = \begin{cases} 1, \text{if order } j \text{ is assigned to driver } i \\ 0, \text{if order } j \text{ is not assigned to driver } i \end{cases} \tag{8}$$

Constraints (1) ensure that all customers are visited exactly once. Constraints (2) and (3) ensure that the driver will eventually return to the depot from the current location. Constraints (4) guarantee the correct flow of vehicles through the arcs, by stating that if a vehicle arrives at a node $i \in N_p$, it must depart from this node. (5) are the subtour elimination constraints. Constraints (6) and (7) are the time window constraints. Constraints (8) are the binary constraints.

2.2.3 Reward

The definition of rewards determines the optimization goal of the whole system. Intuitively, the reward can be defined as the revenue of an order, i. e., $r(S_p, a_p) = r_{ip}$. If no eligible orders are serviced in period p, then $r(S_p, a_p) = 0$.

2.2.4 Post-Decision State

A combination of a state S_p and two actions (a_{p1}, a_{p2}) leads to a known post-decision state S_p^a. The postdecision state contains the updated time pressure T_p^a and new order sets N_p^a.

2.2.5 Transition

The exogenous information provides the set of new customers after Post-Decision State S_p^a. We denote it as w_{p+1}. Based on w_{p+1}, the transition function $W(S_p^a, w_{p+1})$ leads to a new state:

$$S_{p+1} = (V_{p+1}, N_{p+1}, M_{p+1}, T_{p+1}, R_{p+1})$$

2.2.6 Solution

A solution for a Markov decision process is a decision policy π. A policy π assigns a decision $A^\pi(S_p)$ to every state S_p. The overall set of policies is defined as Π. An optimal solution $\pi^* \in \Pi$ maximizes the expected revenue. Formally, the objective function is written as:

$$\pi^* = \operatorname{argmax} \mathbb{E}\Big\{\sum_{p=1}^{P} r\big[s_p, a^\pi(s_p)\big]\Big\}.$$

3　Indicators

The order selection problem is referred to as the appropriate choices among a set of orders to maximize a driver's interests. Three widely perceived factors are delivery time (d), revenue (r), and time pressure (t), whose pairwise combinations are:

$$W_{ij}^1 = \frac{t_j - t_i}{d_{ij}}, W_{ij}^2 = (t_j - t_i)r_{ij},$$

$$\text{and } W_{ij}^3 = \frac{r_{ij}}{d_{ij}} \tag{9}$$

Besides, we propose herein a comprehensive indicator as:

$$W_{ij}^* = \frac{t_j - t_i}{d_{ij}} r_{ij} \tag{10}$$

3.1　Single Indicators

● Delivery time. Denote by i and j two distinct locations, and by d_{ij} the distance to travel from i to j, which can act as an appropriate proxy for delivery time. In general, a driver is willing to serve customers close to him.

● Revenue. Denote by r_{ij} the revenue generated by a driver from location i to customer j. An order with higher r_{ij} seems more appealing.

● Time pressure. Denote by t_i the time when customer i is served, and by t_j the latest time that the customer j expects to be serviced, with which $t_j - t_i$ indicates the time pressure imposed on the driver. In practice, a non-urgent order seems more interesting.

3.2　Composite Indicators

Since the scales of the three indicators are different, min-max normalization is utilized for the composite indicators.

3.2.1　Pairwise Combinations of Indicators

● The ratio of time pressure to delivery time. As defined in (9), it is likely for a driver to choose an order with higher W_{ij}^1, which corresponds to less time pressure (i.e., higher $t_j - t_i$) or less delivery time.

● The product of time pressure and revenue. An order with a higher W_{ij}^2, which corresponds to a higher revenue or less time pressure, seems more attractive.

● The ratio of revenue to delivery time. Obviously, a driver prefers an order with a higher W_{ij}^3, i.e., with a higher revenue or less delivery time.

3.2.2　Comprehensive Indicator

● To achieve a trade-off effect from a driver's perspective, the comprehensive indicator in (10) consists of all three factors. Under the same circumstances, those orders with higher W_{ij}^* are more appealing.

4　Algorithm

The immediate delivery problem can be divided into two stages: the order selection and route optimization. We herein focus on the first stage in real-time logistics delivery, discuss the impacts of various order selection indicators with a common routing stage, and examine the performance of W_{ij}^*.

Algorithm 1describes the order selection algorithm based on comprehensive indicator. It first calculates the descending weights, W_{ij}^*, for all potential orders, after which Algorithm 1 moves to the *while* loop. During each iteration, it selects the order o_{j*} with the highest W_{ij}^* and adds it to the current service route $plan_v$. Thereafter, Algorithm 1 reconfigures the service path using the nearest neighbor algorithm. If feasible, update $plan_v$ with o_{j*}; otherwise simply discard o_{j*}. Finally, Algorithm 1 proceeds to the next round.

Algorithm 1: the Order Selection Problem with W_{ij}^*

Input: current location i, the set of unassigned orders O, and the current service route $plan_v$

Output: updated $plan_v$

$pool \leftarrow \phi$;

For all $o_j \in O$ do

    Calculate W_{ij}^* and $pool \leftarrow pool \bigcup \{o_j\}$

end

While $pool \neq \phi$ do

    $o_{j*} \leftarrow \text{argmax}_{o_j \in pool} W_{ij}^*$ and $plan_v \leftarrow plan_v \bigcup \{o_{j*}\}$;

    set up $plan_v^*$ using the nearest neighbor algorithm;

    If $plan_v^*$ is feasible then

        $plan_v \leftarrow plan_v^*$

    else

        $plan_v \leftarrow plan_v \backslash \{o_{j*}\}$

    end

end

return $plan_v$

---

5 An Illustrative Example

Suppose that we have a set of customers (a,b,c,d,e,f,g). Tab. 1 presents the distances between customers.

Delivery time matrix of potential customers

Tab. 1

Delivery Time	o	a	b	c	d	e	f	g
o	0	2	1	1.5	0.5	2	3	1.5
a	2	0	1.5	2	1	1	1.5	2
b	1	1.5	0	1	2	1.5	2.5	2
c	1.5	2	1	0	1	2	1.5	1.2
d	0.5	1	2	1	0	3	1.5	1.5
e	2	1	1.5	2	3	0	1	1.2
f	3	1.5	2.5	1.5	1.5	1	0	1
g	1.5	2	2	1.2	1.5	1.2	1	0

The time pressure and revenue are shown in Tab. 2.

Delivery time and price of potential customers

Tab. 2

Stage	Stage One		Stage Two		Stage Three		
Order	a	b	c	d	e	f	g
Time Pressure	4.5	2	4	3	4	5	6
Price	6	7	5	14	9	8	3

As shown in Tab. 3, the driver encounters two new orders (a,b) at the initial position o. We calculate their weights. First, order b is added to the current service path since $W_{ob}^* > W_{oa}^*$. By Tab. 1 and Tab. 2, the path is $o \rightarrow b$ since $d_{ob} = 1 < T_b = 2$. Now consider order a. As $d_{ob} = 1 < d_{oa} = 2$ and $d_{ob} + d_{ba} = 2.5 < T_a = 4.5$, the path becomes $o \rightarrow b \rightarrow a$.

The second stage encounters two new orders c and d with $W_{bd}^* = 21 > W_{bc}^* = 20$, $d_{ba} = 1.5 < d_{bd} = 2$ and $d_{ba} + d_{ad} = 2.5 < T_d = 3$. Therefore, order d is preferred. Similarly, after considering order c, a new path $o \rightarrow b \rightarrow c \rightarrow d \rightarrow a$ can be established by calculating the distance between customer b and customer c, a, and d, respectively. The feasible test is $d_{bc} + d_{cd} + d_{da} = 3 < T_a - d_{ob} = 3.5$.

In the third stage, order e and f are successfully incorporated in the path as before with $o \rightarrow b \rightarrow c \rightarrow d \rightarrow a \rightarrow e \rightarrow f$. However, when considering order g, the new path is $o \rightarrow b \rightarrow c \rightarrow d \rightarrow a \rightarrow e \rightarrow f \rightarrow g$ with the nearest neighbor algorithm. Since the overall service time is greater than the service time limit 6, the driver refuses to select order g. Hence, the final path is $o \rightarrow b \rightarrow c \rightarrow d \rightarrow a \rightarrow e \rightarrow f$.

Order selection and routing　　　　　　　　　　**Tab. 3**

Stage	Stage One o	Stage Two b	Stage Three c
New Orders	a, b	c, d	e, f, g
Weight	$W_{ob} > W_{oa}$	$W_{bd} > W_{bc}$	$W_{cf} > W_{ce} > W_{cg}$
Distance	$d_{ob} < d_{oa}$	$d_{ba} < d_{bd}$ $d_{bc} < d_{ba} < d_{bd}$ $d_{cd} < d_{ca}$	$d_{cd} < d_{cf} < d_{ca}$ ...... $d_{ef} < d_{eg}$
Possible Path	$o \rightarrow b$ $o \rightarrow b \rightarrow a$	$o \rightarrow b \rightarrow a \rightarrow d$ $o \rightarrow b \rightarrow c \rightarrow d \rightarrow a$	$o \rightarrow b \rightarrow c \rightarrow d \rightarrow a \rightarrow f$ $o \rightarrow b \rightarrow c \rightarrow d \rightarrow a \rightarrow e \rightarrow f$ $o \rightarrow b \rightarrow c \rightarrow d \rightarrow a \rightarrow e \rightarrow f \rightarrow g$
Time Pressure	$d_{ob} \leqslant T_b$ $d_{ob} + d_{ba} \leqslant T_a$	$d_{ba} + d_{ad} \leqslant T_d$ $d_{bc} \leqslant T_c, d_{bc} + d_{cd} \leqslant T_d$ $d_{bc} + d_{cd} + d_{da} \leqslant T_a - d_{ob}$	$d_{cd} + d_{da} + d_{af} \leqslant T_f$ $d_{cd} + d_{da} + d_{ae} \leqslant T_e$ $d_{cd} + d_{da} + d_{ae} + d_{ef} \leqslant T_f$ $d_{cd} + d_{da} + d_{ae} + d_{ef} + d_{fg} \leqslant T_g$
Service Time	$t_1 = d_{ob}$ $t_2 = d_{ob} + d_{ba}$ $t_1 = 1 < 6$ $t_2 = 2.5 < 6$	$t_1 = d_{ob} + d_{ba} + d_{ad}$ $t_2 = d_{ob} + d_{bc} + d_{cd} + d_{da}$ $t_1 = 3.5 < 6$ $t_2 = 4 < 6$	$t_1 = d_{ob} + d_{bc} + d_{cd} + d_{da} + d_{af}$ $t_2 = d_{ob} + d_{bc} + d_{cd} + d_{da} + d_{ae} + d_{ef}$ $t_3 = d_{ob} + d_{bc} + d_{cd} + d_{da} + d_{ae} + d_{ef} + d_{fg}$ $t_1 = 5.5 < 6 \quad t_2 = 6 = 6 \quad t_3 = 7 > 6$
Final Path	$o \rightarrow b \rightarrow a$	$o \rightarrow b \rightarrow c \rightarrow d \rightarrow a$	$o \rightarrow b \rightarrow c \rightarrow d \rightarrow a \rightarrow e \rightarrow f$

In terms of the total revenue, Tab. 4 compares the average performance between d_{ij} and W_{ij}^* under different scenarios. The first set of experiments is conducted without routing, i.e., a newly received order is directly concatenated to the current path, whereas the second set makes use of the nearest neighbor algorithm to optimize those paths established in the first set to verify the algorithm's effectiveness. It should be noted that a driver's working hours is limited to 6 hours per day. Using only d_{ij}, a driver receives the least set of 4 orders with a total revenue 26 in the entire delivery cycle. Nevertheless, the comprehensive indicator W_{ij}^* increases the number of orders to 5 with a total revenue 40. In addition, the nearest neighbor algorithm effectively improves the driver's service efficiency. In comparison with the case using only d_{ij}, W_{ij}^* brings more orders and higher revenue.

Performance of two indicators on instance

Tab. 4

Order Selection Process	Delivery Route	Number of Orders	Total Revenue
d_{ij} with No Routing	$o \rightarrow b \rightarrow a \rightarrow c \rightarrow f$	4	26
W_{ij}^* with No Routing	$o \rightarrow b \rightarrow a \rightarrow d \rightarrow c \rightarrow f$	5	40

continued

Order Selection Process	Delivery Route	Number of Orders	Total Revenue
d_{ij} with Routing	$o \rightarrow b \rightarrow c \rightarrow d \rightarrow a \rightarrow g$	5	35
W_{ij}^* with Routing	$o \rightarrow b \rightarrow c \rightarrow d \rightarrow a \rightarrow e \rightarrow f$	6	49

6 Experimental Results

To our knowledge, there have been few studies considering the order selection problem from a driver's perspective, and there exist no instances designed to examine the performance of such an algorithm. Consequently, we intend to conduct two sets of experiments in this section. Firstly, we present an overall comparison of the seven indicators by keeping the routing part unchanged with various data distributions. Secondly, we compare the performance among seven indicators in terms of the numbers of potential customers encountered at each stage. Finally, we conduct a sensitivity analysis to evaluate changes in total expected revenue when one parameter changes. We carry out all experiments on a Think-Pad with 1.60 GHz processors and 8 GB RAM. Algorithms are coded in C++.

The first experiment consists of three VRPTWDR test instances. In (Meituan, CFLP, 2019), it is claimed that the local instant retail platform can build a one-hour delivery capability through a powerful delivery network. The data shows that drivers generally receive $5 \sim 10$ Yuan in revenue for the completion of an order. As a result, in instance 1, the distributions for the random variables are assumed as time pressure $T \sim N(65, 40^2)$, the revenue $R \sim N(7, 6^2)$, and the delivery time $D \sim N(100, 10^2)$. Analogously, in instance 2, we assume that T, R, and D are independent random variables following chi-square distributions with 65, 7 and 100 degrees of freedom, respectively. In instance 3, to avoid the data distribution being too dense, we further assume that T, R, and D follow $U(1, 2000)$. The objective is to maximize a driver's total revenue.

Tab. 5 shows the average returns for 200 days, and Tab. 6 exhibits the performance among seven indicators. To ease our presentation, the results of W_{ij}^* are taken as the benchmark 1, and the results of other indicators are ratios to W_{ij}^*. It can be observed that d_{ij} outperforms the other two single indicators. The performance of pairwise combination indicators seems, in general, better than that of any single indicator. In comparsion with others, W_{ij}^* can outperform other indicators in both total revenue and stability. Besides, different distributions (normal, chi-square, uniform) imposed on the instance sets can hardly influence the experimental results.

Performance of seven indicators on three instances with different distributions Tab. 5

Data Sets	Single Index			Composite Index			New Index
	d_{ij}	r_{ij}	$t_j - t_i$	W_{ij}^1	W_{ij}^2	W_{ij}^3	W_{ij}^*
Instance 1	448.97	445.82	425.85	459.55	442.18	463.82	467.94
Instance 2	182.02	156.95	131.70	181.61	159.59	181.86	184.21
Instance 3	14212.26	11607.62	11369.01	14509.41	12114.15	14569.48	14620.67

Performance with different distributions Tab. 6

Data Sets	Single Index			Composite Index			New Index
	d_{ij}	r_{ij}	$t_j - t_i$	W_{ij}^1	W_{ij}^2	W_{ij}^3	W_{ij}^*
Instance 1	0.959	0.953	0.910	0.982	0.945	0.991	1
Instance 2	0.988	0.852	0.715	0.986	0.866	0.987	1
Instance 3	0.972	0.794	0.778	0.992	0.829	0.996	1

With instance 1, the next experiment shows the influence of different potential order sizes at each stage on the results. We assume that the number of new potential customers encountered by a driver at each stage follows a Poisson distribution. Results for seven indicators with $\lambda = 4$ and $\lambda = 20$ are shown in Tab. 7.

The average returns of seven indicators on test instances for 200 days are shown in Tab. 4. It can be observed that as the mean number of potential customers emerging at each stage increases, a driver's total revenue becomes more extensive, and W_{ij}^* significantly outperforms the single indicator. This seems reasonable because the more emerging orders at a stage, the more flexibility a driver has in the order selection problem. In addition, it is common in the real world to have more orders than drivers at least for certain time periods during a day.

Ferformance on modified instances 1 Tab. 7

Data Sets	Single Index			Composite Index			New Index
	W_{ij}^3	d_{ij}	r_{ij}	$t_j - t_i$	W_{ij}^1	W_{ij}^2	W_{ij}^*
$\lambda = 4$	243.79	239.37	237.34	245.44	240.67	243.58	244.08
$\lambda = 20$	448.97	445.82	425.85	459.55	442.18	463.8	467.94

The procedures for sensitivity analysis are summarized as follows:

1) Generate random numbers with the same distribution as in instance 2, but with different Lambda ($\lambda = 10$) as benchmark.

2) Consequently change the freedom degrees of three parameters (order price, time pressure, and distance) by 10% approximately, and then run the model presented above using substitute data. We assessed the effect on the model's accuracy by calculating the percentage difference between the reference total revenue in step 1 and the output values using substitute data in step 2.

3) Repeat step 2 for all the parameters used by the model individually, and observe the trend of the result with the change of the adjustable variables.

Fig. 2 shows the absolute differences between the benchmark values and simulated values with substitute data when one factor changes. Fig. 3 shows the trend of the algorithm results with the dynamic changes of one independent variable. The largest difference were found when the order price changed, and smallest is time pressure. In conclusion, this analysis confirms that order price and distance between two points have significant impact on the results, which consistent with those from the primary experiments to a certain extent.

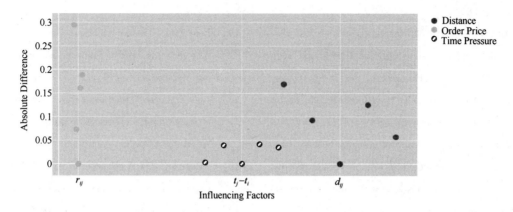

Fig. 2   Sensitivity analysis for one-parameters change

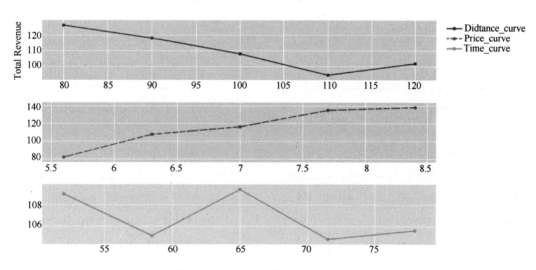

Fig. 3   Sensitivity analysis for one-parameters change

7   Conclusion

With the incorporation of the nearest neighbor algorithm, we propose a comprehensive indicator for the order selection problem in a VRPTWDR model. We model the order dispatch as a sequential decision-making problem and construct several sets of instances so as to examine its performance. Comparative results and statistical analysis demonstrate its effectiveness and robustness.

It remains interesting to investigate the multi-vehicle order matching on the improvement of service completion.

References

[1]   Bendall Gareth, Margot Franc,ois. Greedy-type resistance of combinatorial problems. Discrete Optimization, 2006(3,4): 288-298.

[2]   Bertsimas Dimitris J. A vehicle routing problem with stochastic demand. Operations Research, 1992, 40(3):574-585.

[3]   Cordeau Jean-Francois, Gendreau Michel, Laporte Gilbert, et al. A guide to vehiclerouting heuristics. Journal of the Operational Research society, 2002, 53(5):512-522.

[4]   Dantzig George B, Ramser John H. The truck

dispatching problem. Management Science,1959, 6(1)：80-91.

[5] Dror Moshe, Powell Warren. Stochastic and dynamic modelsin transportation. Operations Research, 1993, 41(1)：11-14.

[6] Flood Merrill M. The traveling-salesman problem. Opera-tions research, 1956, 4(1)：61-75.

[7] Gendreau Michel, Laporte Gilbert, S eguin Ren e. A tabu searchheuristic for the vehicle routing problem with stochasticdemands and customers. Operations Research，1996, 44(3)：469-477.

[8] Glaschenko Andrey, Ivaschenko Anton, Rzevski George, et al. Multi-agent real time scheduling system for taxi companies. 8th International Conference on Autonomous Agents and Multiagent Systems (AAMAS 2009), Budapest, Hungary, 2009：29-36.

[9] Group M D. Announcement of the results for the year ended. December 31, 2020.

[10] Gutin Gregory, Yeo Anders, Zverovich Alexey. Traveling salesman should not be greedy：domination analysis of greedy-type heuristics for the TSP. Discrete Applied Mathematics, 2002, 117 (1-3)：81-86.

[11] IResearch. 2020 China's Just-in-Time Logistics Industry Report. 2020.

[12] Kallehauge Brian, Larsen Jesper, Madsen Oli BG, et al. Vehicle routing problem with time windows. Column Generation, 2005：67-98.

[13] Korte Bernhard, Vygen Jens. Combinatorial optimization：theory and algorithms. 2009.

[14] Lee Der horng, Wang Hao, Cheu Ruey Long, et al. Taxi dispatch system based on current demands and real-time traffic conditions. Transportation Research Record, 2004, 1882（1）：193-200.

[15] Li Yang, Fan Houming, Zhang Xiaonan. A periodic optimization model and solution for capacitated vehicle routing problem with dynamic re-

quests. Chinese Journal of Management Science, 2021：1-12.

[16] Liao Ziqi. Taxi dispatching via global positioning systems. IEEE Transactions on Engineering Management, 2001, 48(3)：342-347.

[17] Liao Ziqi. Real-time taxi dispatching using global positioning systems. Communications of the ACM, 2003, 46(5)：81-83.

[18] Meituan, CFLP. Report on the development of Chinesei mmediate delivery business in 2019.

[19] Psaraftis Harilaos N. Dynamic vehicle routing：Status and prospects. Annals of operations research, 1995, 61(1)：143-164.

[20] Puterman Martin L. Markov decision processes：discrete stochastic dynamic programming. 2014.

[21] Rosenkrantz Daniel J, Stearns Richard E, Lewis Philip M Ⅱ. An analysis of several heuristics for the traveling salesman problem. SIAM Journal on Computing, 1977,6(3)：563-581.

[22] Seow Kiam Tian, Dang Nam Hai, Lee Der-Horng. A collabo-rative multiagent taxi-dispatch system. IEEE Transactionson Automation Science and Engineering, 2009, 7(3)：607-616.

[23] Xu Zhe, Li Zhixin, Guan Qingwen, et al. Large-scale order dispatch in on-demand ride-hailing platforms：alearning and planning approach. Proceedings of the 24th ACM SIGKDD International Conference on Knowledge Discovery & Data Mining, 2018：905-913.

[24] Zhang Jingling, Zhao Yanwei, Wang Haiyan, et al. Modeling and algorithms for a dynamic-multi-vehicle routing problem with Customers' dynamic requests. Computer Integrated Manufacturing Systems,2010, 16(3)：543-550.

[25] Zhang Lingyu, Hu Tao, Min Yue,et al. A taxi order dispatch model based on combinatorial optimization. Proceedings of the 23rd ACM SIGKDD International Conference on Knowledge Discovery and Data Mining, 2017：2151-2159.

Exploring the Knowledge Map of Information Technologies Applications in the Field of Construction Engineering and Management

Liang Wang   Yiming Cheng

(School of Maritime Economics and Management, Dalian Maritime University, Dalian 116026, China)

【Abstract】 Information technologies have been widely applied in the field of construction engineering and management (CEM). This study uses the method of scientometric analysis to reveal the status and development trends of information technologies applications in the field of CEM. 2911 articles published in Automation in Construction, Computer-Aided Civil and Infrastructure Engineering and Journal of Computing in Civil Engineering from 2009 to 2018 are collected from the core collection of the Web of Science to construct the comprehensive information technologies applications in the field of CEM. This study reveals that the academic communities are mainly the USA, China, Korea, Canada and Australia, and the institutions from above main countries/regions. The influential authors from these main countries/regions and institutions make major contributions to information technologies applications in the field of CEM. The hot research topics of information technologies applications in the field of CEM from 2009 to 2018 present evolutionary trends, and the emerging topics between 2014 and 2018 reflect the applications of emerging information technologies in the field of CEM. The research contents of information technologies applications in the field of CEM from 2009 to 2018 can be divided into 12 significant knowledge communities and mainly focus on building information modeling, multidimensional visualization model, optimization and simulation methods, construction information management theory and the applications of information technologies on safety management.

【Keywords】 Information Technologies; Construction Engineering and Management; Academic Community; Evolution Trend; Knowledge Map

1 Introduction

Construction engineering and management (CEM) is an academic field which involves the activities of design, construction, and management activities in construction projects (Aboulezz 2003). As a relatively new discipline, CEM has produced numerous scholarly publications which are published in academic journals (Levitt 2007). Some academic journals have become top journals in the field of CEM, such as Automation in Construction, Computer-Aided Civil and Infrastructure Engineering, Journal of Computing in Civil Engineering, Journal of Construction Engineering and Management, Journal of Management in Engineering International Journal of Project Management and Journal of Civil Engineering and Management (Wing 1997). Based on articles published in above academic journals, scientometric analysis has been used to reveal previous theory, method and application in the field of CEM (Jin, et al., 2019; Yu, et al., 2019).

Information technologies, such as building information modeling (BIM) (Succar, 2009; Belsky et al., 2016; Golparvar-Fard et al., 2012), geographical information systems (GIS) (Aydin, 2014; Borrmann et al., 2015; Bansal, 2010), virtual and augmented reality (VR/AR) (Dong et al., 2013; Behzadan and Kamat, 2010; Lin et al., 2014; Sacks et al., 2013), have been widely applied in the field of CEM. Previous studies of information technologies applications in the field of CEM are mainly published in three academic journals which include Automation in Construction (AC), Computer-Aided Civil and Infrastructure Engineering (CACI) and Journal of Computing in Civil Engineering (JCCE). These three academic journals refer to all aspects for the applications of information technologies in the field of CEM. Although existing studies have reviewed the applications of BIM by scientometric analysis (Zhao, 2017; Santos et al., 2017), there is no study that provides a scientometric analysis for the knowledge map of information technologies applications in the field of CEM.

This study aims to provide a scientometric analysis on the knowledge map of information technologies applications in the field of CEM through reviewing the articles published in AC, CACI and JCCE from 2009 to 2018. The following objectives are targeted in this study: (1) identify the academic communities of articles published in AC, CACI and JCCE, including main countries/regions and institutions and influential authors; (2) compare the evolution trends of keywords in different periods; (3) reveal the knowledge structure of articles published in AC, CACI and JCCE. The contents of above three aspects reveal the knowledge map of information technologies applications in the field of CEM.

2 Methodology

2.1 Analysis Method

Scientometric analysis is used to reveal the development of science based on existing knowledge documents (Chen et al., 2009).

Scientometric analysis mainly contains three analysis methods: academic collaboration analysis, co-occurrence analysis and co-citation analysis; These three analysis methods can be used to demonstrate academic communities, predict the evolution trends of science and detect knowledge structure (Garfield, 1979; Cohen, 1981). CiteSpace is developed by Professor Chaomei Chen for scientometric analysis and integrates academic collaboration analysis, co-occurrence analysis, co-citation analysis and visualization functions as one comprehensive approach for scientometric analysis (Chen, 2004; Chen, 2006).

In this study, we use CiteSpace for constructing a comprehensive analysis framework to produce a scientometric analysis of articles published in AC, CACI and JCCE from 2009 to 2018. This study involves three scientometric analysis methods: (1) collaboration network analysis, including the analysis of country/region and institution and the analysis of author; (2) co-occurrence network analysis for keywords; (3) co-citation network analysis for references. These three scientometric analysis methods are used to construct a comprehensive knowledge map of articles published in AC, CACI and JCCE from 2009 to 2018. Fig. 1 shows the details of analysis framework.

2. 2　Data Collection

The core collection of the Web of Science (WoS) is adopted for data collection on most previous scientometric analysis studies (Li et al., 2017; Wang et al., 2018; Si et al., 2018). This study searches the documents records of AC, CA-

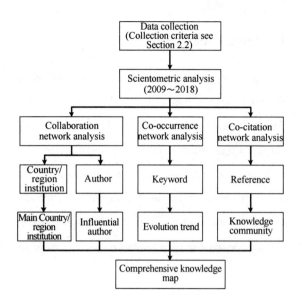

Fig. 1　Analysis framework

CI and JCCE from the core collection of the WoS. We apply the following data collection criteria: [TS= Publication title= ("Automation in Construction" OR "Computer-Aided Civil and Infrastructure Engineering" OR "Journal of Computing in Civil Engineering")]. The time range of documents records retrieval is set from 2009 to 2018. A total of 2911 articles are obtained after the data retrieval process. These 2911 articles are used for the scientometric analysis of articles published in Automation in Construction from 2009 to 2018. The annual number of 2911 articles is shown in Fig. 2.

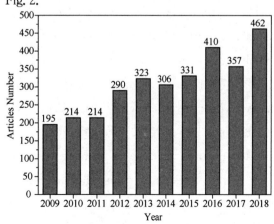

Fig. 2　Annual articles number statistics

3 Results Analysis

In this section, we conduct scientometric analysis on articles published in AC, CACI and JCCE from 2009 to 2018. Scientometric analysis process is conducted according to the analysis method and framework designed in the Section 2.1. In the following sections, analysis results are visualized by CiteSpace.

The different colors of links in visual analysis results represent different years from 2009 to 2018. The transition of colors from blue to yellow represents the time transition from past to present (Chen, 2006). For example, blue indicates that links occurred in 2009, whereas yellow denotes that links occurred in 2018. The detailed expression of different colors is shown in Fig. 3.

| 2009 | 2010 | 2011 | 2012 | 2013 | 2014 | 2015 | 2016 | 2017 | 2018 |

Fig. 3　Colors interpretation of links from 2009 to 2018

3.1 Collaboration Network Analysis

3.1.1 Countries/Regions and Institutions

The collaboration network of countries/regions and institutions reflect the collaborative relationships of countries/regions and institutions on articles published in AC, CACI and JCCE from 2009 to 2018. This collaboration network is produced based on the articles number of countries/regions and institutions and includes 349 nodes and 785 links. The dimension of the nodes indicates the total articles number of countries/regions and institutions published from 2009 to 2018. As shown in Fig. 4, the USA (972 articles), China (776 articles), Korea (290 articles), Canada (243 articles), Australia (191 articles) and England (157 articles) have made major contributions to articles published in AC, CACI and JCCE from 2009 to 2018. The large number of articles denotes that these countries/regions have greater advantages than other countries/regions on articles published in AC, CACI and JCCE from 2009 to 2018. As the most devel-

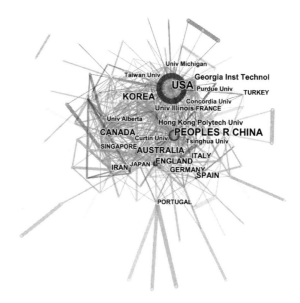

Fig. 4　Collaboration network of countries/regions and institutions

oped country in construction industry, the USA has the greatest contribution to articles published in AC, CACI and JCCE from 2009 to 2018. As the largest market in the construction industry, China also has major contribution to articles published in AC, CACI and JCCE from 2009 to 2018.

The contributions of institutions are also identified in Fig. 4. Some institutions have made major contributions to articles published in AC, CACI and JCCE from 2009 to 2018, such as Georgia Institute of Technology (117 articles), Hong Kong Polytechnic University (90 articles), University of Illinois (88 articles), Purdue University (65 articles), University of Michigan (63 articles), Taiwan University of Science and Technology (63 articles), University of Alberta (63 articles), Yonsei University (57 articles), Tsinghua University (52 articles), Huazhong University

of Science and Technology (51 articles) and Taiwan University (51 articles). These institutions are important centers for the applications of information technologies in the field of CEM. Simultaneously, these institutions are all located in countries/regions with major contributions to articles published in AC, CACI and JCCE from 2009 to 2018. The top 20 most productive countries/regions and institutions are summarized in Tab. 1.

Overview of top 20 most productive countries/regions and institutions　　　　Tab. 1

Countries/Regions			Institutions		
Rank	Name	Articles Number	Rank	Name	Articles Number
1	USA	972	1	Georgia Institute of Technology	117
2	China	776	2	Hong Kong Polytechnic University	90
3	Korea	290	3	University of Illinois	88
4	Canada	243	4	Purdue University	65
5	Australia	191	5~7	University of Michigan	63
6	England	157		Taiwan University of Science and Technology	
7	Spain	124		University of Alberta	
8	Germany	85	8	Yonsei University	57
9	Italy	75	9	Tsinghua University	52
10	Singapore	62	10,11	Huazhong University of Science and Technology	51
11	Iran	67		Taiwan University	
12	France	50	12	Concordia University	50
13	Turkey	44	13	Curtin University	47
14	Japan	43	14	University of Texas at Austin	45
15	India Netherlands	42	15,16	Tongji University	42
				University of Waterloo	
16	Portugal	40	17,18	Carnegie Mellon University	41
17	Switzerland	39		City University of Hong Kong	
18	Poland	37	19	National University of Singapore	37
19	Egypt	31	20	KyungHee University	34

Additionally, nodes with high betweenness centrality are distinguished by purple rings in Fig. 4. These nodes play important roles in connecting other nodes in the collaboration network of countries/regions and institutions. Most of nodes with high betweenness centrality are countries/regions, such as the USA (centrality = 0.34), China (centrality = 0.24), Korea (centrality = 0.21), Canada (centrality = 0.15), Australia (centrality = 0.1) and England (centrality = 0.1). Institutions with high betweenness centrality include Georgia Institute of Technology (centrality = 0.09), Hong Kong Polytechnic University (centrality = 0.08) and University of Illinois (centrality = 0.08).

3.1.2 Authors

According to the articles number pub-lished by authors, we identify the top 10 most productive authors on articles published in AC, CACI and JCCE from 2009 to 2018. As shown in Tab.2, Heng Li (Hong Kong Polytechnic University), SangHyun Lee (University of Michigan) and Hyoungkwan Kim (Yonsei University) occupy the top 3 positions on articles published in AC, CACI and JCCE from 2009 to 2018. Most of top 10 most productive authors are from the USA (4 authors), China (2 authors) and Korea (2 authors).

Overview of top 10 most productive authors　　　　　　　　　　**Tab. 2**

Author	Institution	Country	Count
Heng Li	Hong Kong Polytechnic University	China	43
SangHyun Lee	University of Michigan	USA	42
Hyoungkwan Kim	Yonsei University	Korea	36
Ioannis Brilakis	University of Cambridge	England	29
Min-Yuan Cheng	Taiwan University of Science and Technology	China	29
Changwan Kim	Chung-Ang University	Korea	28
Burcu Akinci	Carnegie Mellon University	USA	26
Carl T. Haas	University of Waterloo	Canada	26
Vineet R. Kamat	University of Michigan	USA	26
Jochen Teizer	Georgia Institute of Technology	USA	26

The collaboration network of authors reflects the collaborative relationship of authors on articles published in AC, CACI and JCCE from 2009 to 2018. This collaboration network is produced based on the collaborative articles between authors. In the collaboration network of authors, nodes represent authors and the links between different authors represent the co-authorship relationships in collaborative articles. The nodes dimension represents the collaborative articles number of authors, and the thickness of the links denotes the collaborative strength between authors. As shown in Fig.5, the collaboration network of authors includes 558 nodes and 600 links.

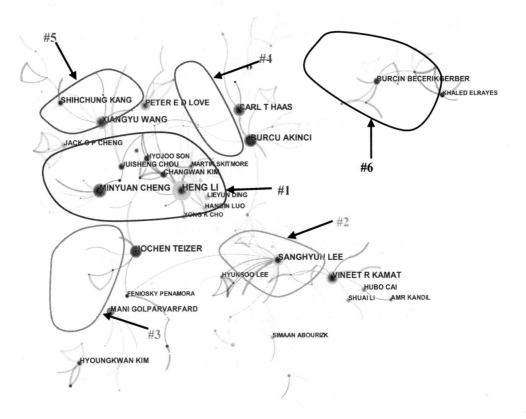

Fig. 5　Collaboration network of authors

3.2　Co-occurrence Network Analysis

Keywords reflect the core contents of articles published in AC, CACI and JCCE from 2009 to 2018. Thus, the change of primary keywords in different periods reflects the evolution trends of articles published in AC, CACI and JCCE from 2009 to 2018. We use keywords from 2911 articles to produce the co-occurrence network of keywords. As shown in Fig. 6, the co-occurrence network of keywords includes 565 nodes and 1401 links. The dimension of crosses in Fig. 6 represents the total frequency of keywords appearing in 2911 articles, and the thickness of the links denotes the connection levels between keywords.

As shown in Fig. 6, the top 20 high-frequency keywords are identified from the co-occurrence network of keywords. The top 20 high-frequency keywords are "building information modeling/BIM" (frequency = 282), "optimization" (frequency = 92), "simulation" (frequency = 85), "construction management" (frequency = 65), "genetic algorithm" (frequency = 62), "construction" (frequency = 54), "automation" (frequency = 53), "scheduling" (frequency = 39), "ontology" (frequency = 39), "computer vision" (frequency = 38), "safety" (frequency = 37), "project management" (frequency = 36), "algorithm" (frequency = 36), "machine learning" (frequency = 34), "interoperability" (frequency = 33), "image processing" (frequency = 32), "construction safety" (frequency = 31), "point cloud" (frequency = 31, "productivity" (frequency = 28) and "visualization" (frequency = 27).

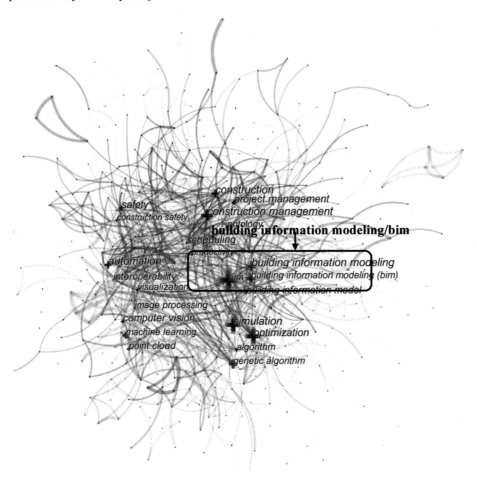

Fig. 6 Co-occurrence network of keywords

The keyword "building information modeling/BIM" has the highest frequency and keeps high connection levels with other keywords, which reflects the applications of building information modeling/BIM in the field of CEM (Succar, 2009; Zhang et al., 2013; Porwal and Hewage, 2013; Jung and Joo, 2011; Borrmann et al., 2015; Lin et al., 2016; Eastman et al., 2009; Beach et al., 2017). The keywords "simulation" "optimization" "algorithm" and "genetic algorithm" refer to apply information technologies for simulation and optimization in the field of CEM

(Granadeiro et al., 2013; Chen et al., 2010; Ghoddousi et al., 2013; Golparvar-Fard et al., 2009a; Son et al., 2011; Gao and Zhang, 2013; Marano et al., 2011). The keywords "construction" "construction management" "ontology" and "project management" reflect the fundamental theories of information technologies applications in the field of CEM (Irizarry et al., 2013; Hartmann et al., 2012; Dave and Koskela, 2009; Park et al., 2013; Lee et al., 2011a; Dossick et al., 2010; Du and El-Gafy, 2012; Martínez—Rojas et al., 2015). Safety problems can be improved by the applications of information technologies in the field of CEM (Teizer et al., 2010; Cheng and Teizer, 2013; Han and Lee, 2013; Lee et al., 2011b; Li et al., 2012; Castillo et al., 2017), which is reflected by the keywords "safety" and "construction safety". In addition to the building information model/BIM, other information technologies, such as point cloud (Wang et al., 2015; Anil et al., 2013; Dimitrov et al., 2016; Sharif et al., 2017), computer vision (Hamledari et al., 2017; Tomé et al., 2015; Gong and Caldas, 2009; Dawood et al., 2017), machine learning (Son et al., 2011; Zhou et al., 2017) and image processing (Liu et al., 2014; Sirazitdinova et al., 2018), are also widely applied in the field of CEM, which is reflected by the keywords "automation" "interoperability" "point cloud" "computer vision" "machine learning" "visualization" and "image processing". The keywords "productivity" and "scheduling" reflect that the final aim of information technologies applications is improving the performance of construction engineering management (Grau et al., 2009; Nath et al., 2015; Ghoddousi et al., 2013; Gong and Caldas, 2009; Lee et al., 2010; Lim et al., 2014; Bügler et al., 2017).

The co-occurrence network of keywords presents the static features of scientific domain which can not reveal the changes of keywords over time (Chen, 2006, Si et al., 2018). Thus, the co-occurrence network of keywords is unable to describe the evolution trends of articles published in different periods. In order to explore the evolution trends of articles

published in AC, CACI and JCCE from 2009 to 2018, the quantitative analysis of top 20 high-frequency keywords in different periods is summarized in Fig. 7.

Fig. 7 shows the evolution trends and interaction of top 20 high-frequency keywords in different periods. The top 20 keywords of different periods have changed in the latest 10 years, which reflects the transformation of important research topics on articles published in AC, CACI and JCCE from 2009 to 2018. Although the keyword "building information modeling/BIM" always occupies the first position from 2009 to 2018, the frequency of "building information modeling/BIM" that occurs between 2014 and 2018 is three times of frequency that occurs between 2009 and 2013. The frequency of other top 20 high-frequency keywords has changed significantly in different periods. There are seven top 20 high-frequency keywords between 2009 and 2013 that disappear from the list of top 20 high-frequency keywords between 2014 and 2018. These disappearing top 20 high-frequency keywords include "project management" "augmented reality" "productivity" "visualization" "rfid" "information management" and "decision support system". Additionally, there are seven new keywords appear in the list of top 20 high-frequency keywords between 2014 and 2018, including "machine learning" "construction safety" "computer vision" "point cloud" "ifc" "support vector machine" and "laser scanning". These new top 20 high-frequency keywords between 2014 and 2018 reflect the applications of emerging information technologies, such as point cloud, laser scanning and machine learning. The ranks of five top 20 high-frequency keywords between 2009 and 2013 decline in the period between 2014 and 2018, including "construction management" "automation" "safety""algorithm" and "productivity". The ranks of keywords "optimization" "simulation" "genetic algorithm" "ontology" and "in-

teroperability" rise in the list of top 20 high-frequency keywords between 2014 and 2018. The upward trends of these five keywords reflect that the issues of optimization, simulation and genetic algorithm get more attentions on articles published in AC, CACI and JCCE between 2014 and 2018.

2009～2013   2014～2018

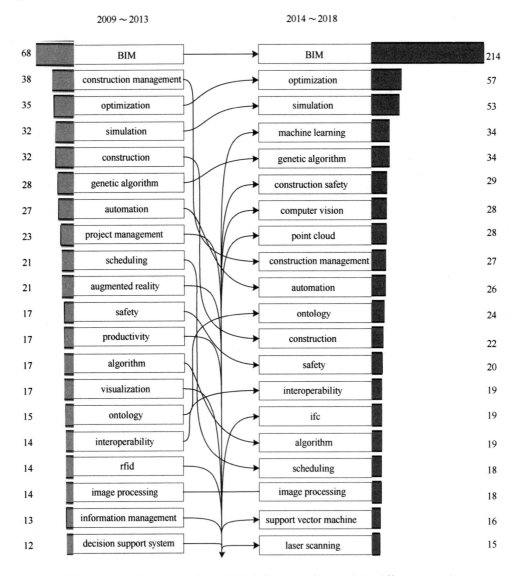

Fig. 7   Evolution trends of top 20 high-frequency keywords in different periods

3.3   Co-citation Network Analysis

We use the references cited by the 2911

articles selected in the section 2.2 to produce the co-citation network of references. As shown in Fig. 8, the co-citation network of references includes 1067 nodes and 1682 links. Each node represents one reference, and the link between nodes represents the co-citation relationships. The size of nodes represents the frequency of references cited by the 2911 articles, and the larger nodes denote the significant of references on articles published in AC, CACI and JCCE from 2009 to 2018.

Fig. 8   Co-citation network of references

The frequency of references cited by the 2911 articles represents the influence of references on articles published in AC, CACI and JCCE from 2009 to 2018. Jiang and Adeli (Jiang and Adeli, 2008b) receives 81 cocitations and occupies the first position, followed by Eastman et al. (Eastman et al., 2011) (frequency = 68), Tang et al. (Tang et al., 2010) (frequency = 67), Jiang and Adeli (Jiang and Adeli, 2008a) (frequency = 62) and Park et al. (Park et al., 2007) (frequency = 61). A total of 14 references are selected from the cocitation network as classical references on articles published in AC, CACI and JCCE from 2009 to 2018. The cocitations frequencies of these 14 classical references are larger than 40, and the details are summarized in Tab. 3.

14 classical references from 2009 to 2018   Tab. 3

No.	Frequency	Author	Title	Source
1	81	Jiang and Adeli，2008b	Neuro-genetic algorithm for non-linear active control of structures	International Journal for Numerical Methods in Engineering
2	68	Eastman et al.，2011	BIM handbook-a guide to building information modeling for owners, managers, designers, engineers, and contractors, second edition	John Wiley & Sons
3	67	Tanget al.，2010	Automatic reconstruction of as-built building information models from laser-scanned point clouds：a review of related techniques	Automation in Construction
4	62	Jiang and Adeli，2008a	Dynamic fuzzy wavelet neuroemulator for non-linear control of irregular building structures	International Journal for Numerical Methods in Engineering
5	61	Park et al.，2007	A new approach for health monitoring of structures：terrestrial laser scanning	Computer-Aided Civil and Infrastructure Engineering
6	58	Adeli and Jiang，2006	Dynamic fuzzy wavelet neural network model for structural system identification	Journal of Structural Engineering-ASCE
7	56	Jiang and Adeli，2007	Pseudospectra, music, and dynamic wavelet neural network for damage detection of highrise buildings	International Journal for Numerical Methods in Engineering
8	55	Eastman et al.，2008	BIM handbook-a Guide to building information modeling for owners, managers, designers, engineers, and contractors, second edition	John Wiley & Sons
9	46	Adeli and Jiang，2008	Intelligent infrastructure：neural networks, wavelets, and chaos theory for intelligent transportation systems and smart structures	Taylor & Francis Group
10	43	Golparvar—Fard et al.，2009b	D4AR-a 4-dimensional augmented reality model for automating construction progress monitoring data collection, processing and communication	Journal of Information Technology in Construction
11	43	Zhang et al.，2013	Building information modeling (BIM) and safety：automatic safety checking of construction models and schedules	Automation in Construction
12	42	Bosché，2010	Automated recognition of 3D CAD model objects in laser scans and calculation of as-built dimensions for dimensional compliance control in construction	Advanced Engineering Informatics
13	41	Jiang et al.，2007	Bayesian wavelet packet denoising for structural system identification	Structural Control and Health Monitoring
14	40	Jiang and Adeli，2005	Dynamic wavelet neural network for nonlinear identification of highrise buildings	Computer-Aided Civil and Infrastructure Engineering

As shown in Tab. 3, Jiang and Adeli publishes 7 classical references (Jiang and Adeli, 2008b; Jiang and Adeli, 2008a; Adeli and Jiang, 2006; Jiang and Adeli, 2007; Adeli and Jiang, 2008; Jiang and Adeli, 2005; Jiang et al., 2007) on articles published in AC, CACI and JCCE from 2009 to 2018. These 14 classical references reflect the research contents of information technologies applications in the field of CEM, which are mainly focus on BIM (Zhang et al., 2013; Eastman et al., 2011; Tang et al., 2010; Handbook, 2008), algorithm (Jiang and Adeli, 2008b; Adeli and Jiang, 2006; Jiang et al., 2007; Jiang and Adeli, 2005), optimization (Jiang and Adeli, 2008a; Jiang and Adeli, 2007; Adeli and Jiang, 2008), laser scanning (Park et al., 2007; Bosché, 2010) and augmented reality (Golparvar-Fard et al., 2009b). The research contents of 14 classical references are consistent with the analysis results of keywords in the section 3.2.

In order to explore the knowledge communities of articles published in AC, CACI and JCCE from 2009 to 2018, cluster analysis is conducted based on the co-citation network of references. The cluster label is identified by the indexing terms (T) of references. The result of cluster analysis of references is shown in Fig. 9.

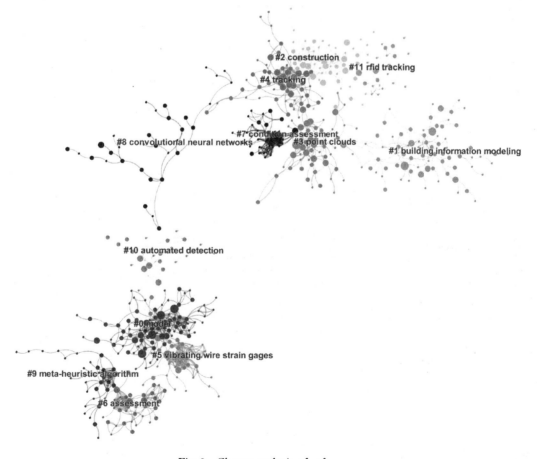

Fig. 9　Cluster analysis of references

Fig. 9 shows the result of cluster analysis. The nodes with the same color are divided in the same cluster, and different clusters present different colors. The number of nodes in a cluster represents the size of cluster. The co-citation clusters are identified based on the indexing terms (T) of references cited in each cluster by the Latent Semantic Indexing (LSI). There are 12 significant co-citation clusters which sizes are larger than 30. These 13 significant co-citation clusters are selected to describe the knowledge communities of articles published in AC, CACI and JCCE from 2009 to 2018. The cluster ID of maximum cluster is #0, whereas the cluster ID of minimum cluster is #11. The details of 12 significant co-citation clusters are described in Tab. 4.

12 significant co-citation clusters from 2009 to 2018　　　　　　　**Tab. 4**

Cluster ID	Size	Label (LSI)	Alternative Labels	Mean Year	Representative Reference
#0	89	model	modeling; radial basis functions	2010	Lee and Wei, 2010
#1	77	building information modeling	discrete event simulation; integrated framework	2010	Ding et al., 2014
#2	65	construction	physiological status; off-duty worker	2010	Zhu and Brilakis, 2010
#3	59	point clouds	as-built BIM; least—squares adjustment	2010	Kim et al., 2016
#4	51	tracking	detection; equipment	2012	Kim et al., 2017
#5	44	vibrating wire strain gages	steel beam-columns; estimation	2009	Aldemir, 2010
#6	37	assessment	multi-criteria model; strategic transportation planning	2005	Sun and Kondyli, 2010
#7	36	condition assessment	sewer pipeline inspection; automated defect detection	2006	Zhu et al., 2010
#8	33	convolutional neural networks	relevance networks; site images	2015	Xue and Li, 2018
#9	32	meta-heuristic algorithm	hybrid laminate composite structures; multi-objective optimal design	2009	Lin, 2011
#10	25	automated detection	using deep learning techniques; closed—circuit television images	2012	Kamaliardakani et al., 2014
#11	22	rfid tracking	function-space assignment; optimization	2006	Moon and Yang, 2009

As shown in Tab. 4, the cluster # 0 "model" is the largest cluster among these 12 significant co-citation clusters. The cluster # 0 has 89 members, and the representative reference is published by Lee and Wei (Lee and Wei, 2010) who design a feature selection method for providing a sequential forecast of accident duration from the time of accident notification to the accident site clearance. The cluster # 1 "building information modeling" has 77 members, and the representative reference is Ding et al. (Ding et al., 2014) which develops a BIM application framework in the process of expanding from 3D to computable nD models. The cluster # 2 "construction" has 65 members, and the representative reference is published by Zhu and Brilakis (Zhu and Brilakis, 2010) who design a novel method of automated concrete column detection from visual data. The cluster # 3 "point clouds" has 59 members, and the representative reference is Kim et al. (Kim et al., 2016) which demonstrates a reasonable system performance in construction site images for recognizes a varying number of construction objects classes in an image. The cluster # 4 "tracking" has 51 members, and the representative reference is published by Kim et al. (Kim et al., 2017) who detect construction equipment using a region-based fully convolutional network and transfer learning. The cluster # 5 "vibrating wire strain gages" has 44 members, and the representative reference is Aldemir (Aldemir, 2010) which proposes a simple integral type quadratic functional as the performance index for the optimal control policy of earthquake excited structures.

The cluster # 6 "assessment" has 37 members, and the representative reference is published by Sun and Kondyli (Sun and Kondyli, 2010) who quantify the vehicle interactions during a lane-changing maneuver on urban streets. The cluster # 7 "condition assessment" has 36 members, and the representative reference is Zhu et al. (Zhu et al., 2010) which proposes a novel method that can detect large-scale bridge concrete columns for the purpose of eventually creating an automated bridge condition assessment system. The cluster # 8 "convolutional neural networks" has 33 members, and the representative reference is published by Xue and Li (Xue and Li, 2018) who propose a fully convolutional network model for shield tunnel lining defects. The cluster # 9 "meta-heuristic algorithm" has 32 members, and the representative reference is Lin (Lin, 2011) which develops a dual variable approximation-based descent method for a bi-level continuous dynamic network design problem. The cluster # 10 "automated detection" has 25 members, and the representative reference is published by Kamaliardakani et al. (Kamaliardakani et al., 2014) who develop an algorithm to automatically detect sealed cracks in pavement surface images. The cluster # 11 "rfid tracking" has 22 members, and the representative reference is Moon and Yang (Moon and Yang, 2009) which provides data communications in the concrete pouring operation using RFID technology.

4 Conclusion

This study focuses on academic communities, research topics evolution trends and knowledge communities covered on articles published in AC, CACI and JCCE from 2009 to 2018 through a scientometric analysis, which contributes to construction the comprehensive knowledge map of information technologies applications in the field of CEM. A total of 2911 articles published in AC, CACI and JCCE from 2009 to 2018 are adopted for scientometric analysis. The results of scientometric analysis explore the knowledge map of information technologies applications in the field of CEM, including main country/region and institution, influential author, evolution trend and knowledge community. The comprehensive knowledge map of information technologies applications in the field of CEM from 2009 to 2018 is summarized in Fig. 10.

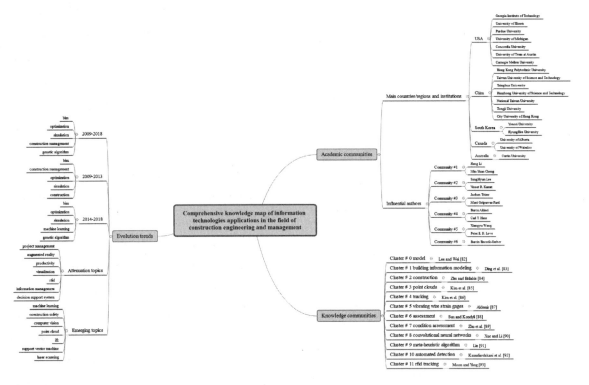

Fig. 10 Comprehensive knowledge map of information technologies applications in the field of CEM

As shown in Fig. 10, the main countries/regions are the USA, China, Korea, Canada and Australia, and the main institutions include Georgia Institute of Technology, Hong Kong Polytechnic University, University of Illinois, Purdue University, University of Michigan, Taiwan University of Science and Technology, University of Alberta, Yonsei University, Tsinghua University, Huazhong University of Science and Technology, Taiwan University, Concordia University, Curtin University, University of Texas at Austin, Tongji University, University of Waterloo, Carnegie Mellon University, City University of Hong Kong and KyungHee University. The influential authors can be divided into six communities according to the collaborative relationship between authors. These influential authors are mainly from main countries/regions and institutions which make major contributions to the articles published in AC, CACI and JCCE from 2009 to 2018.

The hot research topics of articles published in AC, CACI and JCCE from 2009 to 2018 present evolutionary trends, which is reflect by the changes of top 20 high-frequency keywords in different periods. Seven research topics, including "project management""augmented reality""productivity""visualization" "rfid""information management" and "decision support system", present attenuation trend from 2009 to 2018. Seven emerging topics become hot research topics between 2014 and 2018, including "machine learning""construction safety" "computer vision" "point cloud""ifc""support vector machine" and "laser scanning". These emerging topics between 2014 and 2018 reflect the applications of emerging information technologies, such as point cloud, laser scanning and machine learning.

The research contents of AC, CACI and JCCE from 2009 to 2018 mainly focus on building information modeling, multidimensional visualization model, optimization and simulation methods, construction information management theory and the applications of information technologies on safety management. The results of cluster analysis demonstrate that the references of articles published in AC, CACI and JCCE from 2009 to 2018 can be summarized as 12 significant knowledge communities, including model, building information modeling, construction, point clouds, tracking, vibrating wire strain gages, assess-

ment, condition assessment, convolutional neural networks, meta-heuristic algorithm, automated detection and rfid tracking.

References

[1] Aboulezz M A. Mapping the construction engineering and management discipline. 2003.

[2] Adeli H, Jiang X. Dynamic fuzzy wavelet neural network model for structural system identification. Journal of Structural Engineering, 2006 (132): 102-111.

[3] Adeli H,Jiang X. Intelligent infrastructure: neural networks, wavelets, and chaos theory for intelligent transportation systems and smart structures. CRC press,2008.

[4] Aldemir U. A simple active control algorithm for earthquake excited structures. Computer-Aided Civil and Infrastructure Engineering, 2010(25): 218-225.

[5] Anil E B, Tang P, Akinci B,et al. Deviation analysis method for the assessment of the quality of the as-is Building Information Models generated from point cloud data. Automation in Construction, 2013(35):507-516.

[6] Aydin C C. Designing building façades for the urban rebuilt environment with integration of digital close-range photogrammetry and geographical information systems. Automation in Construction, 2014(43):38-48.

[7] Bansal V. Use of GIS and topology in the identification and resolution of space conflicts. Journal of Computing in Civil Engineering, 2010(25): 159-171.

[8] Beach T, Petri I, Rezgui Y, et al. Management of collaborative BIM data by federating distributed BIM models. Journal of Computing in Civil Engineering, 2017(31): 04017009.

[9] Behzadan A H, Kamat V R. Scalable algorithm

for resolving incorrect occlusion in dynamic augmented reality engineering environments. Computer-Aided Civil and Infrastructure Engineering, 2010(25):3-19.

[10] Belsky M, Sacks R, Brilakis I. Semantic enrichment for building information modeling. Computer-Aided Civil and Infrastructure Engineering, 2016(31): 261-274.

[11] Borrmann A, Kolbe T H, Donaubauer A, et al. Multi-scale geometric-semantic modeling of shield tunnels for GIS and BIM applications. Computer-Aided Civil and Infrastructure Engineering, 2015(30): 263-281.

[12] Bosché F. Automated recognition of 3D CAD model objects in laser scans and calculation of as-built dimensions for dimensional compliance control in construction. Advanced Engineering Informatics, 2010(24):107-118.

[13] Bügler M, Borrmann A, Ogunmakin G, et al. Fusion of photogrammetry and video analysis for productivity assessment of earthwork processes. Computer-Aided Civil and Infrastructure Engineering, 2017(32): 107-123.

[14] Castillo E, Grande Z, Mora E, et al. Complexity reduction and sensitivity analysis in road probabilistic safety assessment Bayesian network models. Computer-Aided Civil and Infrastructure Engineering, 2017(32): 546-561.

[15] Chen C. Searching for intellectual turning points: progressive knowledge domain visualization. Proceedings of the National Academy of Sciences, 2004(101): 5303-5310.

[16] Chen C. CiteSpace II: detecting and visualizing emerging trends and transient patterns in scientific literature. Journal of the American Society for Information Science and Technology, 2006 (57): 359-377.

[17] Chen C, Chen Y, Horowitz, et al. Towards an explanatory and computational theory of scientific discovery. Journal of Informetrics, 2009 (3): 191-209.

[18] Chen Y, Okudan G E, Riley D R. Decision support for construction method selection in concrete buildings: prefabrication adoption and optimization. Automation in Construction, 2010 (19): 665-675.

[19] Cheng T, Teizer J. Real-time resource location data collection and visualization technology for construction safety and activity monitoring applications. Automation in construction, 2013 (34): 3-15.

[20] Cohen J. Publication rate as a function of laboratory size in three biomedical research institutions. Scientometrics, 1981(3): 467-487.

[21] Dave B, Koskela L. Collaborative knowledge management—a construction case study. Automation in construction, 2009(18): 894-902.

[22] Dawood T, Zhu Z, Zayed T. Computer vision-based model for moisture marks detection and recognition in subway networks. Journal of Computing in Civil Engineering, 2017 (32): 04017079.

[23] Dimitrov A, Gu R, Golparvar-Fard M. Non-uniform B-spline surface fitting from unordered 3D point clouds for as-built modeling. Computer-Aided Civil and Infrastructure Engineering, 2016(31): 483-498.

[24] Ding L, Zhou Y, Akinci B. Building Information Modeling (BIM) application framework: the process of expanding from 3D to computable nD. Automation in Construction, 2014 (46):82-93.

[25] Dong S, Feng C, Kamat V R. Sensitivity analysis of augmented reality-assisted building damage reconnaissance using virtual prototyping. Automation in Construction, 2013(33): 24-36.

[26] Dossick C S, Mukherjee A, Rojas E M, et al. Developing construction management events in

situational simulations. Computer-Aided Civil and Infrastructure Engineering, 2010 (25): 205-217.

[27] Du J, El-gafy M. Virtual organizational imitation for construction enterprises: agent-based simulation framework for exploring human and organizational implications in construction management. Journal of Computing in Civil Engineering, 2012(26): 282-297.

[28] Eastman C, Jeong Y S, Sacks R, et al. Exchange model and exchange object concepts for implementation of national BIM standards. Journal of Computing in Civil Engineering, 2009(24): 25-34.

[29] Eastman C, Teicholz P, Sacks R, et al. BIM handbook: a guide to building information modeling for owners, managers, designers, engineers and contractors. John Wiley & Sons, 2011.

[30] Gao H, Zhang X. A Markov-based road maintenance optimization model considering user costs. Computer-Aided Civil and Infrastructure Engineering, 2013(28): 451-464.

[31] Garfield E. Is citation analysis a legitimate evaluation tool? Scientometrics, 1979 (1): 359-375.

[32] Ghoddousi P, Eshtehardian E, Jooybanpour S, et al. Multi-mode resource-constrained discrete time-cost-resource optimization in project scheduling using non-dominated sorting genetic algorithm. Automation in Construction, 2013 (30): 216-227.

[33] Golparvar-fard M, Peña-mora F, Arboleda C A, et al. Visualization of construction progress monitoring with 4D simulation model overlaid on time-lapsed photographs. Journal of Computing in Civil Engineering, 2009(23): 391-404.

[34] Golparvar-fard M, Peña-mora F, Savarese S. D4AR-a 4-dimensional augmented reality model

for automating construction progress monitoring data collection, processing and communication. Journal of Information Technology in Construction, 2009(14): 129-153.

[35] Golparvar-fard M, Peña-mora F, Savarese S. Automated progress monitoring using unordered daily construction photographs and IFC-based building information models. Journal of Computing in Civil Engineering, 2012 (29): 04014025.

[36] Gong J, Caldas C H. Computer vision-based video interpretation model for automated productivity analysis of construction operations. Journal of Computing in Civil Engineering, 2009(24): 252-263.

[37] Granadeiro V, Duarte J P, Correia J R, et al. Building envelope shape design in early stages of the design process: integrating architectural design systems and energy simulation. Automation in Construction, 2013(32):196-209.

[38] Grau D, Caldas C H, Haas C T, et al. Assessing the impact of materials tracking technologies on construction craft productivity. Automation in Construction, 2009 (18): 903-911.

[39] Hamledari H, Mccabe B, Davari S. Automated computer vision-based detection of components of under-construction indoor partitions. Automation in Construction, 2017(74):78-94.

[40] Han S, LEE S. A vision-based motion capture and recognition framework for behavior-based safety management. Automation in Construction, 2013(35): 131-141.

[41] Handbook B. A guide to building information modeling for owners, managers, designers, engineers and contractors/ed. by C. Eastman, P. Teicholz, R. Sacks, K. Liston. Wiley. 2008.

[42] Hartmann T, Van Meerveld H, Vossebeld N, et al. Aligning building information model tools and Construction management methods. Auto-

mation in Construction，2012(22)：605-613.

[43]　Irizarry J，Karan E P，Jalaei F. Integrating BIM and GIS to improve the visual monitoring of construction supply chain management. Automation in Construction，2013(31)：241-254.

[44]　Jiang X，Adeli H. Dynamic wavelet neural network for nonlinear identification of highrise buildings. Computer-Aided Civil and Infrastructure Engineering，2005(20)：316-330.

[45]　Jiang X，Adeli H. Pseudospectra，music，and dynamic wavelet neural network for damage detection of highrise buildings. International Journal for Numerical Methods in Engineering，2007(71)：606-629.

[46]　Jiang X，Adeli H. Dynamic fuzzy wavelet neuroemulator for non-linear control of irregular building structures. International Journal for Numerical Methods in Engineering，2008(74)：1045-1066.

[47]　Jiang X，Adeli H. Neuro-genetic algorithm for non-linear active control of structures. International Journal for Numerical Methods in Engineering，2008(75)：770-786.

[48]　Jiang X，Mahadevan S，Adeli H. Bayesian wavelet packet denoising for structural system identification. Structural Control and Health Monitoring：The Official Journal of the International Association for Structural Control and Monitoring and of the European Association for the Control of Structures，2007(14)：333-356.

[49]　Jin R，Zuo J，Hong J. Scientometric review of articles published in ASCE's from 2000 to 2018. Journal of Construction Engineering and Management，2019(145)：06019001.

[50]　Jung Y，Joo M. Building information modelling (BIM) framework for practical implementation. Automation in Construction，2011(20)：126-133.

[51]　Kamaliardakani M，Sun L，Ardakani M K.

Sealed-crack detection algorithm using heuristic thresholding approach. Journal of Computing in Civil Engineering，2014(30)：04014110.

[52]　Kim H，Kim H，Hong Y W. Detecting construction equipment using a region-based fully convolutional network and transfer learning. Journal of Computing in Civil Engineering，2017(32)：04017082.

[53]　Kim H，Kim K，Kim H. Data-driven scene parsing method for recognizing construction site objects in the whole image. Automation in Construction，2016(71)：271-282.

[54]　Lee D E，Yi C Y，Lim T K，et al. Integrated simulation system for construction operation and project scheduling. Journal of Computing in Civil Engineering，2010(24)：557-569.

[55]　Lee D E，Lim T K，Arditi D. An expert system for auditing quality management systems in construction. Computer-Aided Civil and Infrastructure Engineering，2011(26)：612-631.

[56]　Lee H S，Lee K P，Park M，et al. RFID-based real-time locating system for construction safety management. Journal of Computing in Civil Engineering，2011(26)：366-377.

[57]　Lee Y，Wei C H. A computerized feature selection method using genetic algorithms to forecast freeway accident duration times. Computer-Aided Civil and Infrastructure Engineering，2010(25)：132-148.

[58]　Levitt R E. CEM research for the next 50 years：maximizing economic，environmental，and societal value of the built environment. Journal of Construction Engineering and Management，2007(133)：619-628.

[59]　Li H，Chan G，Skitmore M. Multiuser virtual safety training system for tower crane dismantlement. Journal of Computing in Civil Engineering，2012(26)：638-647.

[60]　Li X，Wu P，Shen G Q，et al. Mapping the

knowledge domains of Building Information Modeling (BIM): a bibliometric approach. Automation in Construction, 2017(84): 195-206.

[61] Lim T K, Yi C Y, Lee D E, et al. Concurrent construction scheduling simulation algorithm. Computer-Aided Civil and Infrastructure Engineering, 2014(29): 449-463.

[62] Lin D Y. A dual variable approximation-based descent method for a bi-level continuous dynamic network design problem. Computer-Aided Civil and Infrastructure Engineering, 2011 (26): 581-594.

[63] Lin J R Hu, Z Z, Zhang J P, et al. A natural-language-based approach to intelligent data retrieval and representation for Cloud BIM. Computer-Aided Civil and Infrastructure Engineering, 2016(31): 18-33.

[64] Lin T H, Liu C H, Tsai M H, et al. Using augmented reality in a multiscreen environment for construction discussion. Journal of Computing in Civil Engineering, 2014 (29): 04014088.

[65] Liu Y F, Cho S, Spencer J R B, et al. Concrete crack assessment using digital image processing and 3D scene reconstruction. Journal of Computing in Civil Engineering, 2014 (30): 04014124.

[66] Marano G C, Quaranta G, Monti G. Modified genetic algorithm for the dynamic identification of structural systems using incomplete measurements. Computer-Aided Civil and Infrastructure Engineering, 2011(26): 92-110.

[67] Martínez-rojas M, Marín N, Vila M A. The role of information technologies to address data handling in construction project management. Journal of Computing in Civil Engineering, 2015(30): 04015064.

[68] Moon S, Yang B. Effective monitoring of the concrete pouring operation in an RFID-based environment. Journal of Computing in Civil Engineering, 2009(24): 108-116.

[69] Nath T, Attarzadeh M, Tiong R L, et al. Productivity improvement of precast shop drawings generation through BIM-based process re-engineering. Automation in Construction, 2015 (54): 54-68.

[70] Park C S, Lee D Y, Kwon O S, et al. A framework for proactive construction defect management using BIM, augmented reality and ontology-based data collection template. Automation in Construction, 2013(33): 61-71.

[71] Park H S, Lee H, Adeli H, et al. A new approach for health monitoring of structures: terrestrial laser scanning. Computer-Aided Civil and Infrastructure Engineering, 2007 (22): 19-30.

[72] Porwal A, Hewage K N. Building Information Modeling (BIM) partnering framework for public construction projects. Automation in Construction, 2013(31): 204-214.

[73] Sacks R, Gurevich U, Belaciano B. Hybrid discrete event simulation and virtual reality experimental setup for construction management research. Journal of Computing in Civil Engineering, 2013(29): 04014029.

[74] Santos R, Costa A A, Grilo A. Bibliometric analysis and review of Building Information Modelling literature published between 2005 and 2015. Automation in Construction, 2017 (80): 118-136.

[75] Sharif M M, Nahangi M, Haas C, et al. Automated model-based finding of 3D objects in cluttered construction point cloud models. Computer-Aided Civil and Infrastructure Engineering, 2017(32): 893-908.

[76] Si H, Shi J G, Wu G, et al. Mapping the bike sharing research published from 2010 to 2018: a scientometric review. Journal of Cleaner Pro-

duction, 2018.

[77] Sirazitdinova E, Pesic I, Schwehn P, et al. Sewer discharge estimation by stereoscopic imaging and synchronized frame processing. Computer-Aided Civil and Infrastructure Engineering, 2018(33): 602-613.

[78] Son H, Kim C, Kin C. Automated color model-based concrete detection in construction-site images by using machine learning algorithms. Journal of Computing in Civil Engineering, 2011(26): 421-433.

[79] Succar B. Building Information Modelling framework: a research and delivery foundation for industry stakeholders. Automation in Construction, 2009(18): 357-375.

[80] Sun D J, Kondyli A. Modeling vehicle interactions during lane-changing behavior on arterial streets. Computer-Aided Civil and Infrastructure Engineering, 2010(25): 557-571.

[81] Tang P, Huber D, Akinci B, et al. Automatic reconstruction of as-built building information models from laser-scanned point clouds: a review of related techniques. Automation in Construction, 2010(19): 829-843.

[82] Teizer J, Allread B S, Fullerton C E, et al. Autonomous pro-active real-time construction worker and equipment operator proximity safety alert system. Automation in Construction, 2010(19): 630-640.

[83] Tomé A, Kuipers M, Pinheiro T, et al. Space-use analysis through computer vision. Automation in Construction, 2015(57): 80-97.

[84] Wang C, Cho Y K, Kim C. Automatic BIM component extraction from point clouds of existing buildings for sustainability applications. Automation in Construction, 2015(56): 1-13.

[85] Wang L, Xue X, Zhang Y, et al. Exploring the emerging evolution trends of urban resilience research by scientometric analysis. International Journal of Environmental Research and Public Health, 2018(15): 2181.

[86] Wing C K. The ranking of construction management journals. Construction Management & Economics, 1997(15): 387-398.

[87] Xue Y, Li Y. A fast detection method via region-based fully convolutional neural networks for shield tunnel lining defects. Computer-Aided Civil and Infrastructure Engineering, 2018 (33): 638-654.

[88] Yu D, Xu Z, Antucheviciene J. Bibliometric analysis of the Journal of Civil Engineering and Management between 2008 and 2018. Journal of Civil Engineering and Management, 2019 (25): 402-410.

[89] Zhang S, Teizer J, Lee J K, et al. Building Information Modeling (BIM) and safety: Automatic safety checking of construction models and schedules. Automation in Construction, 2013(29): 183-195.

[90] Zhao X. A scientometric review of global BIM research: analysis and visualization. Automation in Construction, 2017(80): 37-47.

[91] Zhou Y, Su W, Ding L, et al. Predicting safety risks in deep foundation pits in subway infrastructure projects: support vector machine approach. Journal of Computing in Civil Engineering, 2017(31): 04017052.

[92] Zhu Z, Brilakis I. Concrete column recognition in images and videos. Journal of Computing in Civil Engineering, 2010(24): 478-487.

[93] Zhu Z, German S, Brilakis I. Detection of large-scale concrete columns for automated bridge inspection. Automation in Construction, 2010(19): 1047-1055.

Study on the Pricing Decision Level of PPP Project in Urban Comprehensive Corridor Based on Gray-DEMATEL Model

Lin Yang[1]   Ying Li[2]   Yao Liu[2]

(1. Associate Professor, Dept. of Engineering Management,

School of Civil Engineering, Wuhan University;

2. Postgraduate, Dept. of Engineering Management, School of Civil Engineering,

Wuhan University, 430072)

【Abstract】 In this paper, the development of the comprehensive pipe gallery at home and abroad and the literature review of PPP project pricing and government compensation of the comprehensive pipe gallery are studied. It is found that most of the current research results only consider the balance of interests between project participants, but not the quantitative decision-making level between different decision-makers. Therefore, it is proposed to solve the decision-making hierarchy problem among different decision-making bodies of the PPP project of the integrated pipe gallery. To determine the upper and lower decision-makers in the bi-level programming model, firstly, using qualitative analysis, this paper analyzes the influencing factors of pricing decision-making of an urban underground comprehensive pipe gallery PPP project from the perspective of government departments, social capital, and pipeline units. The study selects 13 key influencing factors, and constructs an index system of influencing factors of pricing decision-making in an urban underground comprehensive pipe gallery PPP project. Then, the Gray-DEMATEL model is used to quantitatively study the influence of each participant on the pricing decision-making of the project, and it is concluded that the participant with the greatest influence on the pricing decision-making of the PPP project of the urban underground comprehensive pipe gallery can occupy the leading position, and is a more important decision-maker. On that basis, the pricing model of a PPP project of the comprehensive pipe gallery is designed. Finally, through the case study of the

PPP project of Wuhan Central City, the rationality of the model is verified. At the same time, the paper puts forward suggestions regarding PPP project pricing and government compensation decision-making.

【Keywords】 Urban Integrated Pipe Gallery PPP Project; Gray-DEMATEL Model; Quantitative Decision-making; Hierarchical Research; Case Study

1 Introduction

The urban underground comprehensive pipe gallery project is an important component of an urban infrastructure construction system. There are many problems in the early pipeline burying due to lack of united planning, such as traffic jams caused by repeated excavation, pipeline leakage created by improper excavation, pavement collapse caused by pipeline leakage, and "Street Zippers" "Urban Watching the Sea" etc. The construction of an urban underground comprehensive pipe gallery can effectively solve the existing problems of underground pipeline burying, and promote the construction process of a sponge city through intensive management of the underground pipeline. With the acceleration of China's modernization program, a new upsurge of urban construction has been initiated in China. More and more cities have begun to participate in the construction of urban underground comprehensive pipe galleries, such as Guangdong Pearl River Delta, Southern new city of Nanjing, Central city of Optical Valley in Wuhan and other places [1]. By the end of July 2018, there were 7867 projects stored by the national PPP integrated information platform in China, with a total investment of RMB 11. 8 trillion, including 307 ur-ban underground integrated pipe gallery PPP projects [2].

At present, there are two problems in the construction of the project in China: one is the problem of financing. In the past, the construction of a comprehensive pipe gallery was fully funded by the government. If the urbanization level is 70% in 2050, the annual municipal infrastructure construction needs RMB 800 billion at least [3]. If we only rely on the government's financial funds, it will not only cause an excessive financial burden, but will also be difficult to fill the huge funding shortfall. In the long run, the funding problem will become a hidden danger. Therefore, introducing the cooperation mode of government and social capital, giving full play to the power of social capital, and letting social capital participate in the construction and operation of urban underground comprehensive pipe galleries with PPP can not only reduce the financial pressure on the government, but also broaden the development space of social capital, fill the capital gap of infrastructure construction, and promote the government to improve service quality and supply efficiency. The other concern is the pricing and charging, which is related to the benefits of urban underground comprehensive pipe gallery projects, and the interests of social capital, pipeline units, and the public themselves. Compared with other countries, the construction of integrated pipe galleries in China started later, and the relevant policies and regulatory systems are not mature. Up to now, there is still no unified pricing and charging policy for integrated pipe gallery projects. The main reason is that the initial construction cost of an urban underground comprehensive pipe gallery is much higher than the first laying cost of traditional direct buried pipeline. If the pricing is too high, some pipeline units are not willing to accept too high charging standards, resulting in a lower entrance rate. Further, the regulatory authorities of different pipeline units are also different, so the charging difficulties will be greatly increased, causing the PPP project not to be successfully implemented [4]. If the pricing of a PPP project is too low, the government will have to exchange high compensation for the investment enthusiasm of social capital, in which case the government's financial pressure has been increased instead, so the charging policy needs to be further studied [5].

Therefore, government departments have announced a series of policies, regulations, and technical guidelines to standardize and popularize the application of urban underground comprehensive pipe gallery PPP projects: specifically the guiding ideology, basic principles, policy measures, and development goals of the comprehensive pipe gallery, encouraging enterprises to invest in the construction and operation of urban underground comprehensive pipe galleries, and promote the application of the PPP mode. It is specified

that the commercial operation mode of "the comprehensive pipe gallery shall be used with compensation, and the pipeline unit shall pay the entrance fee and daily maintenance fee to the construction and operation unit of the comprehensive pipe gallery", and it is pointed out that the charging standard for the use shall be consulted between the payers and the chargers together, and the pipe gallery without conditions for negotiation can carry out government pricing and guidance pricing. It is also required that all local governments strictly implement the paid usage system of the urban underground comprehensive pipe gallery and establish a reasonable charging mechanism.

Although the promotion of the comprehensive corridor is widely supported by all parties in society, the relationship among the three pricing subjects of the project is mutually limited, and the primary factor affecting its wide implementation is the fuzzy relationship in the decision-making hierarchy of project pricing, so the quantitative decision-making hierarchy of urban comprehensive pipe gallery PPP projects needs to be studied urgently.

2 Literature Review

2.1 Foreign Research Status

Japan has named the comprehensive pipe gallery as a utility tunnel, which is defined as the appurtenance of the road, and has promulgated the Utility Tunnel Law, which stipulates that the government and users shall each bear half of the construction and daily maintenance costs. In addition, the construction unit of the comprehensive pipe gallery can be given certain subsidies if the budget allows. Among them, the daily maintenance fee includes the main body maintenance fee, auxiliary equipment maintenance fee, daily operation electricity fee, main body inspection fee, and daily maintenance office fee. The main maintenance cost shall be apportioned to each pipeline unit according to the construction cost-sharing proportion, and the other daily maintenance costs except the main maintenance shall be apportioned by the road management unit and pipeline units. The method of average apportionment or proportional apportionment shall be adopted according to the actual situation. The construction cost of the comprehensive pipe gallery in early French construction is entirely borne by the government, while the daily maintenance cost and other costs are mainly subsidized by the government, and a small part is charged to the private pipeline unit. The specific charging standard is agreed in the franchise agreement [6].

Mason and Baldwin evaluated PPP project pricing and government compensation from the perspective of options, and proposed project pricing and government compensation strategies [7]. Ho and Liang used the real option method to construct the option pricing model (BOT option valuation) of PPP project pricing and government compensation, and verified the feasibility of project financing [8]. On the basis of the option evaluation method, Cheah used the Discounted Cash Flow (DCF) Monte Carlo method to simulate quantitative

evaluation and get reasonable pricing [9]. Emil Evenhuis and Roger Vickerman, based on the case of a PPP transportation project, used the marginal cost pricing method to price, and put forward suggestions on the establishment of a pricing mechanism [10]. Driessen and Tijs, from the perspective of a cooperative game, took a multi-functional water conservancy project as a representative, built a game model, and solved the problem of how to share the cost in project pricing [11]. Anthony Chen et al. comprehensively used the network equilibrium model to balance the sensitivity analysis of road network flow and the Monte Carlo simulation method, which improved the reliability of traffic demand prediction of BOT projects in the traffic field, and on this basis, priced the traffic BOT projects [12].

At present, the pricing mechanism under the PPP model abroad is mostly studied using the option evaluation method, marginal cost pricing method, game model, network equilibrium model, etc. However, research on the quantitative decision-making level has not been systematically carried out.

2.2 Research Status in China

The charging and pricing mechanism of urban underground comprehensive pipe gallery projects under PPP mode implemented in China can barely maintain normal operation, but the recovery of construction cost is still far away. Therefore, domestic scholars are further exploring the pricing and government compensation mechanism of urban underground comprehensive pipe gallery PPP projects. Like

PPP projects in other fields, the pricing research of comprehensive pipe gallery PPP projects mostly adopts different pricing methods. Ye Sudong et al. pointed out that the price structure must reflect the complexity of infrastructure project costs, so the price structure is mainly divided into two parts: (1) the fixed price part, including the capital payment and the fixed operation fee payment; (2) the variable price part, including the changing operation management fee and the raw material and energy payments [13]. Ding Xiaomin et al. obtained the current cost pricing situation with analysis of the Guangzhou University City Utility Tunnel project, and concluded that the main costs are the gallery fee and management fee [14]. Song Ding proposed that there was a contradiction between the government's objectives and the social capital objectives under the PPP mode, and suggested that the pricing objectives of both sides should be fully considered, and a reasonable price for the use of the pipe gallery should be recommended [15]. Based on the literature research and project practice, Zhang Xiaoxiao constructed the pricing model of a PPP project of a comprehensive pipe gallery for pipeline unit payment [16]. Wang Xi et al. used the method of a game to study the charge countermeasures of an urban comprehensive pipe gallery. By constructing the game model between the various parties in charge of the operation of the comprehensive pipe gallery, the authors proposed to formulate the charge pricing mechanism to balance the interests of all parties [17]. Wang Shuying et al. used the method of system dynamics to

build the dynamic model of a revenue system, investigating the impact of relevant factors on the project's revenue, and obtained the regulation proportion of the daily maintenance fee of the pipe gallery [18]. Cui Qiming used the multi-objective programming method to establish a charging and pricing model that meets the target planning of social capital, government, and pipeline units [19]. Wang Jianbo et al. established a model based on the principle of a two-part tariff, and obtained the pricing suitable for the management and operation of the comprehensive pipe gallery [20].

Although the application of integrated pipe galleries started late in China, a boom in the construction of integrated pipe galleries commenced with strong support from the government. After years of development, the construction of the comprehensive pipe gallery in China has also gradually improved. At present, the funding problem has been solved through a new model, the cooperation model between the government and social capital. However, there are still no precise research results on the quantitative decision-making level of such projects.

2.3 Current Situation of Research on the Pricing Decision-making Method of Comprehensive Pipe Gallery PPP Projects

After analyzing the present literature on the pricing and government compensation of urban underground comprehensive pipe gallery PPP projects, the main research methods are as follows: literature research and project practice, game theory, incomplete information static game theory and method, system dynamics and multi-objective planning method. Most of the literature has considered the balance of interests of all parties, but not the decision-making level between different decision-makers in the PPP projects of urban underground comprehensive pipe galleries, and the government's dominant position in the project pricing has not been reflected in the pricing. Different decision-makers are at different decision-making levels. They influence each other, and the objective functions they want to achieve are different, which makes it a difficult problem to set up reasonable PPP project pricing and government compensation.

DEMATEL (Decision-making Trial and Evaluation Laboratory) translates as a decision experiment analysis method, which was first proposed by the American Battelle Institute[21]. It is a method that combines the theory of uncertainty gray system with the method of decision-making experiment analysis, which can have the advantages of both methods. It does not only use the flexibility of the gray system theory to solve the uncertain decision-making problems where people's feedback generally consists of fuzzy numerical estimates, but also solves the problem that the traditional gray decision experiment analysis method scores as a certain fraction value without considering the differences between enterprises. The biggest advantage of the Gray-DEMATEL method is that reliable evaluation results can be obtained with less data, so there is no need to select many experts during

the survey, reducing the tediousness of data processing [22].

This method uses graph theory and matrix theory to identify and analyze the complex causality between various factors. The direct influence matrix is established by comparing the interactions between the factors, then changing the matrix to get the comprehensive influence matrix. Finally, the central degree and cause degree of each influence factor are obtained [23].

Based on the Gray-DEMATEL method, this article determines the leading party in project pricing and uses it as the upper-level plan when pricing to reflect its dominant position, which can well solve the decision-level problems between different decision-making subjects.

3 Relationship Analysis among Participating Entities

3.1 Pricing Targets Analysis

The participants of the urban underground comprehensive pipeline gallery PPP project include government departments, social capital, pipeline units, and the public. By analyzing the factors affecting the pricing decision of the comprehensive underground pipe gallery PPP project, it can be seen that although the public is the largest beneficiary of the project, the public is a passive group and can hardly participate in the pricing decision process of the project. Therefore, when researching the pricing decision of the comprehensive pipe gallery PPP project based on double-layer planning, the public is excluded, and only the pricing targets of the government, social capital, and pipeline unit are considered. The details are as follows:

(1) Government departments: The government departments let social capital participate in urban underground comprehensive pipe gallery construction projects by introducing the PPP model. At the same time, the government as a supervisor can urge social capital to maximize its own advantages of operation and management of the urban underground comprehensive pipe gallery, which can improve the efficiency of project construction, promote economic development, and improve the performance of local governments, so as to maximize the social benefits.

(2) Social capital: As a rational investor, social capital usually pursues the maximization of investment returns. Investment income comes from paid use expenses paid by pipeline units and feasible gap subsidies provided by government departments. Paid use expenses and feasibility gap subsidies can cover the cost of construction and operation of the pipe gallery project. The excess part is the investment income of social capital, and social capital hopes that this part of the investment income will reach the maximum.

(3) Pipeline unit: The pipeline unit hopes that the cost of self-consumption is the smallest. The construction cost of the urban underground comprehensive pipe gallery is high, and the initial construction investment is much higher than the first laying cost of the traditional direct burying method. At present,

most pipeline units are less enthusiastic about entering the gallery, hoping to make the payment as low as possible compared to the first laying cost of the traditional direct pipeline burying method, and achieve the goal of consuming the lowest total cost to enter the urban underground comprehensive pipe gallery.

3.2 Pricing Decision Principles

Whether the pricing of a comprehensive pipe gallery PPP project is reasonable is related to the smooth construction and operation of the project. If the pricing is too high, some pipeline units are unwilling to accept the excessive charging standard, which makes the entrance rate too low and PPP projects cannot be successfully implemented; if the pricing of a PPP project is too low, it cannot guarantee the reasonable return of social capital and will force the government to increase subsidies. This will increase the government's financial pressure instead.

By analyzing the roles and functions of government departments, social capital, pipeline units, and the public in the construction of urban underground comprehensive pipe gallery PPP projects, it can be known that the pricing of urban underground comprehensive pipe gallery PPP projects should follow the following principles: (1) principle of user paying: urban underground comprehensive pipe galleries are constructed for use by pipeline units. According to the principle of user paying, pipeline units should bear the corresponding costs. (2) principle of beneficiary paying: the government department is responsible for the construction of the urban underground comprehensive pipe gallery and also obtains the social efficiency, and the government represents the interests of the general public, so the government should bear the corresponding costs. (3) principle of justice and equity: in the strategy of pricing and government compensation of urban underground comprehensive pipe gallery PPP projects, the cost allocation among cost bearers should be fair and reasonable. (4) principle of the importance of upper decision-makers: it means that the pricing should follow the decision level among decision-makers. The upper decision-makers occupy the dominant position in decision-making and have the largest weighting impact on the pricing of urban underground comprehensive pipe gallery PPP projects, so we can utilize a bi-level programming model to reflect the importance of upper-level decision-makers.

3.3　Decision Relationship Analysis

The three pricing decision-making bodies of the urban underground comprehensive pipe gallery PPP project have different interests represented by government departments, social capital, and pipeline units, so the three parties have different pricing goals. Pricing decision-making subjects interact with each other, and there are contradictions among pricing goals.

If purely considering the pricing target of the pipeline unit to minimize the cost of the pipeline unit, it will increase the financial pressure on the government or affect the social

capital's pricing target and reduce the return on investment. If considering the pricing objective of social capital alone to maximize the return on social capital, it will increase the government's fiscal pressure or affect the pipeline unit's pricing target and increase their cost. In addition, changes in the pricing targets of government departments, such as changes in government feasibility gap subsidies, will also affect the pricing targets of social capital and pipeline units. If the government feasibility gap subsidies are increased, it will increase the returns to social capital or lessen the cost of pipeline units; if the government feasibility gap subsidy is reduced, it will decrease the benefits of social capital or increase the cost of pipeline units.

Therefore, it is necessary to establish an effective pricing model of urban underground comprehensive pipe gallery PPP projects to balance the pricing targets of the government, social capital, and pipeline units. Due to the quasi-operating characteristic of the urban underground comprehensive pipe gallery, its price cannot be completely determined by the market. In order to protect the reasonable return of social capital, the government needs to provide a feasible gap subsidy and supervise social capital to prevent it from seeking excessively high returns. At the same time, we should measure the payment capacity of the pipeline units accurately, improve social capital's operation and management capabilities, prevent the government from paying for the losses caused by the mismanagement of social capital, and finally make the three parties' pricing goals reach equilibrium.

As the government, social capital, and pipeline units, the key roles that influence the pricing and government compensation of urban underground comprehensive pipe gallery PPP projects, are at different decision-making levels, we can express the differences in decision-making levels in a mathematical model and quantify their impact. The upper-level decision-maker makes a decision first, and the lower-level subsystems determine the optimal value within the possible range according to the upper-level decision. The optimal value is fed back to the upper layer, and the upper layer makes its own optimal decision accordingly. Therefore, it is necessary to determine the upper and lower decision-makers among the three parties of government, social capital, and pipeline units.

4　Analysis Process Based on Gray-DE-MATEL Model

4.1　Model Construction based on Gray-DE-MATEL

This article needs to determine the upper decision-maker in the pricing and government compensation of an urban underground comprehensive pipe gallery PPP project. Therefore, the Gray-DEMATEL method is used to quantitatively analyze the impact of each participant on the decision-making of project pricing. The specific steps of using the Gray-DEMATEL method to quantitatively analyze the pricing decision-making of an urban underground comprehensive pipe gallery PPP pro-ject are as follows:

Step 1: Identify the influence factors, construct an index system of key factors, and design a questionnaire.

Use the literature reading method, brainstorming method, or expert interview method to identify each influence factor. Screen out the key influence factors to construct an index system, and design a questionnaire according to the constructed index system.

Step 2: Build a direct impact matrix.

Establish direct influence matrix A, and invite experts to compare the influence relationship between each row factor and column factor in the impact matrix and score them according to the key influence factor index system and expert semantic variable table. A score of 0 means no effect, 1 means very little effect, 2 means little effect, 3 means high effect, and 4 means very high effect. This is shown in Tab. 1.

Expert semantic variable　Tab. 1

Value	Description
0	No effect (N)
1	Very little effect (VL)
2	Little effect (L)
3	High effect (H)
4	Very high effect (VH)

Step 3: Construct a direct impact matrix based on the interval gray number.

Use the interval gray number to convert the direct impact matrix constructed with the expert's scores into the gray number matrix B. That is, according to the expert semantic variable of gray number evaluation in Tab. 2, transform the numerical values of the expert scoring into the corresponding interval gray number, where the interval gray number on the diagonal is all $[0,0]$.

Expert semantic variable of gray number evaluation　　Tab. 2

Value	Description	Gray Number
0	N	$[0,0]$
1	VL	$[0,0.25]$
2	L	$[0.25,0.5]$
3	H	$[0.5,0.75]$
4	VH	$[0.75,1]$

Step 4: Consider all aspects to give weights to experts.

According to the expert weight semantic variable in Tab. 3, give the experts with different work experiences and different research levels weights of interval gray number with ambiguity.

Expert semantic variable of weight　Tab. 3

Semantic Variable	Gray Number
Not Important	$[0,0.3]$
Less Important	$[0.3,0.5]$
Important	$[0.4,0.7]$
More Important	$[0.5,0.9]$
Very Important	$[0.7,1]$

Step 5: Standardize and clear the gray number matrix obtained in Step 3.

Use the CFCS (Converting Fuzzy Data into Crisp Scores) method to standardize and clear the gray number matrix B obtained in Step 3. First, use Formula (1) to calculate the upper and lower limits of the standardized gray number. Then, use Formula (2) to obtain the crisp value after clearing the standardized gray number, and obtain the direct impact matrix of every expert's clear value Z^k, where k takes 1 to indicate the direct impact matrix of Expert 1's clear value.

$$\begin{cases} \overline{\otimes}\,\overline{\lambda}_{ij}^k = \dfrac{\overline{\otimes}\,\lambda_{ij}^k - \min \overline{\otimes}\,\lambda_{ij}^k}{\Delta_{\min}^{\max}} \\[2mm] \underline{\otimes}\,\overline{\lambda}_{ij}^k = \dfrac{\underline{\otimes}\,\lambda_{ij}^k - \min \underline{\otimes}\,\lambda_{ij}^k}{\Delta_{\min}^{\max}} \\[2mm] \Delta_{\min}^{\max} = \max \overline{\otimes}\,\lambda_{ij}^k - \min \underline{\otimes}\,\lambda_{ij}^k \end{cases} \quad (1)$$

$$\begin{cases} Y_{ij}^k = \dfrac{\underline{\otimes}\,\overline{\lambda}_{ij}^k(1-\underline{\otimes}\,\overline{\lambda}_{ij}^k)+\overline{\otimes}\,\overline{\lambda}_{ij}^k \times \overline{\otimes}\,\overline{\lambda}_{ij}^k}{(1-\underline{\otimes}\,\overline{\lambda}_{ij}^k+\overline{\otimes}\,\overline{\lambda}_{ij}^k)} \\[2mm] Z_{ij}^k = \min \underline{\otimes}\,\lambda_{ij}^k + Y_{ij}^k\,\Delta_{\min}^{\max} \end{cases}$$

$$(2)$$

Where $\overline{\otimes}\,\overline{\lambda}_{ij}^k$ is the upper limit in gray number matrix B after it is standardized, $\underline{\otimes}\,\overline{\lambda}_{ij}^k$ is the lower limit in gray number matrix B after it is standardized, $\overline{\otimes}\lambda_{ij}^k$ is the upper limit in the expert's original score after it is transformed into an interval gray number, $\underline{\otimes}\,\lambda_{ij}^k$ is the lower limit in the expert's original score after it is transformed into an interval gray number. $x\overline{\otimes}\lambda_{ij}^k$ is the maximum value of the upper limit in the interval gray number scored by the k different experts on the influence factor in row i, column j. $\min \underline{\otimes}\lambda_{ij}^k$ is the minimum value of the lower limit in the interval gray number scored by the k different experts on the influence factor in row i, column j.

Step 6: Clear the assigned expert weights.

Use Formula (1) and (2) to clear the expert weights assigned in Step 4, with the processing method the same as Step 5, and calculate the direct impact matrix Z after considering weights according to Formula (3).

$$\begin{cases} Z_{ij} = \omega_1 Z_{ij}^1 + \omega_2 Z_{ij}^2 + \cdots + \omega_n Z_{ij}^k \\[2mm] \displaystyle\sum_{k=1}^{n} \omega_k = 1 \end{cases} \quad (3)$$

Where ω_k is the experts' weights when cleared, Z_{ij}^k is the direct impact matrix of crisp value scored by No. k expert on influence factors in row i and column j.

Step 7: Transform the direct impact matrix into a normalized matrix.

Use Formula (4) to normalize the direct impact matrix Z and transform the direct impact matrix Z into a normalized matrix M. That is, divide the direct impact matrix Z by the largest value in the all sums of row in the direct impact matrix Z.

$$\begin{cases} M = S \times Z \\ S = \dfrac{1}{\max\sum\limits_{j=1}^{n} Z_{ij}} , i,j = 1,2,\cdots,n \end{cases} \quad (4)$$

Step 8: Use Formula (5) to calculate the comprehensive impact matrix T, where I is the identity matrix.

$$T = M(I-M)^{-1} \quad (5)$$

Step 9: Use Formula (6) to find the cause degree E_i and the central degree P_i according to the comprehensive impact matrix T.

$$\begin{cases} P_i = \{R_i + C_j \mid i = j\} \\ E_i = \{R_i - C_j \mid i = j\} \end{cases} \quad (6)$$

Where R_i is influence degree, the sum of rows in the matrix T; C_j is influence degree, the sum of columns in the matrix T.

The influence factor whose cause degree E_i is greater than 0 is the cause factor of other influence factors, and the lesser one is the result factor of others. The larger the factor's central degree, the closer it is to other influence factors.

Step 10: Use MATLAB software to draw the distribution map of each influence factor. Establish the Cartesian coordinate system taking the cause degree E_i as the ordinate and the central degree P_i as the abscissa.

4.2 Questionnaire Design and Data Processing

This article uses a literature review and expert research interviews to analyze the influence factors of quantitative decision-making for urban underground comprehensive pipe gallery PPP projects, and selects 13 key factors from the perspective of three project participants: government departments, social capital, and pipeline units (see Tab. 4 below).

According to the expert semantic variables in Tab. 2, and comparing the relationship between each influence factor, values from Tab. 2 were selected to fill in Tab. 3. The values in Tab. 3 refer to the influencing degree of row factor on column factor. For example, in row 2, column 1 of Tab. 3, number 2 expresses that the influence factor a2 has a moderate influence on the influence factor a1.

4.3 Pricing Design of the Comprehensive Pipe Gallery PPP Project

As the franchise period of an urban underground comprehensive pipe gallery PPP project is generally longer than 30 years, the external environment will continue to change during this process, such as the economic environment, policy environment, and so on. Once the basis of the quantitative decision-making level is obtained according to the Gray-DEMATEL model mentioned above, it is possible to start from different levels to solve the pricing problem of urban comprehensive pipe gallery PPP projects. Under the conditions of pricing and government compensation of urban underground comprehensive pipe gallery PPP projects, a pricing function can be designed with maximizing social benefits as the government's goal and maximizing investment returns as the social capital's goal. As social capital will take this into account when making decisions on the pricing of integrated pipe gallery PPP projects, the pipeline units may choose not to enter the gallery for its higher consumption cost than the traditional direct laying method, thereby affecting the overall demand of the integrated pipe gallery. If the paid use fee of the pipe gallery is lower, that is, the consumption cost of the pipeline unit is lower, the demand for the pipe gallery will be higher, and the investment income of social capital will subsequently be higher.

Explanatory notes on the factors　　　　　　　Tab. 4

influencing the pricing decision-making in the urban underground comprehensive pipe gallery PPP project

Category	Number	Influence Factor	Description
Government Department	a1	Government Feasibility Gap Subsidy	Including investment subsidy and price subsidy. Due to the huge construction investment and long payback period, charging the pipeline unit alone cannot fully make up the costs for investors. In order to encourage social capital to invest and ensure its reasonable benefits, the government provides a feasibility gap subsidy
	a2	Laws, Gegulations, and Supporting Policies	Laws, regulations, and supporting policies play a vital role in promoting the implementation of the project. Perfect laws and regulations can formulate more reasonable prices, cut down project risks, and reduce uncertainty
	a3	Government Financial Carrying Capacity	The construction of urban infrastructure is inseparable from the support of government financial funds. The better the level of regional economic development, the higher the fiscal revenue and the stronger the government's fiscal carrying capacity
	a4	Government Price Supervision Mechanism	The government supervision mechanism plays a vital role in the construction of the project. The government's conversion to supervisor in the project can improve the service quality and efficiency of the corridor project
Social Capital	a5	Integrated Pipe Gallery Cost	The cost of the integrated corridor is an important basis for determining the pricing of the project. Part of the cost is borne by social capital, and they charge from pipeline units to make up the cost and obtain a reasonable income
	a6	Expected Investment Return Rate	In the construction of the project, social capital is more concerned about its own return on investment. The investors hope that the greater the investment return income the better, but this is affected by the quasi-operating characteristics of the project, so they can only obtain reasonable returns
	a7	Franchise Operation Period	The length of the franchise period determines the amount of social capital investment income. The franchise period is directly proportional to the social capital investment income. The longer the franchise period, the longer the time for social capital to recover capital and obtain benefits, and the higher the natural income
	a8	Capital Cost	Part of the funds that social capital invests are free funds and the other is borrowed funds. The loan interest shall be paid for the loan fund. If the loan interest rate is high and the loan amount is too high, the corresponding original fund is large
	a9	Technical and Operational Management Experience	Social capital with mature technology and rich management experience can build acorridor with good quality and service, which can control project costs, promote technological and economic development, and improve pipeline units' willingness to enter the corridor and psychological endurance capacity to charging

continued

Category	Number	Influence Factor	Description
Pipeline Unit	a10	First Laying Cost of Direct Buried Pipeline	In fact, the comprehensive benefits brought by the corridor construction are much higher than the traditional direct buried pipelines, but the early construction cost of the corridor is high and the short-term benefits are not obvious. In contrast, the traditional one has a small initial investment and obvious short-term benefits, so the pipeline units are not enthusiastic about entering the corridor
	a11	Willingness of Pipeline Unit to Enter the Corridor	Encourage pipeline units to enter the corridor, increase their willingness, and ensure the utilization rate of comprehensive pipeline corridors. Only by ensuring the utilization rate of the comprehensive pipe gallery can the project be profitable and successfully implemented
	a12	Differentiation of Entrance Fees	Various pipelines distributed in urban underground comprehensive corridors include electricity, communications, heating, water supply, and drainage. Different pipelines occupy different pipe gallery resources
	a13	Payment Capacity	The users of different pipelines in the comprehensive corridor are different, and their comprehensive strengths are different. If the project is priced higher than the pipeline unit's own payment capacity, the pipeline unit will not risk the damage of its own interests to cooperate with government entrance

(1)Construction of social capital objective function for lower decision-makers

After judging the degree of influence of multiple participants on pricing decisions based on the Gray-DEMATEL model above, social capital, as a rational investor, usually aims to maximize the return on investment. The investment income comes from the paid use fee from the pipeline unit and the feasibility gap subsidy provided by the government. Among them, the paid use fee includes the entrance fee and daily maintenance fee, and the feasibility gap subsidy is the government compensation. The cost required for the construction and operation of a comprehensive corridor project includes the construction and the annual operating cost, of which the construction cost includes the pipe gallery body construction and the ancillary facilities construction cost. The annual operating cost includes all the expenses required for the operation of the pipe gallery, such as man-

agement costs and labor costs, calculated according to the actual annual operating costs. Social capital covers the cost required for the construction and operation of the pipe gallery project by collecting paid use fees and government compensation, and the excess is the investment income. Social capital aims at maximizing investment income, so the objective function of social capital for lower-level decision-makers is Equation (7).

$$\text{Max} U_f = PQ + Sk_1 - Ik_2 - Ck_1 \quad (7)$$

Where U_f is the investment income of social capital; P is the paid use fee, $P = P_1 + P_2 \times k_1$, P_1 is the gallery fee, and P_2 is the daily maintenance fee. Convert all funds to the beginning of the operating period based on conditional assumptions. As the entrance fee is paid in one lump at the beginning of the operation period, no conversion is required. However, the daily maintenance fee is paid annually during the operation period, so it needs to be multiplied by the discount coefficient k_1 to discount to the beginning of the operating period considering the time value of funds, where k_1 is the present value coefficient of equivalent series in Engineering Economics. $k_1 = (P/A, i, n) = \dfrac{(1+i)^n - 1}{i(1+i)^n}$, where i is the discount rate, n is the years of operation; Q is the demand of the pipe gallery and meet price demand curve. According to the assumption, the price of the pipe gallery is linearly related to its demand, so $Q = a - (b_1 P_1 + b_2 P_2 k_1)$, which means that the higher the price of the pipe gallery, the higher the cost of the pipeline unit, and the lower the demand for the pipe gallery. a, b_1, b_2 is the parameter of the demand curve in Engineering Economics, and the regression analysis method can be used to analyze the price and demand of multiple comprehensive pipeline gallery PPP projects to obtain it. b_1, b_2 reflects the sensitivity of the pipeline unit to the price. Generally speaking, the current expenditure will be more sensitive than the future expenditure. Similarly, the pipeline unit's expenditure on entrance fees is more sensitive than the daily maintenance fee. So, $b_1 > b_2 > 0$; S is the government compensation, and multiply by the discount coefficient k_1 to discount it to the beginning of the operating period when paid annually during operation; I is the construction cost, disbursed at the beginning of the construction period, and it needs to be multiplied by the discount coefficient k_2 to discount to the beginning of the operation period considering the

time value of the funds. k_2 is the final value coefficient of a one-time payment in Engineering Economics, and $k_2 = (F/P, i, n) = (1+i)^N$, where N is the years of construction; C is the annual operating cost, and multiply by discount coefficient k_1 to discount to the beginning of the operation period when paid annually during operation.

The return rate on project investment directly affects the return on investment of social capital, so the return rate on index project investment is used to restrict the return of social capital. As the purpose of social capital investment is to obtain income, the project investment income cannot be lower than the industry's lowest investment return rate. Meanwhile, the urban underground comprehensive pipe gallery has quasi-operational characteristics, and the price of the pipe gallery cannot be completely determined by the market. Thus, the government needs to set down a reasonable investment return rate to limit the maximum return of social capital. Consequently, the constraint of the project investment return rate is Equation (8).

$$R_{\min} \leqslant R \leqslant R_{\max} \qquad (8)$$

Where R is the project investment return rate, and $R = \dfrac{U_f}{I k_2 + C k_1}$; $R_{\min}$ is the industry's lowest investment return rate; $R_{\max}$ is the maximum reasonable investment return rate produced by the government; $[R_{\min}, R_{\max}]$ is the reasonable profit range of social capital.

Compare the pre-investment of the comprehensive pipe gallery PPP project with the traditional direct buried pipeline method. If the paid use fee paid by the pipeline unit is higher than the first laying cost of the traditional direct burying method, the willingness of the pipeline unit to enter the gallery will be reduced. In order to increase the enthusiasm of pipeline units to enter the gallery, the paid use fee of the pipe gallery should be less than the first laying cost of the traditional direct buried pipeline method. So the constraint of the paid use fee of the pipe gallery is Equation (9).

$$0 < P \leqslant CA \qquad (9)$$

Where CA is the first laying cost of the traditional direct buried pipeline method.

According to the "Guidelinesfor Demonstration of Financial Affordability on Government and Social Capital Cooperation Project" issued by the Finance Ministry (Finance [2015] No. 21), the formula for calculating the amount of operating subsidy expenditures, namely government compensation, is as follows:

Amount of operating subsidy for the year =[Total construction cost of the project × (1 +Reasonable profit rate) × (1+ Annual discount rate) ×n]÷Fiscal operation subsidy cycle+ Annual operating cost × (1+ Reasonable profit rate) −Amount of users paid in the year

To simplify the above formula, use the parameter α to represent the part before the minus and get the simplified relationship between government compensation and daily maintenance costs in Equation (4). The government compensation is not less than 0, and $\alpha \geqslant P_2$.

$$S = \alpha - P_2 \qquad (10)$$

Therefore, based on the quantitative hierarchical decision results of Gray-DEMATEL, it is possible to obtain the quantitative impact of each participating entity on the project pricing and finally obtain the price of the project.

5 Case Study

The comprehensive pipe gallery PPP project of Optics Valley Central City in Wuhan East Lake New Technology Development Zone is in the middle of the second independent innovation demonstration area in China, Optics Valley Central City, with a total land area of approximately 36. 8km². The main construction content includes construction engineering, structural engineering, electrical engineering, telecommunications engineering, drainage engineering, ventilation engineering, and firefighting engineering. The comprehensive pipe gallery project is 24. 67km long and will be constructed in two phases. The first phase of the project is divided into two parts. One is a pipe gallery located in the underground space, with a total length of about 4. 61km. This part of the pipe gallery is constructed synchronously with the underground space, and its civil facilities belong to the underground space engineering; the other part is a non-underground space, with a total length of about 13. 29km. The total length of the second phase project is 6. 77km. After the completion of the comprehensive pipe gallery, it will be able to meet the needs of the 400000 residents in the area. It is currently the largest comprehensive pipe gallery project under construction in Wuhan with the most pipe-lines.

5.1 Data Processing based on the Gray-DE-MATEL Model

The design questionnaire is based on the index system of quantitative decision factors for the 13 urban underground pipe gallery PPP projects listed in Tab. 4. Invite four experts to compare between each relationship of row factor on column and score according to the expert semantic variable table, combined with their own viewpoint and the actual project situation. Then use Gray-DEMATEL for data processing. Firstly, convert the four experts' raw data from scoring form into interval gray number, and use CFCS to crisp every expert's gray number matrix. Simultaneously, use CF-CS to clear every expert's gray number weight on the basis of the given weight of their work experience and research level (Tab. 5).

Next, obtain the direct impact matrix Z (Tab. 6) combined with each expert's weight. Then transform Z into standardized matrix M (Tab. 7), and get comprehensive influence matrix T (Tab. 8). Finally, obtain the cause degree E_i and central degree P_i (Tab. 9). Taking the cause degree E_i as the abscissa and the central degree P_i as the ordinate, construct the Cartesian coordinate system, and draw the distribution diagram of each influence factor (Fig. 1).

Expert gray number weight　　Tab. 5

Number	Expert	Interval Gray Number
1	The project construction and operation principal	[0.5,0.9]
2	Consultant A from consulting company	[0.5,0.9]
3	Professor of project management research	[0.4,0.7]
4	Consultant B from consulting company	[0.4,0.7]

Direct impact matrix Z　　Tab. 6

Factor	a_1	a_2	a_3	a_4	a_5	a_6	a_7	a_8	a_9	a_{10}	a_{11}	a_{12}	a_{13}
a_1	0	0.49	0.69	0.31	0.33	0.51	0.65	0.53	0.72	0.37	0.45	0.85	0.43
a_2	0.66	0	0.55	0.78	0.54	0.44	0.59	0.25	0.37	0.45	0.36	0.51	0.65
a_3	0.75	0.46	0	0.64	0.47	0.44	0.47	0.53	0.61	0.29	0.13	0.38	0.36
a_4	0.68	0.62	0.55	0	0.47	0.75	0.76	0.64	0.20	0.20	0.32	0.51	0.92
a_5	0.37	0.16	0.35	0.55	0	0.69	0.54	0.49	0.27	0.11	0.21	0.45	0.61
a_6	0.53	0.39	0.53	0.71	0.63	0	0.76	0.78	0.57	0.48	0.48	0.71	0.52
a_7	0.51	0.55	0.22	0.60	0.48	0.84	0	0.78	0.45	0.31	0.05	0.61	0.92
a_8	0.46	0.37	0.37	0.51	0.38	0.53	0.83	0	0.29	0.72	0.11	0.43	0.76
a_9	0.64	0.30	0.37	0.31	0.03	0.49	0.54	0.40	0	0.45	0.59	0.60	0.43
a_{10}	0.12	0.06	0.25	0.27	0.30	0.64	0.54	0.62	0.55	0	0.37	0.63	0.43
a_{11}	0.60	0.19	0.35	0.06	0.05	0.40	0.04	0.06	0.56	0.31	0	0.37	0.19
a_{12}	0.60	0.48	0.35	0.46	0.38	0.53	0.61	0.46	0.61	0.13	0.39	0	0.52
a_{13}	0.34	0.37	0.34	0.51	0.33	0.71	0.63	0.39	0.54	0.36	0.61	0.63	0

Standardized matrix M   Tab. 7

Factor	a_1	a_2	a_3	a_4	a_5	a_6	a_7	a_8	a_9	a_{10}	a_{11}	a_{12}	a_{13}
a_1	0	0.07	0.10	0.04	0.05	0.07	0.09	0.08	0.10	0.05	0.06	0.12	0.06
a_2	0.09	0	0.08	0.11	0.08	0.06	0.08	0.04	0.05	0.06	0.05	0.07	0.09
a_3	0.11	0.06	0	0.09	0.07	0.06	0.07	0.08	0.09	0.04	0.02	0.05	0.05
a_4	0.10	0.09	0.08	0	0.07	0.11	0.11	0.09	0.03	0.03	0.05	0.07	0.13
a_5	0.05	0.02	0.05	0.08	0	0.10	0.08	0.07	0.04	0.02	0.03	0.06	0.09
a_6	0.07	0.05	0.07	0.10	0.09	0	0.11	0.11	0.08	0.07	0.07	0.10	0.07
a_7	0.07	0.08	0.03	0.09	0.07	0.12	0	0.11	0.06	0.04	0.01	0.09	0.13
a_8	0.06	0.05	0.05	0.07	0.05	0.07	0.12	0	0.04	0.10	0.02	0.06	0.11
a_9	0.09	0.04	0.05	0.04	0.00	0.07	0.08	0.06	0	0.06	0.08	0.08	0.06
a_{10}	0.02	0.01	0.04	0.04	0.04	0.09	0.08	0.09	0.08	0	0.05	0.09	0.06
a_{11}	0.08	0.03	0.05	0.01	0.01	0.06	0.01	0.01	0.08	0.04	0	0.05	0.03
a_{12}	0.08	0.07	0.05	0.07	0.05	0.07	0.09	0.07	0.09	0.02	0.05	0	0.07
a_{13}	0.05	0.05	0.05	0.07	0.05	0.10	0.09	0.06	0.08	0.05	0.09	0.09	0

Integrated impact matrix T   Tab. 8

Factor	a_1	a_2	a_3	a_4	a_5	a_6	a_7	a_8	a_9	a_{10}	a_{11}	a_{12}	a_{13}
a_1	0.32	0.30	0.34	0.34	0.27	0.41	0.44	0.38	0.39	0.26	0.27	0.45	0.40
a_2	0.40	0.24	0.32	0.39	0.30	0.41	0.43	0.34	0.34	0.27	0.26	0.40	0.42
a_3	0.38	0.28	0.23	0.35	0.27	0.38	0.39	0.35	0.34	0.24	0.21	0.36	0.36
a_4	0.43	0.34	0.34	0.32	0.32	0.48	0.48	0.42	0.34	0.26	0.27	0.43	0.49
a_5	0.30	0.21	0.25	0.31	0.19	0.37	0.36	0.31	0.27	0.19	0.20	0.33	0.36
a_6	0.42	0.32	0.35	0.42	0.34	0.39	0.50	0.45	0.40	0.30	0.30	0.47	0.45
a_7	0.39	0.32	0.29	0.39	0.31	0.47	0.37	0.42	0.36	0.27	0.23	0.43	0.48
a_8	0.35	0.27	0.28	0.35	0.27	0.40	0.44	0.29	0.31	0.30	0.21	0.38	0.42
a_9	0.34	0.23	0.26	0.28	0.19	0.35	0.36	0.30	0.24	0.24	0.25	0.36	0.33
a_{10}	0.26	0.19	0.22	0.26	0.22	0.35	0.34	0.32	0.29	0.17	0.21	0.34	0.32
a_{11}	0.24	0.15	0.18	0.16	0.12	0.23	0.19	0.17	0.23	0.15	0.11	0.23	0.19
a_{12}	0.36	0.28	0.27	0.32	0.26	0.38	0.40	0.34	0.34	0.21	0.24	0.30	0.38
a_{13}	0.34	0.27	0.28	0.34	0.26	0.41	0.41	0.33	0.34	0.25	0.27	0.39	0.31

Cause degree and centrality of the influencing factors   Tab. 9

Factor	Row Sum(R)	Column Sum(C)	Centrality (R+C)	Cause Degree(R−C)
a_1	4.5586	4.522	9.0806	0.0366
a_2	4.5077	3.3854	7.8931	1.1223
a_3	4.1348	3.6019	7.7367	0.5329
a_4	4.9158	4.2163	9.1321	0.6995
a_5	3.6319	3.3145	6.9464	0.3174
a_6	5.1006	5.0301	10.1307	0.0705
a_7	4.7259	5.1034	9.8293	−0.3775

continued

Factor	Row Sum(R)	Column Sum(C)	Centrality (R+C)	Cause Degree(R−C)
a_8	4.2812	4.415	8.6962	−0.1338
a_9	3.7385	4.1758	7.9143	−0.4373
a_{10}	3.4998	3.1043	6.6041	0.3955
a_{11}	2.3339	3.0276	5.3615	−0.6937
a_{12}	4.0745	4.8665	8.941	−0.792
a_{13}	4.1847	4.9251	9.1098	−0.7404

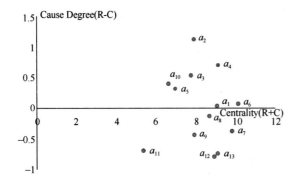

Fig. 1　Distribution map of influence factors

5.2　Result Analysis

5.2.1　Cause Degree Analysis

1. Causal Factor Analysis

The most fundamental factor affecting the pricing decision of the urban underground comprehensive pipe gallery PPP project is the cause. The influence factor of cause degree E_i greater than 0 is the other factors' cause factor, and the larger the value of E_i, the greater its influence on other factors. It can be seen from Fig. 1 that there are seven influence factors whose cause degree E_i is greater than 0, and they are arranged in descending order $a_2 > a_4 > a_3 > a_{10} > a_5 > a_6 > a_1$. Therein, a_2 (laws, regulations, and supporting policies) is the primary influence factor in pricing decision-making of the urban underground comprehensive pipe gallery PPP project.

Therefore, drawing on relevant domestic and foreign advanced systems and combining these with our actual conditions, establishing excellent laws, regulations, and supporting policies for the pricing of urban underground comprehensive pipe gallery projects is the first task in formulating reasonable pricing. The second is a_4 (government price-supervising mechanism). The establishment of a perfect pricing supervision mechanism can prevent social capital from passing on its cost to the pipeline units, and avoid the pipeline units transferring their cost to the public due to excessive pressure. The third is a_3 (the bearing capacity of government finance). Only when the economy of the area develops, where the urban underground comprehensive pipe gallery is located, can the government's financial bearing capacity be improved, and the government's financial bearing capacity is directly related to the subsidy ratio and duration of the urban underground comprehensive pipe gallery project. And finally a_{10} (first laying cost of pipeline unit directly burying), a_5 (the cost of the comprehensive pipe gallery), a_6 (expected return rate on investment), and a_1 (government feasibility gap subsidy) in turn.

From the results, it can be seen that four

of the influence factors in government decision-making are all causal factors, indicating that the government has the greatest influence on pricing in the pricing decision-making of urban underground comprehensive pipe gallery PPP projects, occupying the dominant position, and is the more important decision-maker. Therefore, when establishing a pricing model based on the bi-level planning theory, the government department should be the upper decision-maker.

2. Outcome Factor Analysis

The factor of cause degree E_i less than 0 is others' outcome factor. It is the direct factor influencing the pricing decision-making of urban underground comprehensive pipe gallery PPP projects, greatly influenced by other factors, and will change with other factors, easy to alter in the short term. From Fig. 1, it can be seen that there are six factors with cause degree E_i less than 0, sorted in ascending order $a_8 > a_7 > a_9 > a_{11} > a_{13} > a_{12}$.

5.2.2 Centrality Analysis

The factors with greater centrality P_i are more closely related to other influence factors. It can be seen from Fig. 1 that the order of the centrality from large to small is $a_6 > a_7 > a_4 > a_{13} > a_1 > a_{12} > a_8 > a_9 > a_2 > a_3 > a_5 > a_{10} > a_{11}$. Therein, a_6 (expected investment income return) is most closely related to other influence factors. This factor is highly related to the government feasibility gap subsidy, franchise duration, comprehensive pipe gallery cost, pipeline units' entrance willingness, pipeline units' payment ability, etc. All parties should inspect and judge the impact by the importance of each influence factor of the pricing decision-making in the urban underground integrated pipe gallery PPP project on pricing.

5.3 Solving the Pricing Model

5.3.1 Paid Royalties and Government Compensation

The results obtained by substituting the existing data and parameter estimates into the bi-level planning model are shown in Equation (11).

$$Max\,U_g = (16200 - 0.2\,P_1 - 1.824\,P_2)$$
$$\times (51541 - 0.7\,P_1 - 2.736\,P_2)$$
$$- 5.472S - 121523.47$$
$$\text{s. t. } 0 \leqslant S \leqslant 53530.6$$
$$Max\,U_f = (P_1 - 10.56\,P_2)$$
$$\times (51541 - 0.7\,P_1 - 2.736\,P_2)$$
$$+ 10.56S - 336890.44$$
$$\text{s. t. } 6.5\% \leqslant R \leqslant 10\%$$
$$0 < P_1 + 10.56\,P_2 \leqslant 75875$$
$$S = 17800 - P_2$$

$$(11)$$

The overall entrance gallery fee of the urban underground comprehensive pipe gallery PPP project is calculated to be RMB 275.8375 million. The operation and maintenance cost is RMB 14.18 million per year, and the government compensation is RMB 168.28 million per year. The overall entrance gallery fee is RMB 2612.1 million per year after it is converted into annual expenses.

5.3.2 Comparing the Calculation Results

The plan provided in the "Implementation

Plan of Comprehensive Pipe Gallery Project in Wuhan Optical Valley Central City" is the entrance fee of RMB 28.28 million per year, the annual average daily operation and maintenance fee of RMB 15.71 million per year, and the financial feasibility gap subsidy of RMB 15.778 million per year. Compared with the calculation results obtained above, the entrance fee is RMB 26.112 million per year, the annual average daily operation and maintenance fee is RMB 14.18 million per year, and the government compensation is RMB 16.382 million per year, and it is verified that the calculation results in this paper are similar to those provided in the implementation scheme. Therefore, the pricing and government compensation model of the urban underground comprehensive pipe gallery PPP project established based on the bi-level planning theory is quite reasonable.

6　Conclusion

The construction of an urban underground comprehensive pipe gallery can effectively solve the problems of traditional underground pipelines such as "Urban Watching the Sea" and "Road Zippers". However, currently there are two main problems encountered in the construction of urban underground comprehensive pipe galleries in China. One is the issue of fundraising, and the other is the issue of pricing and charging. Fundraising has been solved by introducing social capital to participate in the construction of urban underground comprehensive pipe galleries by using the PPP model. However, the pricing and charging

mechanism needs to be further improved. Whether the hierarchical relationship between different subjects is clear will affect the smooth implementation of the urban underground comprehensive pipe gallery PPP project. Therefore, this article collects and studies the factors that affect the pricing and charging of current PPP projects in urban underground comprehensive pipe galleries, and classifies them into three categories: government, social capital, and pipeline units. The Gray-DEMATEL method was applied to quantitatively analyze the survey results. From the perspective of government departments, social capital, and pipeline units respectively, the influence factors of the pricing decision-making of urban underground comprehensive pipe gallery PPP projects were qualitatively analyzed, and 13 key factors were screened out. An index system of influence factors for pricing decision-making of urban underground comprehensive pipe gallery PPP projects was established. The results of the quantitative analysis show that the government has the greatest influence on the pricing decision-making of urban underground comprehensive pipe gallery PPP projects, occupying a dominant position, and is a more important decision-maker. This conclusion has been tested in the comprehensive pipe gallery PPP project of Wuhan Optics Valley Central City, and the feasibility, effectiveness, and reasonableness of this method can be verified.

References

[1]　Xu H. The finance ministry: the investment a-

mount of PPP projects in storage exceeds RMB 11mrillion. Tendering & Purchasing Manage, 2018 (10): 8.

[2] Qiang Y, Run W, Shuai B. Study on raising funds for rebuilding of railway goods yard fundamental equipment. Railway Economics Research, 2004(6): 33-34.

[3] He Y, Li Y, Zhang H. Research on PPP pricing and government compensation strategy of integrated pipeline gallery. Journal of Wuhan University of Technology (Information & Management Engineering), 2018(6): 615-621.

[4] Ning Y, Zhao S. Research on the current situation, problems and countermeasures of urban utility tunnel development in China and abroad. Value Engineering, 2018(3): 103-105.

[5] Bai H. A trend study of urban common tunnel development. China Municipal Engineering, 2015(6): 78-81,95.

[6] Cui Q. Study on the urban utility tunnel toll pricing under PPP model. Beijing University of Civil Engineering and Architecture,2017.

[7] Mason P, Bald C Y. Evaluation of government subsidies to large-scale energy projects. Advances in Futures and Options Research, 1988 (3): 169-181.

[8] Ho S P, Liu L. An option pricing-based model for evaluating the financial viability of privatized infrastructure projects. Construction Management & Economics, 2002,20(2): 143-156.

[9] Cheah C Y J, Liu J. Valuing governmental support in infrastructure projects as real options using monte carlo simulation. Construction Management & Economics, 2006,24(5): 545-554.

[10] Vickerman R, Evenhuis E. Transport pricing and Public-Private Partnerships. Research in Transportation Economics, 2010,30(1): 6-14.

[11] Driessen T, and Tijs S H. The cost gap method and other cost allocation methods for multipur-

pose water projects. Water Resources Res, 1985(10): 1469-1475.

[12] Anthony Chen, Yang Hai, Hong K Lo, et al. Capacity reliability of road network: an assessment methodology and numerical results. Transportation Research Part B,2002, 36(3): 225-252.

[13] Ye S, Liu Y, and Wang S. Price design. China Construction News, 2007(6).

[14] Ding X, Zhang J, Pang Y. Discussion on construction and management of the common trench in Guangzhou University Town. Chinese Journal of Underground Space and Engineering, 2010(S1): 1385-1390.

[15] Song D. Research on the operations management of tube gallery based on PPP model. Beijing University of Civil Engineering and Architecture,2014.

[16] Zhang X. The research on the pricing of utility tunnel Public-Private partnership projects. Shandong Jianzhu University,2017.

[17] Wang X, and Zhu F. Research of charge strategy for urban municipal utility tunnel based on game theory analysis. Chinese Journal of Underground Space and Engineering, 2013(1): 197-203.

[18] Wang S, Jin Y. Study on the benefit of PPP project in underground comprehensive corridor based on the system dynamics. Price: Theory & Practice, 2017(5): 143-146.

[19] Wang J, Zhang S, Niu F. Two-part tariff method in application of utility tunnel toll pricing. Architecture Technology, 2017 (7): 772-775.

[20] Wu W W, Lee Y T. Developing global managers' competencies using the fuzzy DEMATEL method. Expert Systems with Applications, 2007, 32(2): 499-507.

[21] Zhang K, Liang D, Liu Y. Analysis of affecting

factors of agricultural products cold chain logistics based on Grey-DEMATEL. Journal of Shenyang University (Nature Science), 2018, 30 (5): 80-84.

[22] Wang S. Study on influencing factors on the green development of the port based on the Grey-DEMATEL method. Dalian University of Technology, 2016.

Practical Application of Safety Management in the Reconstruction and Expansion Project of Expressway

Ying Li   Jinshan Feng

(School of Transportation and Civil Engineering,
Shan Dong JiaoTong University,Shandong JiNan 250357, China)

【Abstract】 During the construction of highway reconstruction and expansion projects, safety management plays an important role in engineering protection. This article takes the reconstruction and expansion project of the Laiwu-Taian section of the Qinglan Expressway as an example, and systematically introduces the project safety management system and the special safety measures taken by the project under the epidemic situation.

【Keywords】 Safety Management; Highway Engineering; Expressway; Reconstruction and Expansion Projects

At present, with the rapid construction of China's highway network, China has built the world's largest road transportation network. The construction technology and construction management of expressways have reached the world's leading level. Entering a new period of development, with the needs of economic development, more roads need to be built. Restricted by project cost, in certain areas, rebuilding and expanding existing roads is the best choice for expressway construction. In the process of expressway reconstruction and expansion, safety management has played an important role[1].

1 Project Overview

Qingdao-Lanzhou Expressway (hereinafter referred to as Qinglan Expressway) is the sixth horizontal line of the 18 east-west horizontal lines of the national expressway network, with a total length of about 1850km, passing through Shandong, Hebei, Shanxi, Shaanxi, Ningxia and Gansu provinces. Qinglan Expressway is connected with Beijing-Shanghai Expressway, Taixin Expressway, Beijing-Taiwan Expressway and the planned Jinan-Taian Expressway, which can effectively realize seamless connection between road networks.

The reconstruction and expansion project of the Laiwu-Taian section of the Qinglan Expressway is an important part of the Qinglan Expressway construction project. The project adopts the two-way eight-lane highway standard.

The total length of the project is 63. 802km, of which the length of the reconstructed and expanded road section is 20. 254km. The engineering design speed is 120km/h. The project contains 8 overpasses, 31 culverts, 2 service areas and 5 toll stations. After the project is completed, it will increase the traffic flow on the road and provide strong support for the conversion of new and old kinetic energy in the areas along the route.

2   Construction Difficulties and Safety Management Goals

The main construction difficulties are as follows. The construction route is long, which involves the construction and transformation of interconnection, overpass, and service area. Road repairs are not allowed during holidays, nights and special service hours. Need to cooperate with the local road construction plan. Subgrade and bridge construction progress needs to be combined with the original road traffic, and a safe and efficient construction organization system and traffic management system are needed[2].

However, the use and maintenance of the roadbed in the construction section is in good condition, and no major settlement or deformation has been seen on the whole line, which reduces the difficulty of construction. Based on this, the goal of safe and civilized construction was determined, requiring no death and serious injury accidents, no major traffic accidents, no major mechanical and electrical equipment accidents, no major fire accidents, and the frequency of industrial accidents less

than 1. 5‰, and that the project passed ISO standards to ensure the requirements of a safe and civilized construction site.

3   Safety Management Measures to Ensure the Successful Completion of the Project

3. 1   Establish a Safety Management Organization Structure System

In order to achieve the project goals such as project duration and quality, the project department has established a management organization and a safety department to carry out special safety management work (Fig. 1). The safety department implements all-round safety management of the project horizontally and vertically.

Vertically, the project manager, as the head of the security department, is responsible for security management within their respective scopes. On the basis of conscientious-

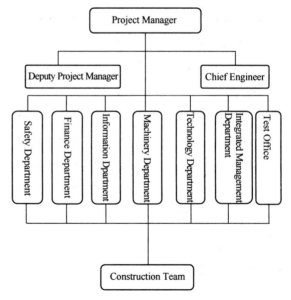

Fig. 1   Management organization chart

ly implementing the relevant national departments and enterprises' safety production responsibility rewards and punishments, the project manager supervises the safety education and technical training of each department and supervises the production team to carry out safety activities; at the same time, it is responsible for the construction management of key dangerous parts of the project, and organize relevant personnel to conduct regular safety inspections to ensure the smooth progress of safe construction. The chief engineer of the project is responsible for the preparation of safety production guarantee plans, the preparation of construction operation plans with higher risk factors, organization of safety technical demonstrations for projects with higher risks and strong professionalism, and supervision and implementation of relevant personnel to implement safety technical preventive measures.

Horizontally, the safety department is responsible for the specific operations of safety management during project implementation. The safety officer is responsible for keeping safety records of all personnel during the entire project construction process, and organizing the operators of various departments to conduct on-site safety knowledge promotion; organizing and participating in safety inspections; grasping safety trends and taking preventive measures against unsafe factors; responsible for carrying out safety training and education and posting safety slogans.

In addition, construction personnel, material workers, data workers, technicians, machine drivers and other technical personnel shall implement safe construction requirements within the scope of their respective duties. The team leader is responsible for the construction site control such as safety technical disclosure, personnel management, hidden danger rectification; and carefully inspect the mechanical equipment, electrical equipment and protective equipment required by each facility before the construction operation. Once an accident occurs, immediately organize personnel to rescue the wounded, protect the scene of the accident, report to relevant departments in a timely manner, participate in accident investigation and analysis, and effectively implement engineering safety technical measures.

3.2　Develop Safety Assurance Measures

Starting from the project situation and construction difficulties, determine the overall plan and deployment. This project adopts parallel operation and flow construction alternately, so that the project can be completed on schedule under the premise of ensuring quality (Fig. 2).

After formulating the overall construction deployment, formulate safety assurance measures that suit the requirements of the project. Carry out necessary safety education and safety training, and conduct safety technical clarifications before starting the job. Strengthen labor safety management to ensure personal safety, and all on-site construction personnel must be listed on duty. During the construction process, the steel bar processing site

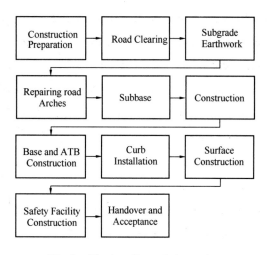

Fig. 2　The key lines of the project

should be level, the workbench should be stable, and the electrical power of each mechanical equipment should be guaranteed to have a grounding protection device; it is strictly forbidden to stand under the lifting arm, the lifting operation shall be uniformly directed, and the safety officer shall stand by to supervise. The installation is in compliance; safety fences must be set up around the transformer, and slogans with a warning effect must be posted. Those involved in high-altitude operations must have regular physical examinations; high-altitude operations should wear seat belts, and avoid outdoor or high-altitude operations during thunderstorms. Establish a safety incident report system, and separately formulate safety technical plans for key safety prevention projects.

3.3　Formulate Civilized Construction Guarantee Measures

The project department strengthens the education of civilized construction for all employees, explains the importance of civilized construction, enables employees to firmly establish the awareness of civilized construction, consciously abide by rules and regulations, standardize construction, and civilized production, and improve the spiritual and civilized quality of construction personnel[3].

Obvious signs should be set up on the construction site, indicating the name of the unit, contact number, supervisor name and position of the supervisor, etc. Slogans of safe production and civilized construction are posted on the construction site. The construction work area, material storage area, office and living area should be clearly demarcated. The construction site is required to be tidy, the residues should be cleaned up in time, and the material yard site should be clean, and it is strictly forbidden to stack them. The machinery and equipment are kept in good condition at all times and placed neatly when not in use. During the construction period, the normal life of the local people shall not be affected, and visitors shall be warmly received and live in harmony with the local residents[4].

Formulate environmental protection guarantee measures. The project department establishes an environmental management system and organizes special personnel to manage environmental issues. The domestic waste and construction waste from the project department and construction site shall be uniformly cleaned and transported on a regular basis. Strengthen the management of materials. Sprinkle water to clean the roads on the construction site regularly. Vehicles entering and leaving the construction site should basically not raise dust. The working and living areas are reasonably arranged, and the equipment with loud noises shall be planned in a centralized manner. Enclosures are set up at the mixing station to prevent materials from flowing into the fish ponds and ponds along the ground and causing water contamination.

3.4　Formulate Safety Assurance Measures in Winter and Rainy Seasons

The project has a long construction period and requires rainy season and winter construction. In order to achieve the project's safety goals, safety assurance measures in the winter and rainy season are formulated.

Before winter construction, prepare winter construction plan and implement it after review by technical quality department. The aggregates for concrete construction in winter should be clean, and should not contain rain, snow, ice and other substances that are easy to freeze and crack. All kinds of machinery used in winter construction should be thoroughly inspected, various lubricating system oils and fuels should be replaced, and problematic machinery and equipment should be repaired in time, and no malfunctioning operation should be allowed. Before pouring concrete, the sundries on the formwork and steel bars must be cleaned, but it cannot be washed with water to prevent the temperature from becoming too low and forming ice. Use sawdust or straw rope for heat preservation treatment for domestic and construction water pipes.

Before the rainy season construction, the rainy season construction plan is compiled and implemented after the technical quality department review.

Drainage measures should be taken. Drainage measures should be taken around the road cutting. An intercepting ditch should be set up on the side slope of the road cutting to prevent the side slope of the road cutting from landslide caused by rain. In the steel bar stacking site, the steel bars should be erected with skids to avoid rainwater soaking the steel bars and causing the steel bars to corrode. Cement materials should be stacked on higher ground, moisture-proof measures should be taken underneath, and drainage work should be done around them. When carrying out concrete and masonry work in the rainy season, the construction specifications should be strictly followed, and the concrete should be poured as far as possible to avoid rainy days.

3.5　Develop Safe Operating Procedures

Formulate rules for the use of mechanical equipment, such as safe operation rules for mixers, safe operation rules for pavers, safe operation rules for road rollers. For example, the driver must be at least 18 years old and have obtained the corresponding certificate, skilled in operation and strong sense of responsibility; understand the relevant construction technology and quality requirements, and have the ability to install and adjust the working devices of different machines; the operator should conduct regular physical examinations without obstruction illnesses and physical defects in the job can continue to work with a certificate; when implementing multi-shift operation machinery, the shift system should be implemented, and written records should be carefully implemented and written records[5]. Improve the safety education of mechanical operators, so that the operators can firmly establish safety awareness, master the mechanical operation methods and construction techniques, and fundamentally prevent the occurrence of safety accidents.

3. 6 Formulate Special Safeguard Measures During the Epidemic

The project started construction in May 2019 and is one of the highway construction projects with the tightest construction period and the heaviest construction task in Shandong Province. The project has been under construction according to the schedule, but the sudden COVID-19 in early 2020 interrupted the construction process. With the gradual control of the epidemic, the project, as a key transportation infrastructure project, gradually resumed work and production, achieving a 100% resumption rate on March 19 2020[6].

The resumption of work is also a critical period for the prevention and control of COVID-19. In order to ensure construction safety during the epidemic, the project mainly adopted the following special safeguard measures.

(1) Establish an organizational system for epidemic prevention and control, and formulate provisions to control COVID-19. Establish and improve the organizational structure for epidemic prevention and control, and form a responsibility system to clarify the size of each person's responsibility to ensure the health of personnel during construction.

(2) Strengthen the awareness of epidemic protection and take specific measures to complete personnel protection. Each project bidding section adopts the "point-to-point" method of rework by constructors chartered and private cars to ensure that personnel arrive at the construction site in a healthy and safe manner. After personnel enter the site, they will be collectively managed. Each residence will designate a special epidemic control personnel. All necessary daily necessities and food will be purchased and distributed by specialized personnel with certificates. At the same time, the personal protection and temperature measurement of the returning personnel are strictly carried out. The above multiple measures ensure the safety of construction personnel.

(3) Adopt new schemes and new technologies to protect the safety of personnel and speed up the construction progress. In response to the epidemic prevention requirements, the project office held a video mobilization meeting for resumption of work, and deployed project officers and personnel from participating units to return to work and epidemic prevention and control work as soon as possible. Actively promote the use of new construction techniques. For the first time in Shandong Province, a fully enclosed steel structure greenhouse production workshop was built as a prefabricated beam field, a mobile pedestal was used for the first time, and an intelligent variable temperature steaming system was applied for the first time. The ap-

plication of these new technologies has greatly improved the prefabrication and maintenance time of the box girder, shortened the production time by nearly 30%, and provided a great help to guarantee the construction period.

In addition, the project office has also taken other measures to speed up the progress of the project. In order to ensure rapid construction after the year, sufficient raw materials were prepared before the epidemic, so that the project was not affected by the shortage of raw materials and provided a strong guarantee for completion on schedule.

4　Conclusion

Under a systematic safety management system, the project has been completed and opened to traffic in December 2020. After the completion of the reconstruction and expansion project of Qinglan Expressway Laitai, the Qinglan Expressway has been connected in the true sense and ensured the smooth flow of the main national corridor. It provides important support for the economic and social development of the areas along the route and even the middle and eastern parts of Shandong Province. The project has been successfully opened to traffic, and the established construction goals and safety and civilized goals have been successfully achieved. Under the management of a sound safety guarantee system and safety measures, the project has achieved the safety management goal of no major accidents and no casualties.

In summary, in the process of road recon-

struction and expansion, a complete safety management system is an important guarantee for the realization of highway reconstruction and expansion. Only by strengthening construction safety management can the quality and safety of construction be further improved, so as to better achieve the overall goal of road construction.

References

[1] Li Bingqian. Study on the safety management and control technology of the open-to-traffic section in the reconstruction and expansion project. Research on Engineering Construction and Control Technology,2018(12):249-250.

[2] Liu Yi. Discussion on the "New Kaiyang Traffic Organization Safety Management" mode of the expressway reconstruction and expansion project. China Construction Metal Structure, 2020.

[3] Liang Chen. Research on traffic safety management of expressway reconstruction and expansion project . Consumer Guide,2019(6):34.

[4] Tao Xuezhou. Analysis of safety management and control of road section open to traffic during reconstruction and expansion project. Engineering Design Construction and Management,2019(17): 101-102.

[5] Yin Liang, Liu Zhengbing, Zhang Dong, et al. How to carry out safety management in reconstruction and expansion of roads . Theoretical Research on Urban Construction (Electronic Edition),2016(6):420.

[6] Jiang Longlong,Li Aiguo,Liu Kun. Running out of qilu new speed——record of the resumption of the reconstruction and expansion project of the Laiwu-Taian section of Qinglan Expressway. China Highway,2020(8):50-51